# Organic Chemistry

## Laboratory Manual
### Second Edition

# GLASSWARE USED IN ORGANIC CHEMISTRY

Round-bottom Flasks with Standard-taper Joints

Distilling Column

West Condenser

Connecting
Adaptor

Claisen
Adaptor

Vacuum
Adaptor

Adaptor
Outlet

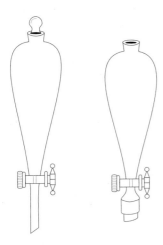  

Separatory Funnels

Büchner
Funnel

Hirsch
Funnel

# Organic Chemistry

# Laboratory Manual
## Second Edition

by

Paris Svoronos
Queensborough Community College

Edward Sarlo
Queensborough Community College

and

Robert J. Kulawiec
Georgetown University

Boston, Massachusetts    Burr Ridge, Illinios    Dubuque, Iowa
Madison, Wisconsin    New York, New York    San Francisco, California    St. Louis, Missouri

# WCB/McGraw-Hill

*A Division of The McGraw-Hill Companies*

**Project Team**

Editor  *Colin Wheatley*
Developmental Editor  *Brittany J. Rossman*
Marketing Manager  *Patrick E. Reidy*
Advertising Coordinator  *Wayne W. Siegert*
Publishing Services Coordinator  *Julie Avery Kennedy*

President and Chief Executive Officer  *Beverly Kolz*
Vice President, Director of Editorial  *Kevin Kane*
Vice President, Sales and Market Expansion  *Virginia S. Moffat*
Vice President, Director of Production  *Colleen A. Yonda*
DIrector of Marketing  *Craig S. Marry*
National Sales Manager  *Douglas J. DiNardo*
Advertising Manager  *Janelle Keeffer*
Production Editorial Manager  *Renée Menne*
Publishing Services Manager  *Karen J. Slaght*
Royalty/Permissions Manager  *Connie Allendorf*

Copyediting and production by Shepherd, Inc.

Cover and interior design by Shepherd, Inc.

Cover photo Images © 1996 PhotoDisc, Inc.

ISBN 978–0–697–33923–2
MHID 0–697–33923–8

Some of the laboratory experiments included in this text may be
hazardous if materials are handled improperly or if procedures are
conducted incorrectly. Safety precautions are necessary when you are
working with chemicals, glass test tubes, hot water baths, sharp
instruments, and the like, or for any procedures that generally require
caution. Your school may have set regulations regarding safety
procedures that your instructor will explain to you. Should you have any
problems with materials or procedures, please ask your instructor for
help.

Printed in the United States of America

10 9

# Contents

**Preface**     **ix**

**1   The Basics**     **1**
1.1   Laboratory Safety     1
1.2   The Laboratory Notebook     2
1.3   Ground-Glass Equipment     3

**2   Stereochemistry and Molecular Models**     **5**
2.1   Constitutional Isomers     6
2.2   Stereoisomers     7
2.3   Fischer Projection Formulas for Chiral Molecules     12
2.4   The R/S Designation for Chiral Carbons     13
2.5   The E/Z Designation of Alkenes     14

**3   Physical Properties**     **19**
3.1   Melting Point     19
3.2   Boiling Point     25
3.3   Refractive Index     26

**4   Distillation**     **29**
4.1   Simple Distillation     31
4.2   Fractional Distillation     32
4.3   Steam Distillation     37
4.4   Micro-Distillation     40

**5   Chromatography**     **42**
5.1   Column Chromatography     42
5.2   Thin-Layer Chromatography     45
5.3   Gas Chromatography     51

**6   Extraction and Recrystallization**     **54**
6.1   Distribution Coefficient     54
6.2   Separation by Extraction     63
6.3   Recrystallization of a Solid     66
6.4   Microrecrystallization of a Solid     72
6.5   Isolation of Caffeine from Tea     75

**7   Equilibrium Constant**     **78**

**8   The Sodium Fusion Test**     **83**
8.1   Sodium Fusion     84
8.2   Test for Sulfur     85
8.3   Test for Halogens     85
8.4   Test for Nitrogen     85

## 9   Spectroscopy     89
9.1   Infrared Spectroscopy     89
9.2   Nuclear Magnetic Resonance     105
9.3   UV-Visible Spectrophotometry     113

## 10   Alkenes     123
10.1   Synthesis of Cyclohexene     123
10.2   Synthesis of 1,2-Cyclohexanediol from Cyclohexene: A Syn Addition     127
10.3   Synthesis of $\alpha,\beta$-Dibromobenzalacetophenone     130
10.4   Oxidation of Cyclohexene to Adipic Acid     133
10.5   Catalytic Hydrogenation of an Alkene     136

## 11   Synthesis and Reactions of Acetylene     141

## 12   Conjugated Dienes: The Diels-Alder Reaction     147

## 13   Alkyl Halides and Nucleophilic Aliphatic Substitution     151
13.1   Synthesis of t-Butyl Chloride     151
13.2   Solvolysis of t-Butyl Chloride: A Kinetic Study     154
13.3   Structural, Solvent, and Temperature Effects on Nucleophilic Substitution Reactions     163

## 14   Alcohols and Ethers     175
14.1   The Grignard Reaction: Synthesis of 2-Methyl-2-Hexanol     175
14.2   Qualitative Tests for Alcohols     180
14.3   The Williamson Ether Synthesis: Preparation of Phenacetin from Acetaminophen     185

## 15   Aromatic Reactions     188
15.1   Nitration of Bromobenzene     188
15.2   Friedel-Crafts Alkylation: Synthesis of 2,5-Di-t-Butyl-1,4-Dimethoxybenzene     192
15.3   Synthesis of 2,4-Dinitrophenylaniline: A Nucleophilic Aromatic Substitution     195
15.4   Oxidation of the Side Chain of an Arene: Synthesis of 2-Chlorobenzoic Acid from 2-Chlorotoluene     197
15.5   Benzyne: Synthesis of Triptycene     200

## 16   Carboxylic Acids and Derivatives     204
16.1   Equivalent Weight of Organic Acids     204
16.2   Hydrolysis of Benzonitrile     211
16.3   Synthesis of Aspirin     214
16.4   Synthesis of Isoamyl Acetate     216
16.5   Imides: Synthesis of N-Phenylphthalimide     219
16.6   Saponification of an Ester     221

## 17   Aldehydes and Ketones     223
17.1   Oxidation of Cyclohexanol to Cyclohexanone     223
17.2   Reduction of Cyclohexanone to Cyclohexanol     226
17.3   Acetal Formation: Synthesis of 4,5-Dimethyldioxolane     229
17.4   The Pinacol Rearrangement: A $^1$H NMR Study     232
17.5   Qualitative Tests for Aldehydes and Ketones     234

## 18   Carbanions and $\alpha,\beta$-Unsaturated Carbonyls     243
18.1   The Aldol Condensation: Synthesis of Dibenzalacetone     243
18.2   A Cyclization Reaction: Synthesis of Cyclopentanone from Adipic Acid     247
18.3   Michael Addition: Reaction of Aniline with Benzalacetophenone     249

**19   Amines**                                                                                **251**
19.1   Reduction of a Nitro Compound to an Amine: Synthesis of m-Aminoacetophenone from m-Nitroacetophenone   251
19.2   Synthesis of Acetanilide                                                                 255
19.3   The Coupling of Aromatic Diazonium Compounds: Azo Dye Formation                          258

**20   Polynuclear Aromatics and Heterocycles**                                                 **260**
20.1   Oxidation of 2-Methylnaphthalene                                                         260
20.2   Synthesis of Benzimidazole                                                               262

**21   Carbohydrates**                                                                          **264**
21.1   The Acid Catalyzed Hydrolysis of Sucrose: A Kinetic Study                                264
21.2   Qualitative Tests for Carbohydrates                                                      273

**22   Amino Acids and Proteins**                                                               **283**
22.1   Synthesis of a Peptide                                                                   283
22.2   Qualitative Tests for Amino Acids and Proteins                                           290
22.3   Decomposition of Hydrogen Peroxide with Catalase                                         299

**23   Lipids: Fats, Oils, and Steroids**                                                       **305**
23.1   Preparation and Properties of a Soap                                                     305
23.2   Qualitative Determination of Unsaturation in Lipids: The Hanus Test                      311
23.3   Quantitative Determination of the Iodine Number of an Unknown Fat or Oil                 315
23.4   The Liebermann-Burchard Test for Cholesterol                                             321

**24   Qualitative Analysis**                                                                   **325**

**Appendix I: How to Balance an Oxidation-Reduction Reaction**                                  **337**

**Appendix II: How to Calculate the Percent Yield of a Reaction**                               **340**

**Appendix III: How to Make Solutions**                                                         **343**

**Appendix IV: Special Reagents**                                                               **346**

**Appendix V: Physical Data of the Liquids Used in This Lab Manual**                            **349**

**Appendix VI: Common Drying Agents for Organic Liquids**                                       **352**

**Index**                                                                                       **354**

# Preface

This manual was written to provide students of organic chemistry with a set of workable laboratory experiments that is comprehensive in scope and that correlates with the theoretical concepts developed in the lecture portion of the course. Although it is primarily intended for a two-semester course, it can easily be adapted to fit a one-semester survey course. In the beginning of a course in organic chemistry, much of the theoretical work concerns the fundamentals: writing structural formulas, acid-base theory, resonance concepts. Similarly, laboratory work also begins with the fundamentals: melting and boiling point determinations, distillation techniques, solvent extractions, and recrystallizations. These experimental procedures, which are probably new to the student, are basic to all laboratory work. The first few experiments should be used to instruct students in **how** to take a melting point, **how** to fractionally distill a mixture and **how** to recrystallize an impure solid. The training the student gets at the beginning greatly influences his or her later performance in the laboratory. Proper technique results in fewer accidents, fewer failed experiments, higher yields, and purer products. In addition, the coordination of lecture and laboratory work is not an issue because the lecture material at this point does not lend itself to laboratory experiments. By the time the lecturer gets to topics that can be correlated with laboratory work (the preparation and reactions of alkenes, for instance), the student will have gained competence in the basic techniques. From there on, he or she will learn to:

- synthesize specific compounds from specific starting materials by reactions that have been described in lecture
- understand why certain precautions should be taken to avoid unwanted side reactions (for instance, the maintenance of absolutely anhydrous conditions for the Grignard reaction, and the use of fractional distillation for the preparation of cyclohexene)
- chemically and/or spectroscopically identify and justify the structure of products
- identify an unknown compound by a qualitative analytical scheme or develop a synthesis for a specific compound using knowledge and initiative.

For these purposes the manual is divided into three main parts:

Part I: *General Techniques* (Chapters 2 through 9)

Part II: *Reactions and Qualitative Analysis* (Chapters 10 through 24). This section contains a variety of synthetic procedures. The synthesis of aromatic carboxylic acids by three different routes (oxidation of 2-chlorotoluene, hydrolysis of methyl benzoate, and hydrolysis of benzonitrile) illustrates the flexibility that is often available in synthetic work. Examples of multistage preparations include the synthesis of adipic acid from cyclohexanol via cyclohexene, and the synthesis of triptycene via a series of reactions involving diazotization, decarboxylation, Diels-Alder cyclization, and anhydride hydrolysis. These preparations require the student to calculate overall yields, and they illustrate how the overall yield depends on the efficiency of each of the reactions involved.

Part II also includes kinetic experiments (the solvolysis of t-butyl chloride and the hydrolysis of sucrose), experiments that require quantitative measurements (of the equilibrium constant of an esterification reaction, of the distribution coefficient of an extraction procedure, of the equivalent weight of a carboxylic acid and of the determination of the iodine number of various fats and oils), and experiments that involve the spectrophotometer (for the depolymerization of plexiglass, for the pinacol-pinacolone rearrangement, and for the distribution coefficient of several dyes). Thus, a wide variety of experiments is available from which an instructor may choose. A complete Qualitative Analysis flowchart is given in Chapter 24.

Part III: The *Appendices* (I through VI). Part III contains three "How to" appendices describing the methods for balancing an equation, the calculation of a percent yield and the preparation of solutions. Students should find these very useful. Additionally this section contains directions for the preparation of special reagents, a very complete table of the physical data for the liquids used in this manual and a listing of the most popular drying agents.

In accord with the tendency toward microtechniques, the manual includes three microscale separation experiments (column chromatography, microrecrystallization, and microdistillation). The small quantities of chemicals involved in microtechniques offer tangible benefits; however, these techniques may require a greater expertise that we can realistically expect a college sophomore to have. Another drawback to the complete conversion to microscale is the expense involved in equipping laboratories with microscale glassware. Thus our inclusion of three microscale techniques is a realistic compromise that offers an introduction to microtechniques while maintaining the traditional semimicroscale methods at which college sophomores are more adept. Several microscale experiments are also described (Friedel-Crafts alkylation dimethoxybenzene, equivalent weight of carboxylic acids).

The importance of laboratory safety cannot be overstressed. All experiments have been carefully screened and selected to minimize the exposure to toxic chemicals. A Safety Tips section before the procedure of every experiment provides important information about the hazardous properties of any toxic chemicals involved and outlines necessary precautions. Efficient fume hoods and the usual safety procedures are mandatory in organic chemistry laboratories. We have found an efficient and valuable exercise for focusing student attention on laboratory safety: After a full discussion on laboratory safety we give each student a sketch of the laboratory room—minus doors, windows, sinks, eye-wash fountains, etc. The student must then identify the location of all these items, plus the fire extinguishers, fire blankets, sand buckets, etc. One of the satisfactions a student gets from organic chemistry is to put theory to work in the laboratory. The experiments in this manual offer just such an opportunity.

Special thanks go to Professor Vaclav Horak of Georgetown University for his constructive criticism, Dr. Chris Waidner for his great help with the spectra, and Dr. Soraya Svoronos and Dr. Suzanne Michel for their patience and support.

# chapter 1

# The Basics

## 1.1  LABORATORY SAFETY

The study of organic chemistry requires working in a chemical laboratory. Laboratory work can be exciting and stimulating, especially when a student successfully completes a difficult synthesis or correctly identifies an unknown. However, there are hazards in laboratory work, and accidents can and do happen. Nevertheless, most accidents can be avoided if the proper procedures are followed. The following safety rules are to a large degree based on common sense.

1. The student should learn the location of the exits, the fire extinguishers, the sand buckets, the first-aid kit, the eyewash fountain, the fire blanket(s), and the safety shower(s). Fume hoods should be amply used, especially when dealing with toxic or flammable solvents, and hoods should *not* be used for storage. The student should be discouraged from keeping purses, briefcases, backpacks, and other personal materials near the experiment area. Books should not be used to raise the heating mantle or the receiving flask during distillation. The address and telephone number of a doctor or nurse responsible for treating persons involved in laboratory accidents should be posted or given to the lab assistants and teaching instructor *and* the students, especially if sessions are held after hours.

2. Acids, bases, and organic solvents are extremely harmful to the eyes, and for this reason, *safety glasses must be worn at all times,* as required by state law. Wearing contact lenses is not advisable since gases formed during an experiment might be entrapped between the eye and the lens. If a chemical gets into the eye, the eye should be immediately flushed with water. It is important that this be done at once. Either an eyewash fountain or distilled water in special eyewash bottles (if available), or even sink water should be used to flush the eye *for at least 10 to 15 minutes.* Medical assistance should then be sought.

3. Suitable clothing should be worn, preferably with a laboratory apron. Avoid loose-fitting shirts and sweaters that may come into contact with a Bunsen burner. For the same reason, long hair should be tied back. Shoes (not sandals) are mandatory to protect the feet from broken glass and spilled chemicals that may be on the floor.

4. The students should wash their hands thoroughly with soap and water before leaving the laboratory.

5. The presence and consumption of any food and drink in the lab is strictly forbidden, because many chemicals are toxic if ingested.

6. When diluting an acid, the student should add the acid *slowly* to the water with stirring to avoid splattering. Concentrated hydrochloric

acid (HCl), nitric acid (HNO$_3$), and ammonia (NH$_3$) emit irritating vapors; therefore, bottles of these chemicals should be opened and their contents poured in a fume hood. The use of a funnel when transferring liquids is highly recommended.

7. Chemicals spilled on a laboratory bench should be cleaned immediately. Concentrated acids can be neutralized with sodium bicarbonate and then flushed with water. Chemicals that come into contact with the skin should be washed away with copious amounts of water.

8. All injuries must be reported to the instructor. *Minor* burns can be treated with either cold water or ice. Wrapping the burned part of the body (usually fingers) with a cloth soaked with 95% ethyl alcohol provides cooling and temporary local anesthesia. A burn ointment from the first-aid kit may be applied. Minor cuts should be flushed with water before a pressure bandage is applied. Severe burns and cuts require medical attention.

9. To reduce the incidence of fires, heating mantles or hot plates should be used instead of Bunsen burners. If Bunsen burners are necessary (for instance, in experiments where large volumes of water are to be evaporated, or during the sodium fusion test), they should be used *only* in the presence of nonflammable liquids. Highly flammable liquids should be boiled away on a steam bath in the hood or, preferably, via simple distillation. Laboratory fires, if they occur, are usually small and easily extinguished. A watch glass held with a pair of crucible tongs on top of the burning contents of a beaker or Erlenmeyer flask usually extinguishes the fire. Larger fires can be put out with a chemical or a CO$_2$ fire extinguisher. Therefore, students should know where the laboratory fire extinguishers are and how to operate them. In case of a serious fire, the laboratory should be evacuated. If a student's clothing catches fire, the student should *walk (not run)* to the safety shower and stay under the water until the fire is out. If a safety shower is not available, the student should be rolled in a safety blanket until the fire is extinguished. The student should then be kept quiet and warm until medical assistance arrives (if required).

10. Many common solvents are both flammable and toxic to varying degrees. Diethyl ether is a very useful but highly flammable solvent, and its use in the laboratory should be limited. No open flames are permitted in the lab when diethyl ether is used. Although most toxic organic liquids (benzene, chloroform, carbon tetrachloride) have been replaced by less toxic materials (ethyl acetate, ethanol, acetone), students should not be exposed to large volumes of any solvent. In this manual, large volumes of solvents are removed by distillation, and not by simple boiling off. Organic substances, including solvents, should not be disposed of by pouring them in the sink, but by placing them in clearly marked waste bottles.

## 1.2   THE LABORATORY NOTEBOOK

Laboratory work requires record keeping, and for this purpose a laboratory notebook is necessary. The instructor's specific instructions on keeping a laboratory notebook should be followed; in addition, the write-up for each experiment should contain the following information:

- the title of the experiment
- the date the experiment was performed
- balanced equations for all reactions, and detailed mechanisms

- a brief description of the procedure, including diagrams of the experimental set-up, as well as any special precautions that were taken. The procedure should be written in a way that would allow someone else to reproduce the results obtained.
- a complete reagent table listing the names, quantities (mass, volume, moles), and all relevant physical properties (molecular weight, melting point [MP], boiling point [BP], density, refractive index) of all chemicals used; the tables should also include the properties and theoretical yield of the expected product
- all data and observations (e.g., melting points, boiling points, refractive index, color of liquid distilled, type of crystals formed)
- all calculations, including the determination of the percent yield (see Appendix II)
- answers to all questions that follow the Experimental part.

Some experiments involve the synthesis of an organic compound that is then handed in and graded. The product should be placed in a tightly capped vial, bottle or test tube (preferably sealed with parafilm) and labeled appropriately. The label should contain the following information:

- the student's name and class section
- the name and structure of the product
- the physical constants of the product (melting or boiling point, refractive index)
- the mass of the product
- the percent yield of the reaction.

Other experiments involve analysis of an unknown. A report form (data sheet) similar to the type included at the end of some of the procedures in this manual should be handed in at the end of the laboratory period. A copy of the data sheet should be included in the final report submitted to the instructor.

The instructor should have a clear understanding of the experimental details. It is up to the instructor to help the student get satisfactory results. He or she should bear in mind that the beginning student has very little knowledge of organic chemistry, therefore, it is important to give a detailed analysis of each experiment. One suggestion is to discuss the *next* scheduled experiment at the end of each experiment and give a reading assignment. To ensure that the student is properly prepared, a five-minute quiz highlighting the basic principles and hazards of the day's experiment may be administered at the beginning of the laboratory period.

## 1.3  GROUND-GLASS EQUIPMENT

Organic laboratories use ground-glass equipment with standard tapered joints. This virtually eliminates the use of corks and rubber stoppers in glassware assembly. When properly greased, standard tapered joints join together easily and make an airtight seal (Figure 1.1). This allows a rapid assembly, and the interchangeable parts enable the student to set up the same equipment quickly for different procedures, for example, a refluxing procedure followed by a distillation. The student must properly support the glassware with clamps and, in some instances, with rubber bands. Laboratory setups are constructed piece by piece and all parts should fit together snugly and be free from strain. A piece of glassware should never

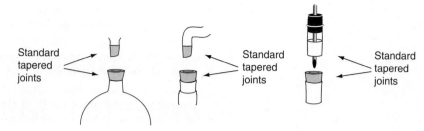

**Figure 1.1**   Standard tapered joints making an airtight seal

be left dangling, even if it seems to fit tightly. When the experiment is over, each piece of glassware should be cleaned and dried before it is put away. Glassware should be washed thoroughly with warm water and a detergent using a test tube brush. Small amounts of organic materials can be removed with 5 to 10 mL of an organic solvent. Acetone is particularly convenient because it dissolves many organic substances, is water soluble, and volatile (boiling point 56°C). Glassware should be separated and never left joined together after use.

# 2

# Stereochemistry and Molecular Models

Unlike many inorganic molecules, organic compounds contain covalent bonds, which are formed by the sharing of electron pairs by two atoms. Usually, but not always, each atom contributes one electron to the bond. Table 2.1 lists the number of covalent bonds formed by elements most commonly found in neutral stable organic molecules.

**Table 2.1** Number of covalent bonds of common elements found in organic molecules

| Element | Symbol | Number of Bonds |
|---------|--------|-----------------|
| Carbon | C | 4 |
| Hydrogen | H | 1 |
| Oxygen | O | 2 |
| Nitrogen | N | 3 |
| Halogens | F, Cl, Br, I | 1 |

Organic compounds are represented by chemical formulas. The **molecular formula** indicates the kinds of atoms and the exact number of each atom in the molecule. This formula describes the molecule's composition but not its structure. For instance, it is immediately evident that $C_4H_{10}$ represents a different compound than $C_5H_{12}$, but the molecular formulas themselves do not tell us about the arrangement of the atoms in the molecules. **Structural formulas,** on the other hand, describe the actual arrangement of these atoms in a molecule. Usually more than one structural formula is possible for a single molecular formula. Compounds with identical molecular formulas, but with different structural formulas, are called **isomers.**

The phenomenon of isomerism and the nature of organic structures can best be understood by constructing molecular models which clarify what might be misunderstood by two-dimensional drawings on paper. In the most common commercially available molecular model kits, balls represent atoms, and sticks or pegs, and springs represent bonds. Each ball has one or more holes to correspond to the total number of bonds that the atom can form. Carbon atoms (black) are connected by wooden pegs or springs which represent single bonds; double and triple bonds like those of ethylene and acetylene, respectively, are represented by curved pegs or springs.

As an exercise, match the color of the balls in your kit with the elements that appear in Table 2.2 by comparing the number of holes with the number of bonds that those elements normally form in neutral compounds.

**Table 2.2**    The molecular model kit

| Atom | Number of Holes | Color |
|---|---|---|
| Carbon | 4 | Black |
| Hydrogen | 1 | |
| Oxygen | 2 | |
| Nitrogen | 3 | |
| Chlorine | 1 | |
| Bromine | 1 | |
| Iodine | 1 | |

There are two general classes of structural isomers: constitutional isomers and stereoisomers. Each of these is further subdivided into several categories.

## 2.1   CONSTITUTIONAL ISOMERS

Constitutional isomers differ in the sequence in which atoms are connected in a molecule. Among these are skeletal, positional and functional group isomers.

### Skeletal Isomers

**Skeletal isomers** have identical molecular formulas, but differ in the arrangement of the carbon skeleton. *n*-Butane and isobutane have identical molecular formulas ($C_4H_{10}$), but different carbon skeletons.

$$CH_3CH_2CH_2CH_3$$

$$CH_3 \diagdown \atop CH_3 \diagup CHCH_3$$

*n*-butane             isobutane

Use the molecular model kit to construct both molecules. Note the difference in the carbon skeletons. The four carbons of *n*-butane form a continuous chain. The longest continuous chain of isobutane consists of only three carbons, and the fourth carbon is attached in the middle. Also note the tetrahedral geometry around each carbon atom. This arrangement is formed via $sp^3$ hybridization.

### Positional Isomers

Two compounds are **positional isomers** if their molecular formulas and carbon skeletons are identical, but the position of a non-carbon atom or group is different. For example, 1-chlorobutane and 2-chlorobutane are positional isomers. Construct models of both by replacing the appropriate hydrogen atoms in *n*-butane with a chlorine atom.

$$CH_3CH_2CH_2CH_2 \atop | \atop Cl$$
        
$$CH_3CH_2CHCH_3 \atop | \atop Cl$$

1-chlorobutane          2-chlorobutane

Use your model kit to answer the following questions: How many positional isomers of $C_4H_9Cl$ are possible? of $C_4H_8Cl_2$?

## Functional Group Isomers

Organic compounds are classified in families of **functional groups** according to their structure and chemical reactivity. Such families are alkanes (R—H), alcohols (R—OH), ethers (R—OR), amines (R—NH$_2$, R—NHR and R—NR$_2$), and alkyl halides (R—X, where X=F, Cl, Br, or I). Also included are compounds that have multiple-bond groups, such as the carbonyl group, >C=O, in aldehydes (R—CHO), ketones (R—COR), carboxylic acids (R—COOH), esters (R—COOR), and amides (R—CONH$_2$, R—CONHR and R—CONR$_2$). Multiple carbon-carbon bond families include alkenes (>C=C<), alkynes (—C≡C—), and aromatic compounds. Functional group isomers are illustrated by ethanol and dimethyl ether.

CH$_3$CH$_2$OH           CH$_3$OCH$_3$
ethanol             dimethyl ether

Use the model kit to build both molecules. Note that the oxygen atom of ethanol is attached to a hydrogen and a carbon, while the oxygen atom of dimethyl ether is bonded to two carbons.

**Question**

1. What is the relationship between
   a. *o*-dichlorobenzene and *p*-dichlorobenzene

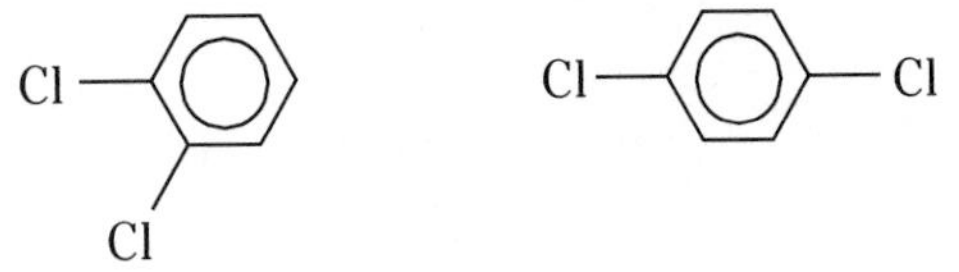

   b.       propanal    and     acetone

   c. cyclohexane and methylcyclopentane

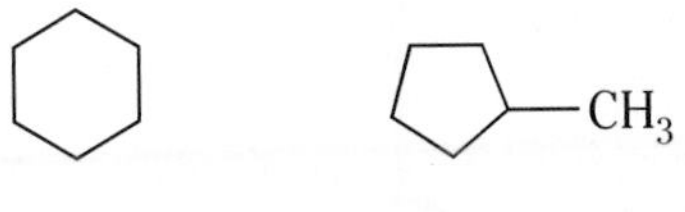

**Answer** The compounds *o*- and *p*-dichlorobenzene (molecular formula C$_6$H$_4$Cl$_2$) are positional isomers; propanal and acetone (molecular formula C$_3$H$_6$O) are functional group isomers; cyclohexane and methylcyclopentane (molecular formula C$_6$H$_{12}$) are skeletal isomers. In all three cases, the *sequence* of atoms is different and each pair of compounds represents a set of constitutional isomers.

## 2.2 STEREOISOMERS

More subtle differences are found in **stereoisomers** which are compounds that have the same sequence of atoms, but different three-dimensional arrangements of those atoms. In other words, the atoms' spatial orientation differs, not the order in which they are connected to one another. **Stereochemistry** is the study of the three-dimensional shapes of

molecules. There are three types of stereoisomers: conformational, geometric and optical isomers. (Geometric and optical isomers are also considered configurational isomers.)

## Conformational Isomers

**Conformational isomers** arise by rotation around a carbon-carbon single
bond and are interchangeable. Rotation results in an alteration of the spatial positions of the atoms. Since only a small amount of energy is required
for rotations around the carbon-carbon single bond, these are regarded as
essentially **free rotations.**

As a specific example, consider an ethane molecule held in such a way
that one can look directly down the carbon-carbon bond. This is illustrated in Figure 2.1a with **Newman projections.** In these drawings, the carbon atom nearest to the observer is represented as a dot (where the three
lines converge), and the second carbon atom is represented as the large
circle. Three hydrogens emanate from each carbon. Figure 2.1b represents
the same molecule with **perspective formulas.**

Rotation around the carbon-carbon bond results in an infinite number
of spatial orientations of the hydrogen atoms. The two pictured in Figure
2.1 are called the **eclipsed** form and the **staggered** form. Since rotation of

a) Newman projections

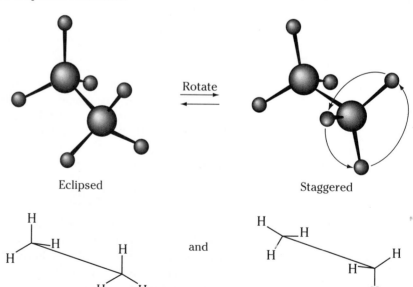

0°, 120°, 240°
Eclipsed

60°, 180°, 300°
Staggered

b) Perspective formulas

Eclipsed          and          Staggered

**Figure 2.1**    The ethane molecule

this kind requires only a small amount of energy, interconversion of these forms is rapid and continuous. Some conformational forms have lower potential energy than others. Of all the possible arrangements of the ethane molecule, the staggered form is the least crowded and requires the lowest potential energy. On the other hand, in the more crowded eclipsed form, the hydrogen atoms and their bonding electrons are closer together. Since electron pairs tend to repel each other, the eclipsed form requires the highest potential energy. Therefore, at any one time, most of the molecules in a sample of ethane will probably be in the staggered form.

Construct models of both the eclipsed form and the staggered form of ethane and satisfy yourself that they are intercovertible by rotation. Make a model of *n*-butane and rotate around the bond between C$_2$ and C$_3$ so that the methyl groups of C-1 and C-4 are *anti* to each other (i.e., 180° apart). Take note of the sawtooth or zigzag arrangement of carbon atoms that accompanies this arrangement. This is a low-potential-energy conformation. Rotate slowly around the bond between C-2 and C-3. The conformations that result from this rotation are called **anti, eclipsed, gauche** and **fully eclipsed** (Figure 2.2). Considering the anti arrangement as the least crowded, and therefore as the lowest potential energy conformation, rank the others in terms of potential energy, lowest to highest.

Cyclohexane is not a planar molecule (why not?) and can exist in two conformational forms, the more stable **chair** form and the less stable **boat** form (Figure 2.3). In the chair structure, the hydrogens are arranged in a

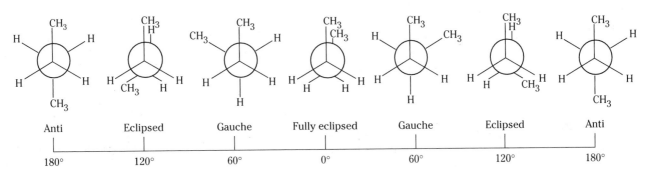

| Anti | Eclipsed | Gauche | Fully eclipsed | Gauche | Eclipsed | Anti |
|------|----------|--------|----------------|--------|----------|------|
| 180° | 120° | 60° | 0° | 60° | 120° | 180° |

**Figure 2.2**    Conformations of *n*-butane

H$_e$ = equatorial hydrogen
H$_a$ = axial hydrogen

Newman projection    Newman projection

chair form    boat form

**Figure 2.3**    The chair and boat conformations of cyclohexane

*staggered* conformation, while in the boat form, some hydrogens are in the higher-energy *eclipsed* conformation. The chair form does not have the eclipsed interactions and is consequently more stable than the boat form. A closer examination of the chair form reveals that six of the twelve hydrogen atoms of the molecule point outward from the ring. These six are called **equatorial** hydrogens. The other six hydrogens are arranged so that they alternate directly up and directly down from the ring. These are called **axial** hydrogens. Make molecular models of the boat and chair forms and compare them with the structures Figure 2.3.

Examine the boat conformation of cyclohexane and identify the eclipsed arrangements (there are two of them). Identify the axial and equatorial hydrogens in the chair form. By carefully rotating the carbon-carbon bonds of the chair form, you can convert it into the boat form. Continue rotating until you have another chair form. This second chair form is the mirror image of the one you started with (Figure 2.4). All of the hydrogens have changed places; those that were equatorial are now axial, and those that were axial are now equatorial. Replace one of the equatorial hydrogens with a colored halogen (G) and again rotate from one chair form to the other. The colored halogen will now be in an axial position.

## Configurational Isomers

Stereoisomers that cannot be interchanged by rotating a carbon-carbon bond are referred to as **configurational isomers.** There are two types: **geometric** and **optical** isomers.

*Geometric Isomers*   Free rotation is not possible with the cyclic compounds and with the alkenes, and this rigidity gives rise to stereoisomers called **geometric isomers** (also called *cis* and *trans* isomers). This is best illustrated with the example of 1,2-dibromocyclopentane. If, for this discussion, we consider the five-membered cyclopentane rings as flat, the bromine atoms may be arranged spatially in two ways: both on the same side of the ring (*cis*), or on opposite sides of the ring (*trans*).

These isomers are not conformational isomers since they cannot be interchanged by rotation. They are also **non-superimposable** (i.e., one does not spatially correspond exactly with the other); therefore, they represent two distinct compounds, each with its own properties (Figure 2.5).

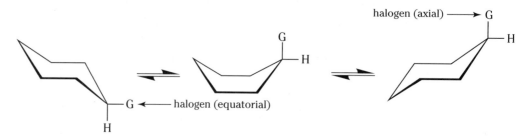

**Figure 2.4**   Interconversion of the chair forms of cyclohexane

*cis*-1, 2-dibromocyclopentane          *trans*-1, 2-dibromocyclopentane

**Figure 2.5**   *Cis*- and *trans*-1,2-dibromocyclopentane

Construct models of the two geometric isomers to satisfy yourself that they are not superimposable and not interchangeable by rotation.

The situation is similar with suitably substituted alkenes, since they also display restricted bond rotation. With alkenes, geometric isomerism can occur only if the two groups attached to each double-bonded carbon are different from each other. Thus, compounds $(AB)C = C(AB)$ and $(AB)C = C(DE)$ can show geometric isomerism since A is different from B, and D is different from E; however, compound $(AB)C = C(DD)$ cannot exist as a *cis/trans* pair, since one of the double-bonded carbons is attached to two D atoms.

Construct models of *cis*- and *trans*-1,2-dibromoethene, $BrCH=CHBr$, using two springs to form the double bond. Take note of the flat geometry and the $120°$ bond angles characteristic of the $sp^2$ hybridized carbons. Satisfy yourself that the two isomers are not superimposable. Now make a model of 1,1-dibromoethene, $Br_2C=CH_2$. Why isn't geometric isomerism possible here?

*Optical Isomerism*   A pure solution of an **optical isomer** can rotate plane-polarized light in either a clockwise (dextrorotary, +) or a counterclockwise (levorotary, −) direction. Structures of optical isomers are distinguished by the fact that they contain no plane, point, or axis of symmetry. Objects that are optically active possess a "handedness" and are said to be **chiral.** Usually chirality in a molecule is associated with one or more carbon atoms that are centers of **asymmetry.** These carbon atoms are called **chiral carbons** or **stereocenters** and they are recognized by the fact that they are bonded to **four different groups or atoms.** For example, compound C(XYZW) contains a chiral carbon and can exist as two stereoisomers that are mirror images of each other (mirror images have the same spatial relationship as a left and right hand).

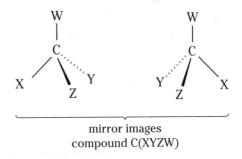

mirror images
compound C(XYZW)

In the simplest cases, the total number of optical isomers can be calculated from the $2^n$ rule, where $n$ is the number of chiral atoms. Thus, if a molecule contains three chiral carbons, it can exist as four pairs of mirror images, for a maximum of $2^3 = 8$ stereoisomers. Mirror images are called **enantiomers.** Molecules that have a plane of symmetry are not optically active and are **achiral.** Achiral molecules are superimposable with their mirror images (i.e., they are identical).

Lactic acid (2-hydroxypropanoic acid) is an example of an optically active molecule with one chiral carbon (Figure 2.6), and it exists as one pair of enantiomers.

Make models of lactic acid that are mirror images of one another and try to superimpose one on the other. As with other chiral objects, lactic acid does not possess a plane of symmetry.

Now construct a model for 2-propanol, $CH_3CH(OH)CH_3$. Is there a plane of symmetry in the molecule? Is the compound optically active?

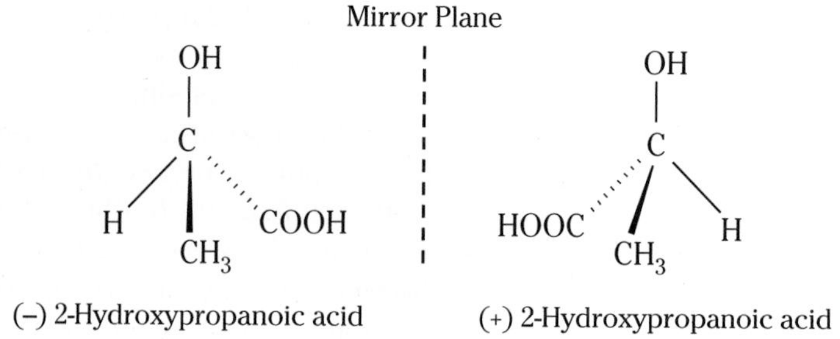

Mirror Plane

(−) 2-Hydroxypropanoic acid          (+) 2-Hydroxypropanoic acid

**Figure 2.6**   Enantiomers of lactic acid (2-hydroxypropanoic acid)

Organic molecules can contain more than one chiral carbon. Remember that a chiral carbon must have four different attachments (a methyl group is different from an ethyl group, for instance; and an isopropyl group is different from an *n*-propyl group).

In the case of 2,3-dibromopentane, there are two chiral carbons (marked with asterisks) and two pairs of enantiomers.

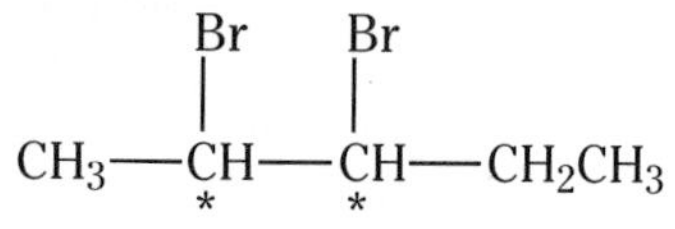

Construct the two pairs of mirror images. Take either of the models from one pair of enantiomers and compare it with either of the models of the second pair of enantiomers. Are these two models mirror images? Save these models for parts 2.3 and 2.4. Stereoisomers that are *not* mirror images are called **diastereomers** (*cis* and *trans* isomers are diastereomers). Enantiomers (mirror images) have identical physical and chemical properties except for the way in which they rotate polarized light (one is [+] and the other is [−]) and the way in which they react chemically with other chiral molecules. The latter is extremely important in biological reactions. On the other hand, diastereomers have different physical and chemical properties.

## 2.3   FISCHER PROJECTION FORMULAS FOR CHIRAL MOLECULES

Three-dimensional structures are difficult to draw and difficult to visualize. For this reason, flat, two-dimensional structures are often used to represent the three-dimensional structures. In flat structures, the molecule is represented by vertical and horizontal lines with the chiral carbon at the intersection. In Figure 2.7, the enantiomers of lactic acid are represented in this way. The horizontal lines represent the bonds of the atoms or groups that would be projecting *outward* in the three-dimensional structure. To test whether the structures are or are not superimposable, either structure can be rotated 180° (i.e., we can invert them) but they *cannot* be taken out of the plane of the paper (i.e., we cannot flip them over). These formulas are called **Fischer projection formulas.**

The four stereoisomers of 2,3-dibromopentane (*n* = 2) exist as pairs of enantiomers. The Fischer projections are shown in Figure 2.8.

Examination of the four structures shows that I and II are mirror images, as are III and IV. However, a comparison of structures I or II with either III or IV shows that they are not mirror images. As mentioned, stereoisomers that are not mirror images are called **diastereomers.** Refer to your models of 2,3-dibromopentane and verify these statements that provide the relationship between the different structures.

a. Three-dimensional drawings.

$$
\begin{array}{c}
COOH \\
HO - \!\!\!\! - H \\
CH_3
\end{array}
\qquad
\begin{array}{c}
COOH \\
H - \!\!\!\! - OH \\
CH_3
\end{array}
$$

b. Flat representations of the three-dimensional molecules (Fischer projection formulas).

$$
\begin{array}{c}
COOH \\
HO - \!\!\!\! - H \\
CH_3
\end{array}
\qquad
\begin{array}{c}
COOH \\
H - \!\!\!\! - OH \\
CH_3
\end{array}
$$

(+)-lactic acid          (−)-lactic acid

**Figure 2.7**  Enantiomers of lactic acid

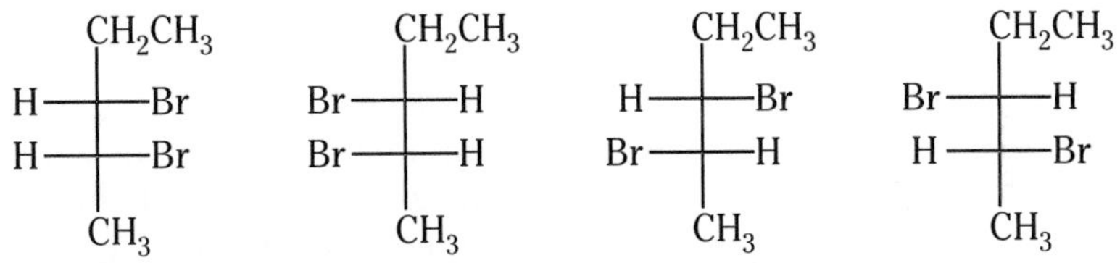

$$
\begin{array}{cccc}
CH_2CH_3 & CH_2CH_3 & CH_2CH_3 & CH_2CH_3 \\
H-\!\!\!-Br & Br-\!\!\!-H & H-\!\!\!-Br & Br-\!\!\!-H \\
H-\!\!\!-Br & Br-\!\!\!-H & Br-\!\!\!-H & H-\!\!\!-Br \\
CH_3 & CH_3 & CH_3 & CH_3 \\
(I) & (II) & (III) & (IV)
\end{array}
$$

enantiomers                    enantiomers

**Figure 2.8**  The four stereoisomers of 2,3-dibromopentane

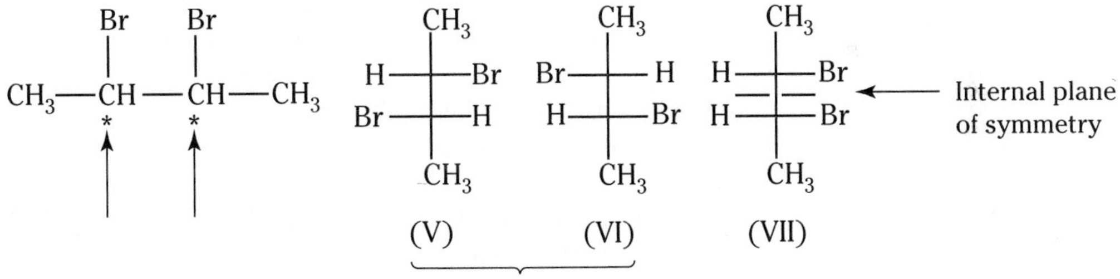

$$
\begin{array}{c}
Br \quad Br \\
CH_3 - CH - CH - CH_3 \\
\ast \qquad \ast
\end{array}
\qquad
\begin{array}{c}
CH_3 \\
H-\!\!\!-Br \\
Br-\!\!\!-H \\
CH_3 \\
(V)
\end{array}
\begin{array}{c}
CH_3 \\
Br-\!\!\!-H \\
H-\!\!\!-Br \\
CH_3 \\
(VI)
\end{array}
\begin{array}{c}
CH_3 \\
H-\!\!\!-Br \\
H-\!\!\!-Br \\
CH_3 \\
(VII)
\end{array}
\quad \longleftarrow
\begin{array}{l}
\text{Internal plane} \\
\text{of symmetry}
\end{array}
$$

chiral carbons          mirror images          meso compound

**Figure 2.9**  Stereoisomers of 2,3-dibromobutane

In some cases the $2^n$ rule does not work. Make models of 2,3-dibromobutane, $CH_3CH(Br)CH(Br)CH_3$. Check to see if any are superimposable. One of the models will contain an internal plane of symmetry, and thus it will be optically inactive. Stereoisomers such as this one, which contain chiral carbons but are optically inactive, are called **meso compounds.** Compare your models to the structures shown in Figure 2.9.

## 2.4  THE *R/S* DESIGNATION FOR CHIRAL CARBONS

There is one chiral carbon in 2-butanol, and therefore, this compound exists as a pair of mirror images. The name *2-butanol* alone does not distinguish one isomer from the other. A set of rules to distinguish these isomers was introduced by the chemists R. S. Cahn, C. K. Ingold, and V. Prelog. This system was subsequently accepted as part of the IUPAC nomenclature rules. In this system, a chiral carbon is identified as having an *R* (Latin *rectus,* mean-

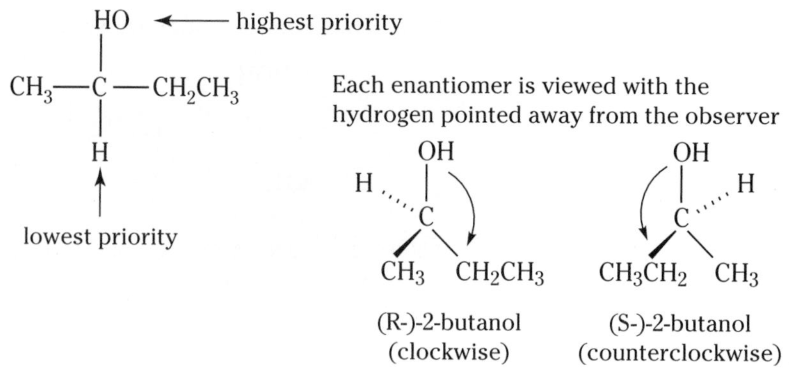

HO ◀——— highest priority

$$CH_3—\overset{\displaystyle |}{\underset{\displaystyle |}{C}}—CH_2CH_3$$

Each enantiomer is viewed with the
hydrogen pointed away from the observer

lowest priority

(R-)-2-butanol
(clockwise)

(S-)-2-butanol
(counterclockwise)

**Figure 2.10**   Assignment of the $R/S$ nomenclature in the enantiomers of 2-butanol

ing right) or an $S$ (Latin *sinister,* meaning left) configuration. The designation is determined by assigning priorities to the four groups attached to the chiral carbon. The rules for assigning priorities are based on the atomic number of the atoms directly bonded to the chiral carbon. The higher the atomic number, the higher the priority. Thus, hydrogen (atomic number 1) has the lowest priority, while oxygen (atomic number 8) has a higher priority than nitrogen (atomic number 7) or carbon (atomic number 6). To apply the rules, a model of the molecule in question is held so that the attachment with the *lowest* priority (often hydrogen) is pointed directly away from the observer. The other three attachments are viewed in terms of their priorities. If the arrangement from highest to lowest priority is in a *clockwise* direction, the chiral carbon is assigned the $R$ designation. If the arrangement is in a *counterclockwise* direction, the chiral carbon is assigned the $S$ configuration. The assignment for 2-butanol is shown in Figure 2.10.

The ethyl group has a higher priority than the methyl group because it has two carbon atoms. Thus, the order of priority is — OH > — $CH_2CH_3$ > — $CH_3$ > — H. This sequence is clockwise in the $R$ isomer, and counterclockwise in the $S$ isomer.

Construct the model for 2-butanol and assign it an $R$ or $S$ designation. Again examine your models of 2,3-dibromopentane and assign $R$ or $S$ designations to both chiral carbons for every stereoisomer. How many stereoisomers are possible for this compound? Assign the $R/S$ designations to each of them.

## 2.5   THE *E/Z* DESIGNATION OF ALKENES

The terms *cis* and *trans* are used to identify the stereochemistry of particular alkenes. However, these terms are often ambiguous. The Cahn, Ingold, and Prelog sequence rules are used to unambiguously identify geometric isomers. The four attachments to the double bond are assigned priorities as is done when determining $R/S$ designations. If the two highest priority groups are on the same side of the double bond, the letter **Z** (German *zusammen,* meaning together) is used. If they are on opposite sides of the double bond, the letter **E** (German *entgegen,* meaning opposite) is used. As an example, consider the geometric isomers of 3-bromo-4-methyl-3-heptene (Figure 2.11).

When all four groups are different, the *cis/trans* designations are ambiguous, but the $E/Z$ designations are clear. The highest priorities are the Br atom and the propyl (— $CH_2CH_2CH_3$) group. In structure A they are on opposite sides, and that structure is therefore ($E$). For isomer B, the high priority groups are on the same side and that isomer is labelled ($Z$).

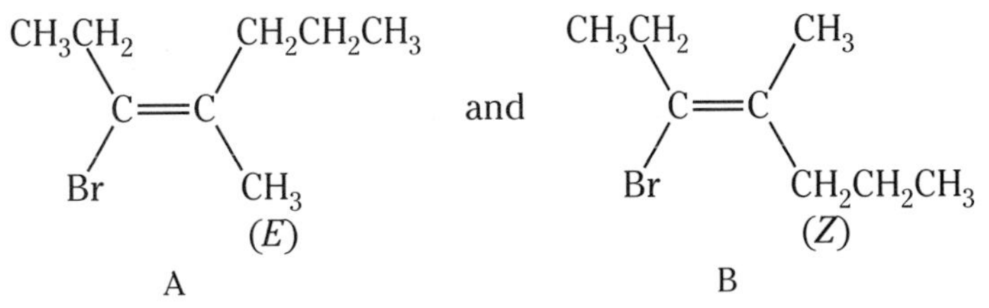

**Figure 2.11** Isomers of 3-bromo-4-methyl-3-heptene: $E/Z$ designation

The correct nomenclature for A is $(E)$-3-bromo-4-methyl-3-heptene and for B is $(Z)$-3-bromo-4-methyl-3-heptene.

Make a model of 2-bromo-2-butene and assign it the $E$ or $Z$ designation. Make a model of 2-bromo-3-methyl-2-butene. Does the $E/Z$ designation apply to it? Explain.

**Definitions**

**Configurational isomers:** Stereoisomers that are not interchangeable by rotation. Examples are geometric and optical isomers.

**Conformational isomers:** Stereoisomers that arise and are interchangeable by rotation around a carbon-carbon single bond. The staggered and eclipsed forms of ethane are examples.

**Diastereomers:** Stereoisomers that are not mirror images and as a result have different physical properties. *Cis-* and *trans-* isomers are examples.

**Enantiomers:** Non-superimposable mirror images. The (+) and (–) isomers of lactic acids are examples.

**Geometric isomers:** Stereoisomers in which groups are on the same side or on opposite sides of a double bond or ring.

**Optical isomers:** Stereoisomers that interact with plane-polarized light and rotate the plane of polarization.

**Racemic mixture:** An equimolar mixture of enantiomers that has a net optical rotation of zero.

**Stereoisomers:** Molecules with the same sequential arrangement of atoms but with different molecular shape.

**Superimposable structures:** Structures that are spatially identical.

**Suggested References**

Carey, F. A., *Organic Chemistry*. 3rd ed., 26–29, 94–103, 172–76, 265–308. New York: McGraw-Hill, 1996.

McMurry, J., *Organic Chemistry*. 4th ed., 59–62, 75–80, 95–98, 106–16, 187–92, 294–335. Pacific Grove, CA: Brooks/Cole, 1996.

Mislow, K., *Introduction to Stereochemistry*. Menlo Park, CA: W. A. Benjamin, 1966.

Morrison, R. T., and R. N. Boyd, *Organic Chemistry*. 6th ed., 125–64. Englewood Cliffs, NJ: Prentice-Hall, 1992.

Solomons, T. W. G., *Organic Chemistry*. 6th ed., 142–220. New York: John Wiley, 1996.

Vollhardt, K. P. C., and N. E. Schore, *Organic Chemistry*. 2nd ed., 136–174. New York: W. H. Freeman, 1994.

Wade, L. G., *Organic Chemistry*. 3rd ed., 97–122, 174–227, 305–08. Englewood Cliffs, NJ: Prentice-Hall, 1995.

**Questions**

*Note:* Models will help you answer the following questions.

1. Give the structural formulas and names of all possible (exclude stereoisomerism)
   a. pentanes $C_5H_{12}$ (three structures)
   b. hexanes $C_6H_{14}$ (five structures)
   c. chlorobutanes $C_4H_9Cl$ (four structures)
   d. difluoroethanes $C_2H_4F_2$ (two structures)
   e. tribromopropanes $C_3H_5Br_3$ (five structures)
   f. trichloro derivatives of isobutane, $C_4H_7Cl_3$ (five structures).

2. Identify the relationship between the structures in each part of question 1.

3. Which of the following compounds contain chiral carbons? Using the $2^n$ rule, calculate the maximum number of optical isomers possible for these compounds:

   a. $CH_3CH_2CH(OH)CH_3$

   b. $CH_3CH(Br)CH_3$

   c. 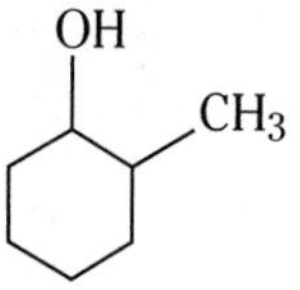  OH
   
   d. $HOCH_2CH(OH)CH(OH)CH(OH)CHO$

   e. 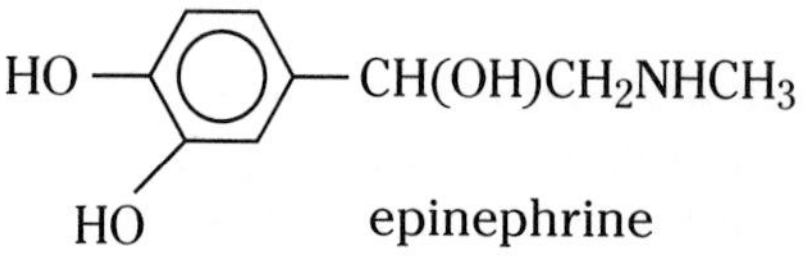 HO — ◯ — $CH(OH)CH_2NHCH_3$

   HO        epinephrine

   f. 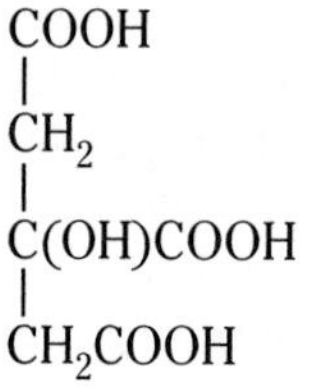 $COOH$
      $|$
      $CH_2$
      $|$
      $C(OH)COOH$
      $|$
      $CH_2COOH$

      citric acid

   g. $CH_3CH(Br)CH{=}CH_2$

   h. 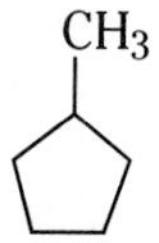  $CH_3$

   i. 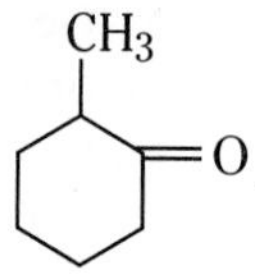  $CH_3$

4. Assign the correct $R$ or $S$ designation to each of the following structures:

   a. 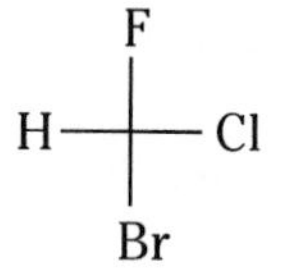
   
   b. 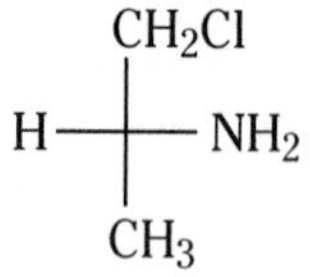
   
   c. 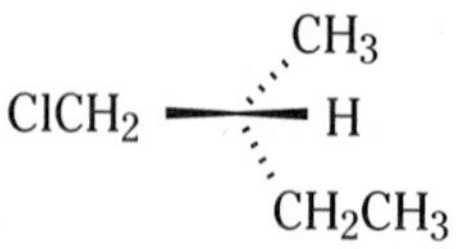
   
   d. 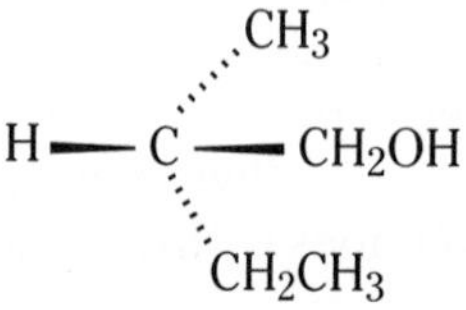
   
   e. 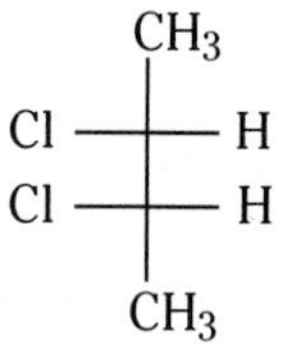
   
   f. 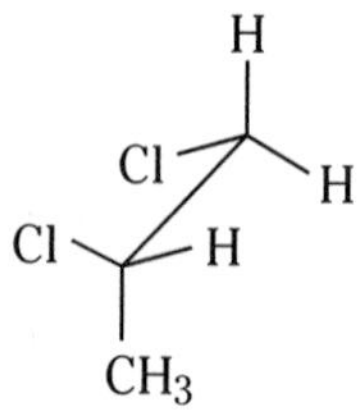
   
   g. 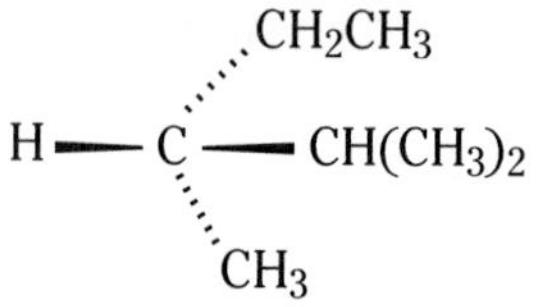
   
   h.

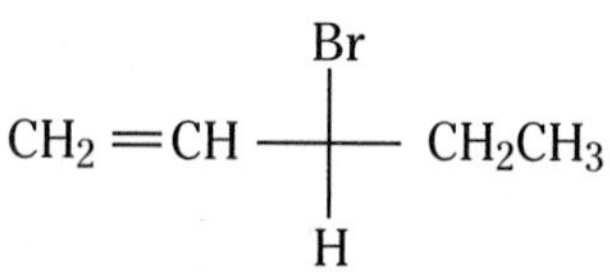

5. Match the stereoisomers that are mirror images and identify any *meso* compounds.

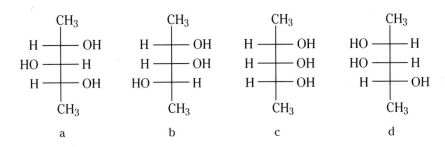

       a                 b                  c                  d

6. What is the relationship between the following structures?

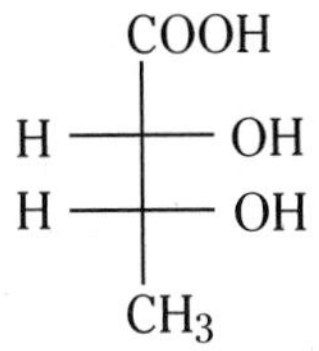

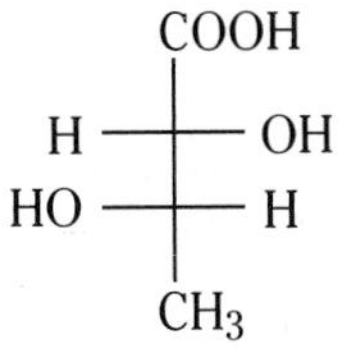

```
      COOH                    COOH
  H ──┼── OH            H ──┼── OH
  H ──┼── OH           HO ──┼── H
       CH3                   CH3
```

7. For each case, decide whether the following pairs are skeletal, positional, or functional group isomers:

a. 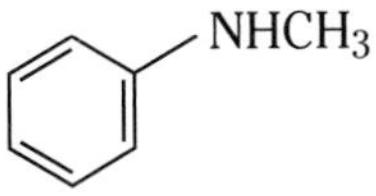 ⬡—NHCH₃    and    ⬡—NH₂ , CH₃

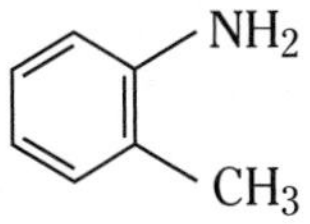

b. 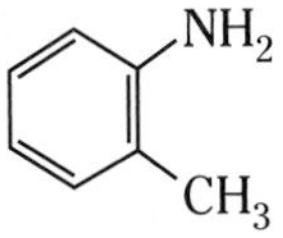 ⬡—NH₂ , CH₃    and    CH₃—⬡—NH₂

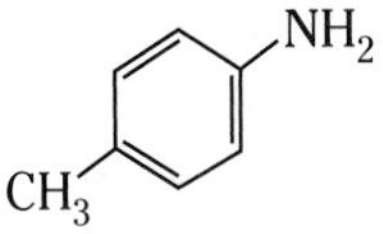

c. $CH_3CH_2CH_2OH$    and    $CH_3OCH_2CH_3$

d. 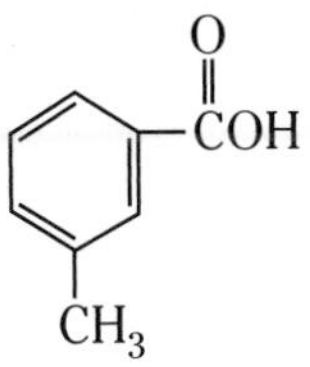 ⬡(—COH, O)(CH₃)    and    ⬡(—CH, O)(CH₂OH)

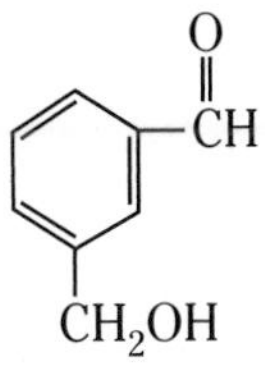

e. 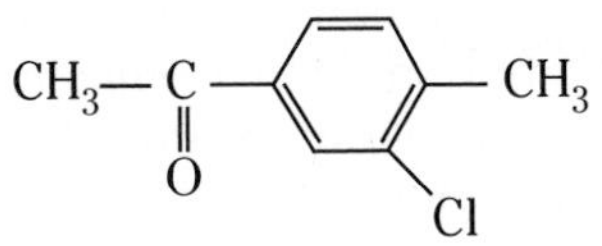 $CH_3$—C(=O)—⬡(—CH₃)(Cl)    and    $CH_3$—C(=O)—⬡(—CH₃)(Cl)

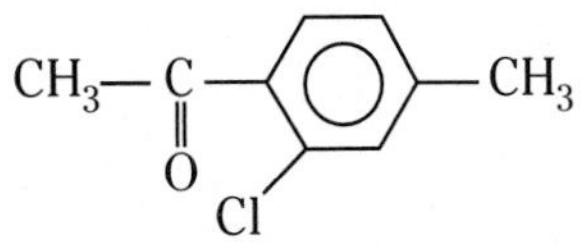

f. 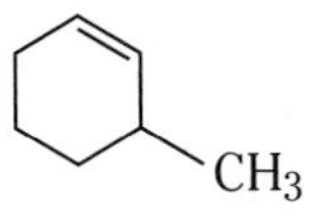 ⬡—CH₃    and    ⬡—CH₃

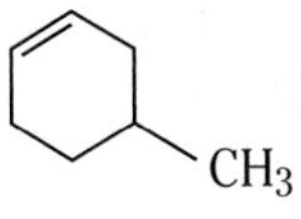

g.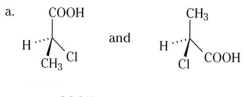

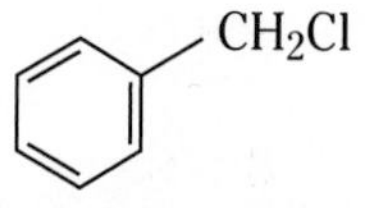

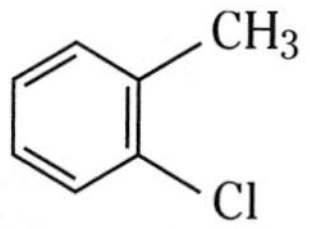

8. For each case, decide whether the following pairs are enantiomers, diastereomers, or identical molecules:

a.

$$COOH \qquad\qquad CH_3$$

H⸴⸴⸴⧹ (CH₃, Cl)   and   H⸴⸴⸴⧹ (Cl, COOH)

b.

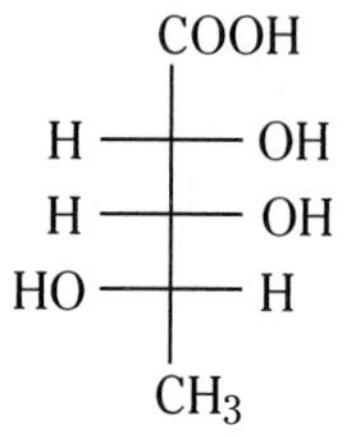

| COOH | | |
|---|---|---|
| H — | — OH | |
| H — | — OH | |
| HO — | — H | |
| | CH₃ | |

and

| CH₃ | | |
|---|---|---|
| HO — | — H | |
| HO — | — H | |
| H — | — OH | |
| | COOH | |

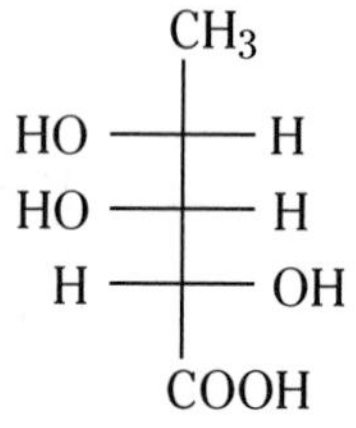

c.

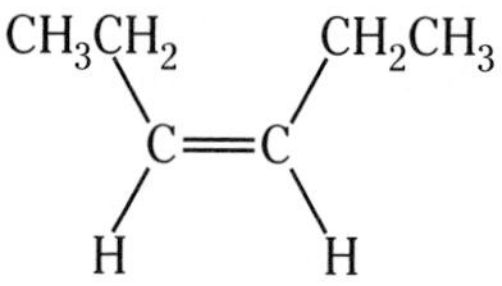

CH₃CH₂ , CH₂CH₃

C=C

H , H

and

CH₃CH₂ , H

C=C

H , CH₂CH₃

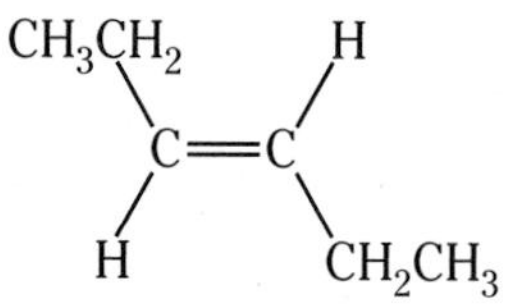

d.

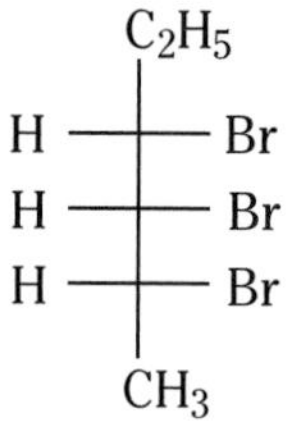

| C₂H₅ | | |
|---|---|---|
| H — | — Br | |
| H — | — Br | |
| H — | — Br | |
| | CH₃ | |

and

| C₂H₅ | | |
|---|---|---|
| Br — | — H | |
| Br — | — H | |
| Br — | — H | |
| | CH₃ | |

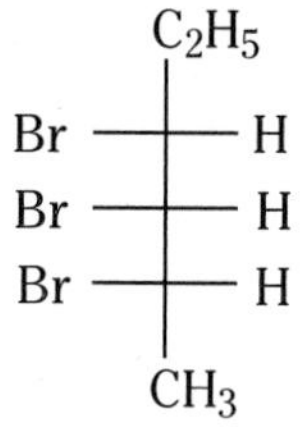

e.

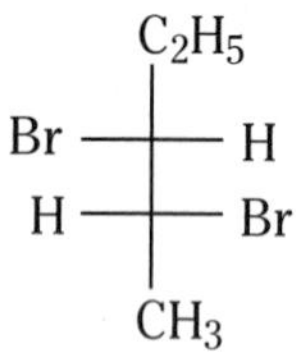

| C₂H₅ | | |
|---|---|---|
| Br — | — H | |
| H — | — Br | |
| | CH₃ | |

and

| C₂H₅ | | |
|---|---|---|
| H — | — Br | |
| Br — | — H | |
| | CH₃ | |

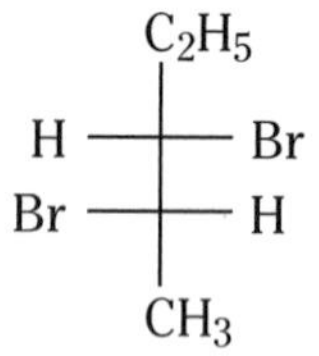

9. Draw the structure of 1,2-dimethylcyclohexane in the chair form
   a. with both methyl groups in the equatorial positions.
   b. with both methyl groups in the axial positions.
   c. with one methyl group in the axial and one in the equatorial position. Identify each as *cis-* or *trans-*.

10. How many configurational isomers are possible for each of the following compounds?
    a. 2,4-hexadiene
    b. 4-hexen-2-ol
    c. 1,4-pentadien-3-ol

# chapter

# 3

# Physical Properties

## 3.1  MELTING POINT

The **melting point** (MP) of a compound is defined as the temperature at which the solid and liquid phases are in equilibrium with each other. Most organic solids melt between 40°C and 300°C, a range that can be conveniently measured. Thus, the melting point is a useful physical constant that is routinely used to assist in the identification of a substance. In practice, the melting point of a compound is actually a narrow range of temperatures over which the compound melts. This range depends on the purity of the sample and on the technique employed to determine the melting point. Impurities tend to both lower and broaden the range of the melting point. Compounds of acceptable purity will melt over a very narrow temperature range (generally within 0.5°C to 2.0°C).

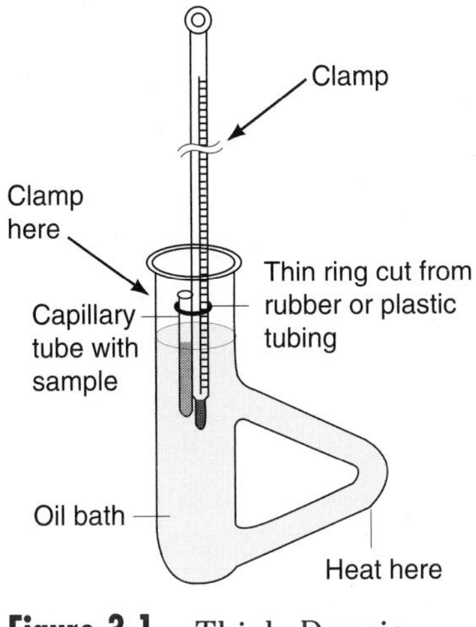

Melting points were traditionally measured using the Thiele-Dennis tube (Figure 3.1). The tube is partially filled with paraffin or mineral oil (if temperatures are not to exceed 150°C to 200°C), or with silicone fluid (if temperatures up to 300°C are required). It is essential that the bath liquid circulates so that the temperature remains uniform. A few crystals of the sample are placed in a capillary sealed at one end, which is tied to a thermometer with a thin ring cut from rubber or plastic tubing, or with a rubber band. The thermometer is suspended in the oil bath as shown in Figure 3.1. Heat is then applied to the sidearm with a Bunsen burner, and the convection currents produced provide adequate circulation for a uniform temperature.

Many electrically heated melting-point devices are commercially available. There are generally two types:

1. the metal-block apparatus, and its variations (Figure 3.2). The dials on these machines are rheostats that control the voltage and thus the rate of heating of the sample. The sample is placed in a capillary and is viewed through an eyepiece equipped with a magnifying glass. This provides a clear view of the melting crystals and the mercury thread in the thermometer, and it enables the accurate measurement of the melting point of small samples
2. the hot-stage apparatus, such as the Fisher-Johns apparatus and the microscope with hot-stage apparatus (Figure 3.3). In this type of device, the sample is placed between two glass plates on an electrically heated metal block and is carefully observed through the magnifying lens or microscope during the melting process.

The rate of heating should be gradual in all cases of melting-point determination. If the temperature is increased too rapidly, it will overshoot

**Figure 3.1**  Thiele-Dennis tube for melting point determination

Clamp

Clamp here

Capillary tube with sample

Thin ring cut from rubber or plastic tubing

Oil bath

Heat here

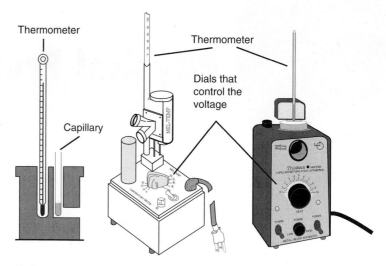

**Figure 3.2**  Examples of the metal-block apparatus

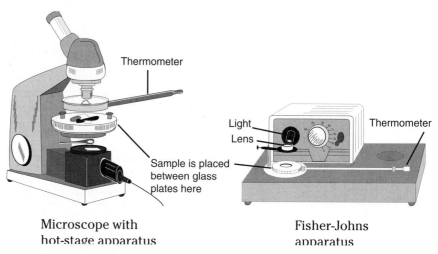

Microscope with
hot-stage apparatus

Fisher-Johns
apparatus

**Figure 3.3**  Examples of melting point apparatus

the correct melting point because the temperature within the capillary or between the plates will not have had the opportunity to attain thermal equilibrium. Rapid heating also tends to give a wider melting point range because the mercury rises very quickly. It takes time for the sample to absorb heat and melt (especially in the Thiele tube), and if the temperature increases too fast, the range will be broad.

There are many instances of two or more compounds having the same (or nearly the same) melting point; for example, benzamide and maleic acid both melt at about 130°C. However, a mixture of both compounds will melt at a temperature lower than 130°C and over a broad melting range. This is because each compound in the mixture acts as an impurity to the other.

In the following experiment, the melting points of pure maleic acid and pure benzamide will be determined, along with the melting points of several mixtures of the two. The melting point temperature will be plotted as a function of the sample's composition (Figure 3.4). This graph, the **melting-point diagram,** will illustrate the way in which "impurities" in a sample (with rare exceptions) will cause the mixture to melt over a broad range of temperatures that are lower than the melting temperatures of the pure compounds.

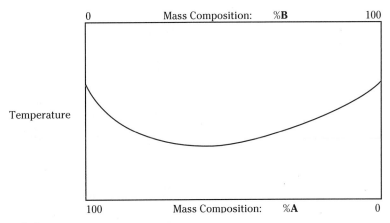

**Figure 3.4** Melting-point diagram for a mixture of solids *A* and *B*

This behavior of mixtures allows the identity of a compound to be firmly established by a procedure called **mixed melting.** For example, if a sample is suspected to be benzamide and is mixed with pure benzamide, the melting point of the mixture will be identical to that of benzamide *only* if the sample really was benzamide. If the sample is anything other than benzamide, the melting point of the mixture will be lower and broader.

Some compounds chemically decompose when melted. Some others are **polymorphic;** that is, they exist in different crystalline forms that have different melting points. To avoid possible errors due to decomposition or polymorphism, samples should be discarded and not re-used after the melting point is determined. To avoid the problem of decomposition, it is often necessary to preheat the bath to a temperature just below the (suspected) melting point, introduce the sample, and quickly obtain the melting point before any appreciable decomposition can occur. Decomposition is evidenced by a significant color change of the solid, or sometimes by a release of gas during melting.

**Suggested References**

Brown, T. L., H. E. Le May, Jr., and B. E. Bursten, *Chemistry: The Central Science.* 5th ed., 383–84, 462–66. Englewood Cliffs, NJ: Prentice-Hall, 1990.

Fieser, L. F., and M. Fieser, "Diels Alder Solvents." In *Reagents for Organic Synthesis,* 241–43. New York: John Wiley, 1967.

"Melting Point." In *McGraw-Hill Encyclopedia of Science and Technology,* 7th ed., vol. 10, 616–17. New York: McGraw-Hill, 1992.

Umland, J. B., *General Chemistry.* 503–06. Minneapolis: West, 1993.

**Experimental**

### A. Using a Thiele-Dennis tube

1. Place a small amount of the sample whose melting point is to be determined on a piece of weighing paper. Insert some of the sample into the capillary tube by tapping the open end of the tube into the sample. Invert the tube and tap it on the bench top until the sample falls to the bottom of the tube (approximately 1/8 to 1/4 inch high). Avoid using too large a sample because a large sample requires more time to melt and often results in a higher, broader melting point range.

2. Attach the capillary tube to the end of the thermometer with a rubber band and suspend the thermometer in the oil bath of the Dennis-Thiele tube (see Figure 3.1). Keep the rubber band out of the oil.

3. Heat the oil and allow its temperature to rise rather rapidly until it is about 20°C below the suspected melting point of the sample. Then

decrease the rate of heating. The temperature of the oil should rise at a rate of approximately 1–2°C/min as the melting point of the sample is approached.

4. Record the temperature at which liquid first appears, and then record the temperature at which the sample is entirely converted to a liquid. These two temperatures are the melting point range.
5. Determine the melting point of the following samples:
   a. benzamide
   b. maleic acid
   c. mixture of maleic acid: benzamide (75:25)
   d. mixture of maleic acid: benzamide (50:50)
   e. mixture of maleic acid: benzamide (25:75)
   f. three of the available unknowns.
6. Plot the melting points of the samples of parts (a) through (e) on the graph paper on page 23.

## B. Using a metal-block apparatus
Use a capillary as in part A and carefully determine the melting point range as described above (step 4).

## C. Using a hot-stage apparatus
Place a few crystals between two melting point plates and carefully determine the melting point range as described in part A, step 4.

**Questions**

1. When determining the melting point of a substance, should one use a large or a small sample in the capillary? Should the sample be firmly or loosely packed? Explain.
2. Does atmospheric pressure affect the melting point of a compound?
3. What effect would rapid heating of a sample have on the observed melting point?
4. A student has an unknown that could be either benzamide or maleic acid. How could he/she decide which one it is, using only melting point determinations?
5. A student determined the melting point of an unknown solid to be 115°C. He allowed the sample to cool and solidify and then repeated the melting point determination using the same sample. He found the melting point to be 125°C. Suggest an explanation.
6. Molecular weights can be determined by freezing point depression measurements. A 6.0 g sample of a compound whose empirical formula is $(CH_2)_x$ is dissolved in 200 g of benzene. The freezing point of the solution is 1.83°C below that of pure benzene. The molal freezing point constant for benzene is 5.12°C/m. Determine
   a. the molecular weight of the compound
   b. the correct molecular formula of the compound.
   Suggest a possible structure using your answer from (b) above. (*Hint:* Use the freezing point depression formula from any general chemistry textbook—see suggested references).

# GRAPH PAPER MELTING POINT

Mass Composition: Maleic Acid (%)

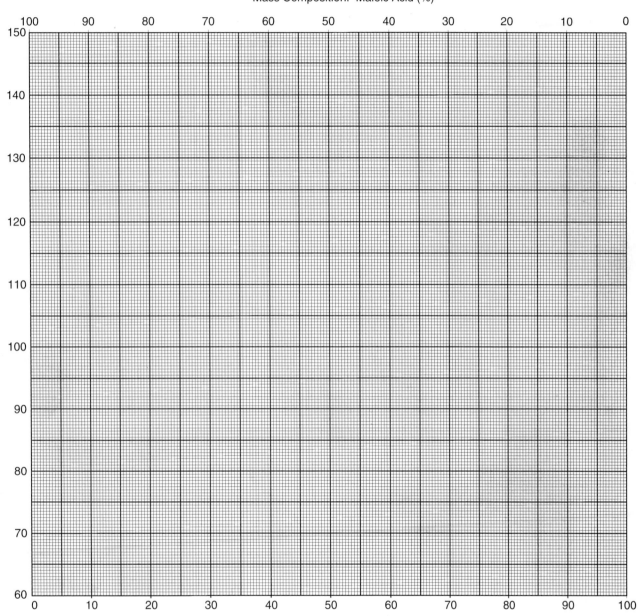

Mass Composition: Benzamide (%)

## 3.2    BOILING POINT

The **boiling point** (BP) of a liquid is defined as the temperature at which the vapor pressure of the liquid equals the external (normally the atmospheric) pressure. The boiling point is a physical property that is characteristic of the compound and is a function of the external pressure. Thus in a vacuum, the boiling point drops dramatically—a phenomenon that is exploited in vacuum distillations which are performed when the compound to be distilled is unstable at its normal atmospheric boiling point.

In this experiment, the boiling point of an unknown liquid will be determined. A water bath can be used if the boiling point is less than 95°C, otherwise an oil bath is necessary. This experiment may be combined with Experiment 3.3 (refractive index) to identify an unknown liquid.

**Suggested References**  Barrow, G. M., *Physical Chemistry.* 413–16. New York: McGraw-Hill, 1988.
"Boiling Point." In *McGraw-Hill Encyclopedia of Science and Technology,* 7th ed., vol. 2, 702–03. New York: McGraw-Hill, 1992.
Brown, T. L., H. E. Le May, Jr., and B. E. Bursten, *Chemistry: The Central Science.* 381–82, 461–62. Englewood Cliffs, NJ: Prentice-Hall, 1990.

---

**Experimental**

1. Place 1 mL of the unknown liquid in a regular-size test tube.
2. Carefully cut off about 1 inch from the sealed end of a melting-point capillary tube and insert it in the test tube, closed end up (Figure 3.5a).
3. Attach the test tube to the end of a thermometer with a rubber ring or a rubber band.
4. Slowly heat the assembly in a water (or oil) bath (Figure 3.5b) until a constant, rapid stream of bubbles starts coming from the bottom of the capillary tube.
5. Remove the heat and record the temperature at which the *last* bubble comes out and the liquid just starts to rise into the capillary tube. This is the boiling point of the liquid.

**Questions**

1. What will be the effect on the normal boiling point of a pure liquid if:
   a. a nonvolatile impurity is added?
   b. a volatile impurity is added?
2. When determining the boiling point, why should the water (or oil) bath be heated slowly?

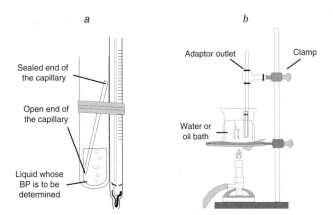

**Figure 3.5**    Microboiling point set-up. *a,* The capillary is placed inside the test tube containing the liquid. A rubber band is wrapped around the thermometer and the test tube. *b,* Set up (*a*) is placed in a beaker of boiling water

## 3.3   REFRACTIVE INDEX

The **refractive index** is an intrinsic property of the compound—as intrinsic as the boiling point and melting point. It is defined as the ratio of the speed of light through a vacuum to the speed of light through the liquid compound. Thus, no refractive index is expected to have a value under 1.000. It can be readily determined to a few parts in 10,000; therefore, the refractive index provides a very accurate physical constant for identification. The value is usually reported as $n^{25}$, in which the superscript represents the temperature (at °C).

The speed of light through a medium is directly related to the structure of the molecule and, in particular, to the types of bonds or functional groups present. The lowest common refractive index is that of water (1.333), while those of most organic compounds lie in the range between 1.35 and 1.70 (Table 3.1).

**Table 3.1**   Refractive indices of common substances

| Compound | Refractive Index | Compound | Refractive Index |
|---|---|---|---|
| Acetone | $1.359^{25}$ | Methylene chloride | $1.424^{25}$ |
| Acetonitrile | $1.344^{25}$ | Tetrahydrofuran | $1.407^{25}$ |
| Acetic acid | $1.372^{25}$ | Dimethylformamide | $1.431^{25}$ |
| Chloroform | $1.445^{25}$ | Carbon tetrachloride | $1.460^{25}$ |
| Benzene | $1.500^{25}$ | Methyl ethyl ketone | $1.378^{25}$ |
| Ethyl acetate | $1.372^{25}$ | Dimethyl sulfoxide | $1.478^{25}$ |
| 2-methyl 1-propanol | $1.377^{25}$ | Diethyl ether | $1.351^{25}$ |
| Toluene | $1.496^{25}$ | Ethanol (anhydrous) | $1.361^{25}$ |

The design of most refractometers follows the general diagram shown in Figure 3.6. The hinged prisms are enclosed within a compartment that is maintained at a constant temperature and the liquid is placed between the prisms. Monochromatic light (usually the sodium D-line) passed through the sample can then be seen as a partially shaded circle (Figure 3.7). The hinged prisms can be rotated so that the circle's horizontal dividing line is placed in the center of the diagonal cross-hairs. The refractive index can then be read directly from the scale on the instrument.

One of the most popular refractometers is the Abbe-Spencer refractometer (Figure 3.8). The apparatus should always be kept clean, and special care should be taken to prevent scratching the prism. A good low-boiling solvent (such as ethanol or acetone), as well as soft tissues, are recommended for cleaning.

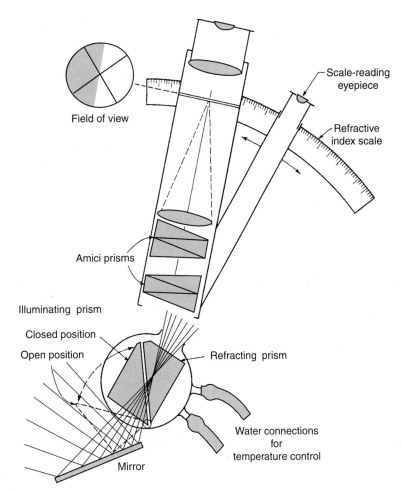

**Figure 3.6**   General diagram of a refractometer

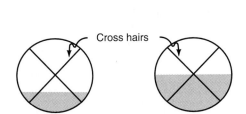

**Figure 3.7**   Adjustment of a refractometer.
*Left*, view when the refractometer is out of
adjustment. *Right*, view when adjusted

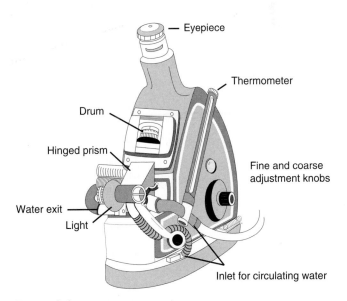

**Figure 3.8**   Abbe-Spencer refractometer

**Suggested References**   Barrow, G. M., *Physical Chemistry*. 668. New York: McGraw-Hill, 1988.

George, S. W., and J. A. Campbell, "Refractive Indices of Some Carbon Compounds as a Function of Temperature." *J. Chem. Educ.* 44 (1967): 393.

Hartkopf, A., R. R. Schroeder, and C. H. Meyers, "Quantitative Analysis of Xylene Mixtures by Refractometry." *J. Chem. Educ.* 51 (1974): 405.

Maley, L.E., "Refractometers." *J. Chem. Educ.* 45 (1968): A467.

"Refractometric Analysis." In *McGraw-Hill Encyclopedia of Science and Technology*, 7th ed., vol. 15, 253–54. New York: McGraw-Hill, 1992.

Shoemaker, D. P., and C. W. Garland, *Experiments in Physical Chemistry*. 445–49. New York: McGraw-Hill, 1967.

**Experimental**

1. Place a few drops of ethanol or acetone on the prism using a disposable pipette. *Avoid touching the prism with the pipette.* Wipe off the prism with a soft tissue.
2. Place a drop or two of the sample on the lower prism so that the entire width of the prism plate is covered. Bring down the upper prism into contact with the lower prism.
3. Turn on the lamp and manipulate the controls so that the light/dark interface bisects the cross-hairs visible in the eyepiece. Read and record the refractive index found on the scale built into the instrument.
4. Repeat this procedure twice.
5. Find the average of the results. This is the refractive index of the sample at the temperature at which the readings were taken. The refractive index varies with temperature: lower values are recorded at higher temperatures. As a rule of thumb, the refractive index changes by 0.0004 units per degree.
6. Identify an unknown that has been selected by your instructor from Table 3.2, after determining its BP (Experiment 3.2) and refractive index. Provide the IUPAC names of all compounds in Table 3.2.

**Table 3.2**   List of unknown liquids and their BP and refractive indices

| Compound | | | |
|---|---|---|---|
| **Common Name** | **IUPAC Name** | **BP (°C)** | **Refractive Index ($n^{25}$)** |
| Cyclopentane | | 49 | 1.406 |
| Acetone | | 58 | 1.359 |
| Methyl alcohol | | 65 | 1.323 |
| Ethyl acetate | | 77 | 1.372 |
| Cyclohexane | | 81 | 1.344 |
| Acetonitrile | | 82 | 1.426 |
| Isopropyl alcohol | | 82 | 1.377 |
| *t*-Butyl alcohol | | 82 | 1.387 |
| *sec*-Butyl bromide | | 91 | 1.437 |
| *n*-Propyl alcohol | | 97 | 1.385 |
| Toluene | | 110 | 1.496 |
| *n*-Butyl alcohol | | 117 | 1.399 |

**Questions**

1. A student obtains a refractive index reading of 1.4684 at 20°C. Calculate the refractive index at 25°C and at 18°C.
2. Would a liquid's refractive index measured at a high altitude be different from that measured at sea level? Explain.

# Distillation

**Distillation** is the chief technique used to separate and purify liquids. In this process, a liquid in one vessel is vaporized, and the vapor is subsequently condensed in another vessel. There are four types of distillation: **simple, fractional, steam,** and **vacuum** distillation. The first three types and relevant experiments are described in this chapter.

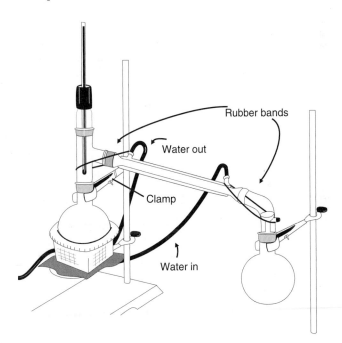

**Figure 4.1** Simple distillation apparatus

 **Safety Tips**

During any form of distillation the following should not be overlooked:
- ✔ The round-bottomed flask from which the liquid is distilled should be firmly clamped on a ring stand. This will provide the setup with a degree of security (see Figure 4.1).
- ✔ Ground-glass joints should be lightly greased to prevent fusion of the glass pieces. Avoid using excessive grease, which might contaminate the distillate.

✔ Rubber bands should be used to hold the connecting adaptor and the condenser together, and to hold the condenser and the vacuum adaptor together. This will prevent slipping and subsequent breaking of the glassware.

✔ Boiling chips should be added to the distilling flask. This promotes homogeneous heating and inhibits the sudden appearance of large bubbles that can burst with enough energy to splatter the liquid and cause "bumping" of the overheated solution. Boiling chips (which are generally nonreusable) have many pores that contain air. The solution is agitated by the warm stream of air released by the chips during the heating period. Thus the chips prevent superheating. If by mistake boiling chips are not added, one should **never** add them directly to the hot solution, since a violent liquid eruption may occur. Instead, the solution should be cooled down to room temperature, the boiling chips added, and the solution reheated.

✔ Hoses should be secured onto the condenser tube. The hoses convey cool water to the jacket that surrounds the condenser tube. Figure 4.1 demonstrates the correct placement of the water intake and outtake hoses. Water should be inserted in a slow stream from a faucet (*do not use the vacuum line!*) and the water jacket should be filled slowly while the condenser is rotated to remove all air bubbles.

✔ Use of an excessive number of clamps should be avoided. Too many clamps can lead to leaks and glassware breakage. Clamps are recommended on the distilling round-bottomed flask (as described previously) and on the collecting flask.

✔ The position of the thermometer is of paramount importance. The top of the mercury bulb should be at the same level as the bottom part of the connecting adaptor (Figure 4.2). If the thermometer is placed lower, a higher temperature of the overheated liquid will be recorded. On the other hand, temperatures that are much lower than the true boiling point will be recorded if the thermometer is placed higher.

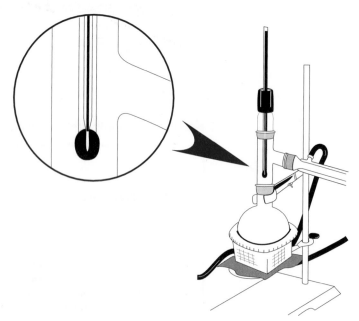

**Figure 4.2**   Setup showing the proper position of the thermometer

# 4.1    SIMPLE DISTILLATION

The apparatus for a simple distillation is shown in Figure 4.1. As the liquid is heated, its vapor rises and comes into contact with the thermometer. The vapor then passes through a water-cooled condenser, where it cools, reliquifies, and drips into a collecting flask. For a pure liquid compound, the boiling point read on the thermometer will remain constant throughout the distillation. During the distillation of an ideal mixture of two liquids, the lower-boiling component will vaporize first at a constant temperature (its boiling point). The higher-boiling component will follow when its boiling point is reached. Mixtures of liquids with boiling points within 30°C or less of each other are not easily separated by simple distillation. In these cases fractional distillation is a much more efficient technique.

**Suggested References**    "Distillation." In *McGraw-Hill Encyclopedia of Science and Technology,* 7th ed., vol. 5, 364–68. New York: McGraw-Hill, 1992.

Ghayourmanesh, S., "Distillation and Evaporation." In *Magill's Survey of Science: Applied Science,* 682–88. Pasadena: Salem Press, 1992.

---

**Experimental**

**Part I.  Simple distillation of the solvent in a dye solution**

1. Set up a simple distillation apparatus (see Figure 4.1) using a 50-mL round-bottomed flask. Add 30 mL of a solution containing 1% methylene blue dissolved in one of the organic solvents listed in Table 3.2 (*use a glass short-stemmed funnel*) and a few boiling chips, and begin heating. Methylene blue can be replaced by any other commercially available dye.

2. Collect the distillate in a round-bottomed flask and record the constant temperature at which the main fraction of the distillate is collected. Identify the solvent by comparing the constant temperature with the BP of the compounds listed in Table 3.2.

3. This part may be combined with the refractive index described in Section 3.3.

**Part II.  Simple distillation of a two-liquid mixture**

1. Perform a simple distillation as described in part I using 30 mL of a 50:50 acetone-water mixture. Use a 10-mL graduated cylinder to collect the distillate.

2. Collect the distillate in a graduated cylinder and record the temperature for every 1 mL of distillate. Once the 10-mL mark is reached, rotate the vacuum adaptor upwards, discard the first 10 mL of the distillate as instructed and continue the distillation process. Discontinue heating once the temperature reaches 100°C.

3. Plot the various temperatures (*y*-axis) versus the number of milliliters collected (*x*-axis) on the graph paper (p. 35). Connect the points to produce a smooth curve.

## 4.2    FRACTIONAL DISTILLATION

A fractionating column is placed between the distilling flask and the connecting adaptor (Figure 4.3) to promote an efficient separation of components that have small differences in boiling point. Inert packing material in the column, such as glass beads, glass wool, steel wool, or copper wire, continuously subjects the upward-moving mixture to many vaporization-condensation cycles. As the vapor rises in the column, it cools and condenses. The condensate runs down the column until it is reheated and vaporized again. Each time the condensate is vaporized, the vapor becomes richer in the component with the lower boiling point. Each cycle, therefore, represents a simple distillation, and the process continues until only vapor of the lower boiling point component reaches the thermometer and the condenser. The result is that the mixture undergoes many simple distillations in one operation.

**Experimental**

1. Clamp a 50-mL round-bottomed flask on a ring stand, and place 30 mL of a fresh 50:50 acetone-water mixture in this flask through a funnel.
2. Assemble the fractional distillation apparatus (see Figure 4.3). Pack the fractionating column with copper mesh or glass beads. If the beads are too small to be held by the condenser, first place a piece of lumped aluminum foil in the bottom of the condenser and then add the beads. Wrapping the fractionating column with an insulating material such as aluminum foil is recommended, but is not usually necessary. Use a 10-mL graduated cylinder to collect the distillate as in part II, p. 31.
3. Distill slowly and record the temperature for every 1 mL collected. Discontinue heating once the temperature reaches 100°C.
4. Plot the temperature (*y*-axis) versus the number of milliliters distilled (*x*-axis) on the graph paper provided (p. 35). Connect the points to produce a smooth curve. Compare the graphs obtained from simple distillation and fractional distillation of the acetone-water mixture.

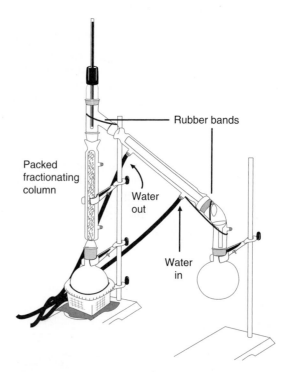

**Figure 4.3**   Fractional distillation apparatus

**Questions**

1. What is the purpose of the following in a distillation experiment?
   a. greasing the joints
   b. using rubber bands
   c. properly positioning the thermometer
   d. using boiling chips
   e. rotating the condenser while slowly filling the jacket with water.

| Compound | BP (°C) | MW |
|---|---|---|
| Dimethyl ether | −24 | 46 |
| Carbon disulfide | 46 | 76 |
| n-Hexane | 69 | 86 |
| n-Heptane | 98 | 100 |
| Water | 100 | 18 |
| Ethanol | 80 | 46 |

2. Plot boiling point (BP) (y-axis) versus molecular weight (MW) (x-axis) for the first four compounds listed in the table above.
   a. Do all points lie on a straight line?
   b. Include water and ethanol in your plot. Do they fall on the line plotted in part (a)? Why?

3. **Raoult's law** is a law of ideal solutions: For a solution composed of components A and B,

$$P_A = (P^°_A)(x_A)$$

   where $P_A$ is the partial vapor pressure of $A$ arising from the solution, $P^°_A$ is the vapor pressure of pure $A$ independent of $B$, and $x_A$ is the mole fraction of A.

   A solution consists of 360 g of sucrose (table sugar) and 144 g of water. Assume that the solution follows Raoult's law and that the vapor pressure of sucrose is negligible. The normal boiling point of water at atmospheric pressure is 100°C. The molecular weight of sucrose is 180.
   a. What is the mole fraction of water in solution?
   b. What is the mole fraction of sucrose in solution?
   c. If the solution is heated to 100°C, what will be the vapor pressure of the water rising from the solution?
   d. Will the solution boil at this temperature? (*Hint:* For a solution to boil at atmospheric pressure, the vapor pressure must be equal to 760 mm Hg). What does this tell you about the boiling point of a sucrose solution? Is it going to be higher or lower than 100 °C?

4. Sodium chloride has a greater effect on the boiling point of a solution than an equimolar amount sucrose. Explain. (*Hint:* Sucrose is a covalent compound, while sodium chloride is an ionic compound.)

**GRAPH PAPER: SIMPLE AND FRACTIONAL DISTILLATION OF A 50-50% ACETONE-WATER MIXTURE**

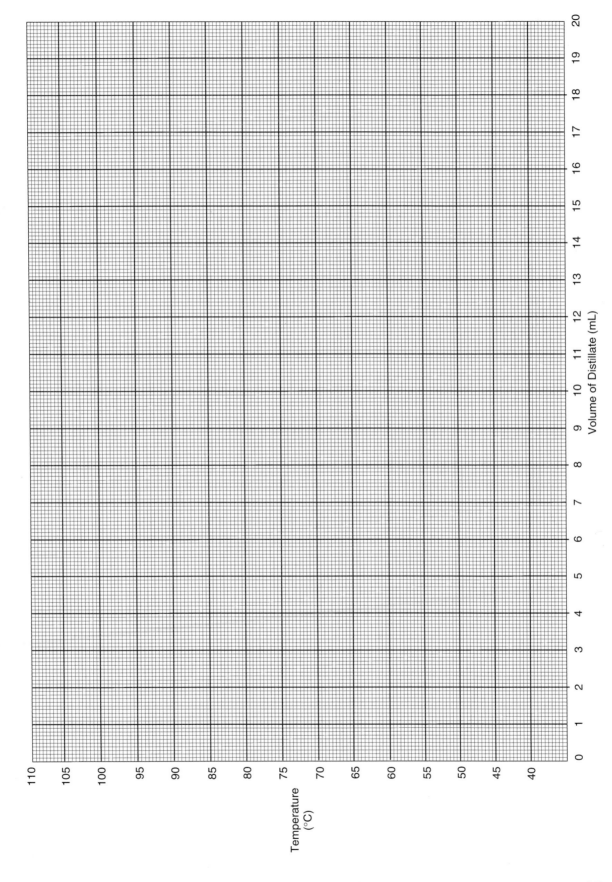

Temperature
(°C)

Volume of Distillate (mL)

## 4.3    STEAM DISTILLATION

Many organic compounds decompose near their boiling points. For these compounds, co-distillation with steam is useful because non-ideal mixtures deviate from Raoult's law and allow distillation at temperatures lower than the boiling point of either component. Distillation with water is a special kind of **azeotropic** distillation. An azeotropic solution is defined as a solution of two or more components; the composition of which does not change upon distillation. Such minimum-boiling azeotropic distillations are often used to separate volatile organic compounds from non-volatile inorganic salts, tars, or leaves and seeds of plants.

Steam distillation depends on the mutual insolubility of many organic compounds with water. It also requires that these compounds should not react with water and have an appreciable vapor pressure at approximately 100°C. When heated, the total vapor pressure above the two-phase mixture is equal to the sum of the vapor pressures of the pure components. When the sum is equal to 760 mm Hg (or 1 atm) co-distillation of both components will occur. For example: in a water-benzene mixture, steam distillation will not occur before the temperature of 69.3°C is reached. At that temperature, the partial pressure of benzene is 533 mm Hg and that of water is 227 mm Hg (note that the sum is equal to 760 mm Hg). The mole fraction in this case can be calculated as 0.701 for benzene (533/760) and 0.299 for water (227/760). The weight percentages in the vapor phase will be different, since water and benzene have different molecular weights.

The setup for a steam distillation resembles that of a simple or fractional distillation. The source of steam can be either internal (a large amount of water in the distilling flask) or external (superheated steam passed into the distillation flask directly from a steam line).

In this experiment, citral (molecular formula $C_{10}H_{16}O$) will be separated from lemon grass oil. Citral is the most important member of the acyclic monoterpenes, and it constitutes about 80 percent of East Indian lemon grass oil, the essential oil of *cymbopogon flexucosus*. It is also found in the skins of lemons, oranges, and some other citrus fruits. Although to humans citral tastes and smells pleasant, it is secreted by certain insects to ward off potential predators. It is commercially important in the synthesis of vitamin A, and is used as a constituent of perfumes in which a lemon-like essence is desired.

After it is steam distilled, citral is dissolved in diethyl ether (ether) or methylene chloride to separate it from water. Both solvents are water-immiscible, but traces of water will dissolve in them. The small amount of water in the solvent will be removed by adding a small amount of an anhydrous salt such as calcium chloride, sodium or magnesium sulfate, or sodium carbonate. These salts form hydrates with water. Calcium chloride, for example forms a dihydrate, and magnesium sulfate a heptahydrate:

$$CaCl_2 + 2H_2O \longrightarrow CaCl_2 \cdot 2H_2O$$
calcium chloride                    calcium chloride dihydrate

$$MgSO_4 + 7H_2O \longrightarrow MgSO_4 \cdot 7H_2O$$
magnesium sulfate                  magnesium sulfate heptahydrate

These **_drying agents_** are very commonly used in organic chemistry. However, the hydrated salts will release water when heated at temperatures slightly above 100°C; therefore, they are filtered from the organic solvent before distillation. A list of the most commonly used drying agents appears in Appendix VI.

**Suggested References**   Bruno, T. J., and P. D. N. Svoronos, *CRC Handbook of Basic Tables for Chemical Analysis.* 470–71. Boca Raton, FL: CRC, 1989.

"Distillation." In *McGraw-Hill Encyclopedia of Science and Technology,* 7th ed., vol. 5, 364–68. New York: McGraw-Hill, 1992.

"Extraction." In *McGraw-Hill Encyclopedia of Science and Technology,* vol. 6, 572–74. New York: McGraw-Hill, 1992.

Ghayourmanesh, S., "Distillation and Evaporation." In *Magill's Survey of Science: Applied Science.* 682–88. Pasadena: Salem Press, 1993.

Ghayourmanesh, S., "Solvent Extraction." In *Magill's Survey of Science: Applied. Science.* 2391–96. Pasadena: Salem Press, 1993.

Solomons, T. W. G., *Organic Chemistry.* 5th ed., 1053–62. New York: John Wiley, 1992.

Wade, L. G., *Organic Chemistry.* 3rd ed., 1224–27. Englewood Cliffs, NJ: Prentice-Hall, 1995.

---

 **Safety Tip**

   Ether is a volatile and flammable solvent. All flames should be extinguished in the laboratory when ether is in use. Ether has a high vapor pressure and the separatory funnel should be vented often to prevent a pressure buildup.

   Methylene chloride is harmful when ingested or when its vapors are inhaled (recommended level RL = 350 mg/m$^3$). Therefore, it should be used with adequate ventilation.

---

**Experimental**

1.  Set up a simple distillation apparatus using a 250-mL round-bottomed flask as the distilling flask. Using a funnel, place in the flask 80–100 mL of water, 5 g of lemon grass oil, and a few boiling chips.

2.  Distill the mixture rapidly until a total of 50 mL has been collected.

3.  Transfer the mixture to a separatory funnel, add 20 mL of ether, swirl the funnel gently, and let it stand on a ring support or clamp it on a ring stand. As a precaution in case the funnel leaks, place a beaker or Erlenmeyer flask under it. Methylene chloride may be used instead of ether.

4.  Separate the water layer from the organic layer (which layer is which?) and place the organic layer in a conical flask. Do not discard the second layer until the end of the experiment since there is a possiblity of mislabeling the layers.

5.  Add 0.5 g of anhydrous sodium or magnesium sulfate to the conical flask, cover the flask with a watch glass, and let the mixture stand for about 5 minutes.

6.  Gravity filter the solution into a dry conical flask, and remove the organic solvent (ether or methylene chloride) by heating the solution on a steam bath in the hood or, preferably, by simple distillation. The distillate should be disposed of properly by the instructor.

7.  Weigh the dried citral in a preweighed vial and determine the percent recovery from the original sample of lemon grass oil.

**Questions**

1.  What three requirements must be satisfied before a compound can be steam distilled?

2.  The molecular weights of benzene and water are 78 and 18, respectively. Calculate the weight percent of each of the components distilled at 69.3°C in the example discussed in the introduction.

3. What is the purpose of anhydrous sodium or magnesium sulfate in this experiment?

4. Citral is classified as a terpene. In one paragraph, describe what a terpene is and give several examples. Refer to any of the suggested references.

5. Citral is actually a mixture of two isomers: geranial and neral. Draw the structures of these two compounds and show how they differ.

6. After drying a sample with anhydrous sodium or magnesium sulfate, the liquid is separated from the drying agent by filtration before it is distilled. Explain why the drying agent must be removed by filtration *before* distillation.

## 4.4   MICRO-DISTILLATION

The need to distill small amounts of liquid often arises in laboratory experiments. The main problem that is encountered in this situation is the small volumes to be distilled. A regular distillation setup that uses 19/22 joints is inadequate because the vapors must travel long distances to be isolated in the collecting flask. The Hickman flask is useful because of the short path the vapors have to travel and the absence of joints or connections. The long neck of the Hickman flask (Figure 4.4) serves as the condenser, and the collection collar serves as the receiving flask. The disadvantage of the Hickman flask lies in the inefficient isolation of low-boiling fractions. However, it is very well suited for high-boiling fractions, including some that are solids at room temperature. The flask can also be used for vacuum distillations. In all cases, the Hickman flask can be swirled gently during the distillation process to increase the surface area and to reduce any foaming that may occur.

**Experimental**

1. Clamp the Hickman flask in an oil bath or heating mantle as seen in Figure 4.4a. Insert a long-stem disposable pipette until it reaches the bottom of the flask. Using another disposable pipette or dropper, introduce the liquid to be distilled through the long-stem disposable pipette (Figure 4.4b). Remove the long-stem pipette *without letting it touch the walls of the Hickman flask*. The volume recommended depends on the size of the Hickman flask; it generally should not exceed 30% of the volume of the lower bulb of the Hickman flask. Examples of liquids to be distilled are cyclohexanol (BP 161°C), *n*-amyl acetate (BP 148°C) cyclohexanone (BP 156°C), 1-pentanol (BP 138°C), 2-hexanone (BP 127°C) and *n*-butyl acetate (BP 126°C).

2. Insert a thermometer through an adaptor outlet or a rubber stopper, and clamp it on top of the Hickman flask. The thermometer bulb should be at the same level as the neck of the flask. *It should neither be immersed into the liquid, nor be placed too high above it.*

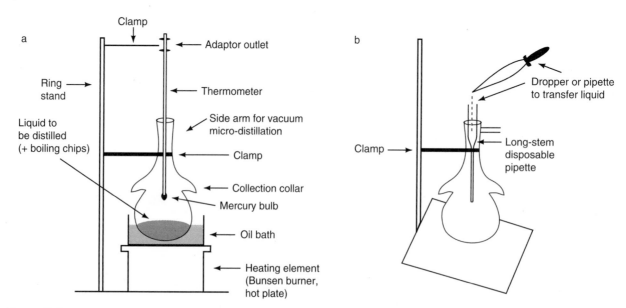

**Figure 4.4**   Micro-distillation setup. *a,* General setup for Hickman flask micro-distillation; *b,* Liquid transfer into the Hickman flask

3. Add a few boiling chips and start heating the setup gently. It is important that only the bottom part be immersed in the oil bath and not the collection collar. If a vacuum distillation is required, attach the vacuum line to the side arm. The vacuum can be provided by the water aspirator or by an oil pump.

4. Condensation will first be noted on the upper walls of the flask, and the droplets of the distillate will accumulate at the collection collar. Heating should be constant, and a temperature drop avoided.

5. When the distillation process has been completed, allow the Hickman flask to cool and withdraw the collected purified liquid from the collection collar using a bent long-stem pipette. If a vacuum was applied, it should be disconnected before the cooling and collection of the distillate. If the distillate solidifies, use an infrared lamp to melt it, and then withdraw the liquid using a warm, bent, long-stem pipette that has been previously placed in an oven.

**Questions**

1. Explain the purpose of the following:
   a. using a long-stem disposable pipette before inserting the liquid to be distilled
   b. making sure the thermometer is not immersed in the lower part of the Hickman flask
   c. making sure that only the bottom part of the Hickman flask is immersed in the oil bath during distillation.

2. Discuss the advantages and disadvantages of the Hickman flask distillation and compare with those of the conventional distillation setup.

# chapter 5

# Chromatography

**Chromatography** (chroma, greek for color) is the technique used to separate the components of a mixture by distributing them between two phases. It received its name from the first experiments in which colored compounds were separated. Three types of chromatography are used extensively in organic chemistry: column, thin-layer, and gas chromatography. Other types, such as high-pressure liquid (HPLC), gel-permeation and ion-exchange chromatography, are other popular types. Every chromatographic analysis depends on the distribution of substances between two phases: a **stationary** phase and a **mobile** phase. The sample is passed through the stationary phase via a (gas or liquid) mobile phase, and the compounds separate according to their extent of adsorption (clinging) to this stationary phase. The compounds that adhere more tightly need larger volumes of mobile phase to be washed through to the far end of the stationary phase. Thus, in **adsorption chromatography,** the components' adsorption and subsequent desorption (detachment) from a solid (or liquid) makes separation possible.

## 5.1   COLUMN CHROMATOGRAPHY

In **column chromatography,** a finely ground solid (the stationary phase) is placed in a glass tube, and a solution containing a solvent and the substances to be separated is placed on top of it. In operation, fresh solvent (the mobile phase) is continually run through the column (Figure 5.1). As the mobile phase proceeds down the column, the compounds of the sample are selectively adsorbed and readsorbed on the stationary phase many, many times.

Column chromatography is used to separate nonvolatile mixtures. It can be carried out with large samples (100 mg to 100 g), and takes between 30 minutes and several hours to perform. Two commonly used stationary-phase adsorbents used for the column material are aluminum oxide (otherwise known as alumina, $Al_2O_3$) and silica gel ($SiO_2 \cdot xH_2O$). The degree of adsorption to the stationary phase depends on van der Waals and dipole forces, as well as stronger hydrogen bonding, complexation and coordination interactions. Many organic liquids may be used as the mobile phase; however, the solvent chosen must not react with the components of the sample. Generally, polar compounds are eluted (washed out) with polar solvents, and non-polar components with non-polar solvents. Table 5.1 lists the approximate increase in eluting power of different solvents for alumina and silica.

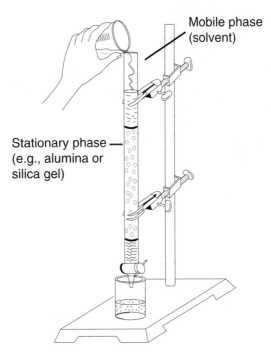

Mobile phase (solvent)

Stationary phase — (e.g., alumina or silica gel)

**Figure 5.1**   Column chromatography setup

**Table 5.1**   Eluotropic series of solvents for alumina and silica

|  | | **Alumina** | **Silica** |
|---|---|---|---|
| increasing eluting power | ↓ | Pentane | Cyclohexane |
| | | Petroleum ether | Petroleum ether |
| | | Hexane | Pentane |
| | | Cyclohexane | Ethyl ether |
| | | Ethyl ether | Ethyl acetate |
| | | Methylene chloride | Ethanol |
| | | Ethyl acetate | Water |
| | | Isopropyl alcohol | Acetone |
| | | Ethanol | Acetic acid |
| | | Methanol | Methanol |
| | | Acetic acid | |
| | | Water | |

The construction of the column is crucial to the success of chromatography. Poor separation of the mixture results if the adsorbent has cracks or channels, or if it has a nonhorizontal or irregular surface.

The column must not be allowed to dry out once it has been prepared and is in operation. The column can dry if the meniscus of the solvent is allowed to fall below the upper level of the adsorbent. Drying forces the adsorbent to pull away from the walls of the glass tubing, and causes channels to form.

Column chromatography is especially suited for purifying gram-sized quantities of materials. It effects separation far more completely than distillation or recrystallization techniques. Its disadvantage is that several trials are often required to establish optimal conditions of stationary and mobile phases. It is also not suited for very large quantities of material.

**Suggested References**    Christian, G. D., *Analytical Chemistry.* 137–40. New York: John Wiley, 1977.

"Chromatography." In *McGraw-Hill Encyclopedia of Science and Technology,* 7th ed., vol. 3, 594–602. New York: McGraw-Hill, 1992.

Gordon, A. J., and R. A. Ford. *The Chemist's Companion: A Handbook of Practical Data, Techniques, and References.* New York: John Wiley, 1972.

Harris, D. C., *Quantitative Chemical Analysis.* 602–4. San Francisco: W. H. Freeman, 1982.

Skoog, D. A., and J. J. Leary. *Principles of Instrumental Analysis.* 4th ed., 580. Fort Worth: Saunders, 1992.

**Experimental**

1. With a thin glass rod, insert a small uniform plug of glass wool into the bottom of the microcolumn or a disposable pipette.
2. Prepare a suspension of alumina in ethanol in a 100-mL beaker, and pour it slowly into the top of the column as seen in Figure 5.1. Avoid overflowing the column.
3. Place a finger on top of the column, invert it, and shake it vigorously until the alumina is suspended in the solvent and all air bubbles are eliminated. If the column has a stopcock, make sure it is closed. If a disposable pipette is used, make sure that the alumina is completely covered by the ethanol.
4. Turn the column right-side up and clamp it on a ring stand over a beaker or conical flask. If the column is too narrow to clamp, insert it through a rubber stopper or the adaptor outlet and clamp it by the stopper. *Note: Use the hood if the eluting solvent is toxic (e.g., chloroform, carbon tetrachloride, benzene).*
5. Rinse down any traces of alumina adhering to the column with a few drops of the solvent. Make sure that the top of the column is covered by ethanol.
6. When the solvent is just 2–3 mm above the top of the alumina, place a few drops of the sample: a dye solution containing 5% methylene blue and fluorescein (1:1 mixture) in 95% ethanol directly on the top of the column.
7. Continue to elute the mixture with ethanol. Collect the eluate in a test tube at the bottom of the column. *Do not allow the column to dry at any time.* When the first component, methylene blue, has been eluted, (the next few drops should be colorless) replace the ethanol with water. Collect the second component, fluorescein, in another test tube.

**Questions**

1. Give a reason for the following in the above experiment:
   a. using glass wool (step 1)
   b. not allowing the column to dry.
2. Which component, methylene blue or fluorescein, is:
   a. adsorbed more strongly to the alumina?
   b. more polar?

## 5.2  THIN-LAYER CHROMATOGRAPHY

Column chromatography is useful for separating gram-sized quantities, but it is not as helpful or economical for isolating very small quantities. **Thin-layer chromatography (TLC)** is simple, inexpensive, fast, and efficient. It is used primarily to determine the number of components in a mixture, and to establish whether two unknowns have identical or different structures. It can also be used to identify a compound by comparing its mobility on a stationary phase to that of an authentic sample. In this exercise, the $R_f$ values (see below) of three organic compounds in two different solvent mixtures will be measured, and via this information the identification of two of these compounds in a "unknown" solution will be achieved.

TLC uses small glass or plastic slides (or plates) coated with a very thin layer of a stationary phase (such as silica gel, $SiO_2$). However, since the adsorbent layer is very thin, volatile samples can evaporate. This limits the application of TLC to only nonvolatile solids and liquids. To obtain a thin-layer chromatogram, a tiny amount of a solution containing one (or more) organic compound(s) is applied to the bottom of the TLC slide, using a micropipette or a thin capillary, in a procedure called "spotting." After allowing the solvent to evaporate, the slide is carefully placed in a developing chamber with a small amount of developing solvent (Figure 5.2). The developing solvent is adsorbed onto the silica gel, and capillary action causes the solvent to travel up the TLC slide, carrying the organic compound with it. **More polar compounds bind more strongly to the polar silica gel (primarily by hydrogen bonding) and thus travel a shorter distance up the slide. Less polar compounds bind less strongly to silica gel (primarily by van der Waals forces) and thus travel a longer distance up the slide.** Thus, differences in polarity cause different compounds to have different mobilities on silica gel and therefore allow compounds to be separated. After the solvent has nearly reached the top of the slide, the TLC plate is removed from the chamber and the solvent is allowed to evaporate. The separated spots can then be visualized with a chemical "stain," such as iodine vapors. Alternatively, if the stationary phase has been treated with a fluorescent indicator, the spots will be visible if the developed slide is viewed under ultraviolet light, presuming that the molecules absorb in the UV range. Such molecules should carry a chromophore such as an aromatic ring or other conjugated system.

Under a specific set of experimental conditions, a given compound will always travel a fixed distance relative to the distance traveled by the solvent front. The mobility of a compound on silica gel is therefore measured by the $R_f$ value, defined as the ratio of the distance traveled by the spot to the distance traveled by the solvent front, with measurements made from the center of each spot (Figure 5.3).

$R_f$ values are expressed in decimal notation (i.e., values range from 0.00 to 1.00)

$$R_f = \frac{\text{distance traveled by the compound (in cm)}}{\text{distance traveled by solvent front (in cm)}}$$

As described previously, the $R_f$ value depends on the polarity of the compound—**more polar compounds show smaller $R_f$ values, while less polar compounds show greater $R_f$ values.** However, solvent polarity also plays an important role. Since more polar solvents also bind more strongly to silica gel than less polar solvents, **a given compound will have a**

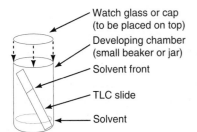

Watch glass or cap
(to be placed on top)

Developing chamber
(small beaker or jar)

Solvent front

TLC slide

Solvent

**Figure 5.2**  Thin-layer chromatography apparatus

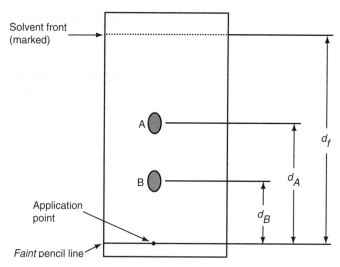

**Figure 5.3**  A thin-layer chromatogram

**greater $R_f$ value when the slide is developed with a more polar solvent than with a less polar solvent.** You can think of the more polar solvent as "displacing" the compound from the surface of the silica more efficiently, thus allowing it to travel up the plate more rapidly. Organic chemists frequently use mixtures of less toxic solvents such as hexane (nonpolar) and ethyl acetate (polar) to obtain developing solvents with varying polarities. Two different solvent mixtures will be used in this experiment: 1:1 hexane-ethyl acetate (more polar because it has a higher percentage of the more polar ethyl acetate) and 9:1 hexane-ethyl acetate (less polar because it has a lower percentage of ethyl acetate).

**Suggested References**

Christian, G., *Analytical Chemistry.* 174–75. New York: John Wiley, 1977.

"Chromatography." In *McGraw-Hill Encyclopedia of Science and Technology,* 7th ed., vol. 3, 594–602. New York: McGraw-Hill, 1992.

Flaschka, H.A., A.J. Barnard, Jr., and P.E. Sturrock, *Quantitative Analytical Chemistry.* 546–49. Boston: Willard Grant, 1980.

Harris, D. C., *Quantitative Chemical Analysis.* 584–85. San Francisco: W. H. Freeman, 1982.

Skoog, D. A., and J. J. Leary, *Principles of Instrumental Analysis.* 4th ed., 663–67. Fort Worth: Saunders, 1992.

 **Safety Tips**

✔  The compounds used are not highly toxic, but avoid skin contact or unnecessary inhalation.

✔  Hexane and ethyl acetate are both highly flammable; no open flames should be used.

✔  UV light can cause skin burns and eye damage. *DO NOT look directly into the UV lamp! DO NOT shine it on your skin!*

**Experimental**

1. Obtain three TLC slides and a TLC micropipette or capillary. Choose one of the two developing solvent mixtures available: either 9:1 hexane-ethyl acetate (less polar) or 1:1 hexane-ethyl acetate (more polar). Be sure to *record* which solvent mixture you use. In your

screw-cap jar or small beaker equipped with a watch glass cover, place sufficient solvent to cover the bottom of the chamber to a depth of about 5 mm (1/4 inch).

2. Mark a TLC slide as shown in Figure 5.4 by *gently* drawing a faint pencil line horizontally, about 1 cm (1/4–1/2 inch) from the bottom of the slide. *Lightly* draw three small pencil marks vertically on this line, about 1 cm apart. At the top of the slide, above the small vertical pencil marks, write **S** (for *trans*-stilbene), **D** (for 1,2-dimethoxybenzene), and **A** (for acetanilide), as shown.

3. "Spot" solutions of the three "known" compounds available in the hood. The three known compounds, in order of *increasing* polarity, are: *trans*-stilbene, 1,2-dimethoxybenzene, and acetanilide, as 1% solutions in ethyl acetate. Spot a solution by dipping the tip of your micropipette or capillary in the solution and then *gently but rapidly* touching the pipette to the correct pencil mark (as labeled—either **S, D** or **A**) on your TLC slide. You should only dispense a very tiny drop of the solution onto your slide in this manner. Your instructor will demonstrate this procedure. After spotting one solution, rinse out your pipette by dipping it into acetone and touching it to a paper towel or filter paper to remove the acetone. Then, in the same way, spot the other solutions onto the other pencil marks on the same TLC slide. *It is vital that the "known" solutions do not become cross-contaminated!*

4. Develop the TLC slide by gently placing it in your jar or beaker and covering the chamber with a cap or watch glass. The spots should *not* be below the level of solvent in the chamber, and *the chamber should be left undisturbed during the development process.* When the solvent front rises to about 1 cm (1/4–1/2 inch) from the top of the slide, carefully remove the slide, gently mark the level attained by the solvent front (i.e., the highest point reached by the rising solvent), and allow the slide to dry for a minute or so.

5. "Visualize" the spots by exposing the plate to UV light; you may also try visualizing them with iodine vapors. This is accomplished by placing the developed slide in a jar that contains iodine crystals and is marked "iodine chamber." The spots may be clearly seen after a few (3–5) minutes.

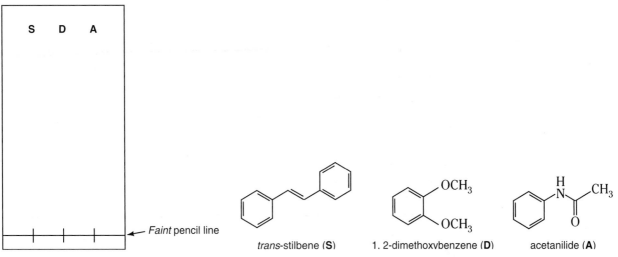

**Figure 5.4** A marked TLC slide

6. Circle the spots gently with a pencil, and calculate the $R_f$ values for the three "known" compounds using the formula above; record them in your notebook. Measurements should be made from the center of each spot.

7. Repeat steps 2–6 using the other developing solution. At this point, you should have measured and recorded $R_f$ values for all three of the "known" compounds in both solvent mixtures.

8. Mark a third TLC slide as described in step 2, except with only one mark on the baseline. Spot the "unknown" solution assigned to you (making sure to record the number of the "unknown" you used) as described in step 3.

9. Develop the slide in one of the developing solvents used previously, and visualize it as described above. Based on the results of this test, you should be able to identify which two of the three "known" compounds are present in your "unknown" solution. If you cannot, then repeat the experiment using the other solvent mixture. Record you observations in your notebook.

10. When you have identified both of the compounds present in your "unknown," record your conclusions on the data sheet. Label **all** of your developed TLC slides with a pencil, noting the compound(s) spotted, the developing solvent, visualization method(s) used, and calculated $R_f$ values.

11. Dispose of the organic solvents as instructed. If you used plastic-backed TLC slides, tape them to your lab report. If you used glass-backed TLC slides, dispose of them in the glass waste container.

**Questions**

1. A spot has traveled 2.4 cm on a thin-layer chromatogram; and the solvent front, 6.4 cm. Calculate the $R_f$ of the spot.

2. Arrange the following compounds in order of increasing $R_f$ value on silica gel:
   a. benzaldehyde ($C_6H_5CHO$)
   b. $n$-decane ($CH_3(CH_2)_8CH_3$)
   c. benzoic acid ($C_6H_5CO_2H$).

3. Which of the compounds in question 2 will be eluted first in column chromatography?

# DATA SHEET: THIN LAYER CHROMATOGRAPHY

| | Hexane-Ethyl Acetate (1:1) | | Hexane-Ethyl Acetate (9:1) | |
|---|---|---|---|---|
| | Distance Travelled (mm) | $R_f$ | Distance Travelled (mm) | $R_f$ |
| *Trans*-stilbene | | | | |
| 1,2-dimethoxybenzene | | | | |
| Acetanilide | | | | |
| Solvent front | | — | | — |

Based on my results, unknown #_____ contains (circle all applicable)

    *trans*-stilbene        1,2-dimethoxybenzene        acetanilide

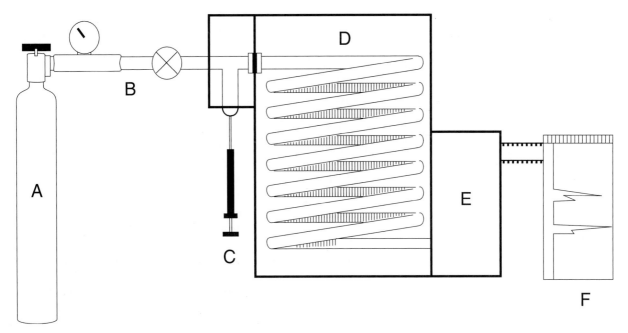

**Figure 5.5** General diagram for a gas chromatograph. *A,* Gas cylinder; *B,* Regulator; *C,* Injection port; *D,* Column; *E,* Detector; *F,* Recorder

## 5.3   GAS CHROMATOGRAPHY

**Gas Chromatography (GC)** is one of the most valuable techniques in organic analysis. Since its introduction in the early 1950s, it has been applied to the separation, identification, and analysis of virtually all gas and volatile liquid mixtures—air, drugs, gasoline, moon dust, and natural products.

The general diagram of a gas chromatograph is seen in Figure 5.5. The flow of the carrier gas (usually an inert gas such as helium, nitrogen, argon) found in the gas cylinder, *A,* is controlled by the regulator *B*. The sample is inserted through the injection port, *C,* and the carrier gas carries it through the column, *D*. After the components have been separated, they are analyzed by detector, *E,* and the results are registered on the recorder, *F*.

In gas chromatography, the mobile phase is an inert gas, and the stationary phase can be either a liquid (**gas-liquid chromatography,** or **GLC**) or a solid (**gas-solid chromatography** or **GSC**). Since the mobile phase is a gas, the system must be confined to a pressure-tight volume; usually the apparatus consists of a metal column in which the stationary phase is placed, the ends of which are closed with porous plugs. The sample is introduced at the head of the column in a small area and is then "carried" away by the carrier gas. Although the column is packed, it contains a considerable amount of space in which the mobile phase flows.

Because the carrier gas flows continuously through the column, the molecules that make up the components of the sample are carried by it but only at those times when they are not attracted to and immobilized by the stationary phase. The longer a component is immobilized, the longer it will take to come out of (be eluted from) the column. With a properly chosen column, the components of the mixture will be immobilized for different lengths of time, and will be eluted one at a time.

A detector at the column exit provides evidence for the separation of the mixture. The detector is attached to a recorder that graphically documents each component as it is eluted. The graph is called a **chromatogram** (Figure 5.6), and it allows the investigator to measure retention times as well as relative concentrations.

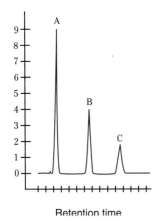

**Figure 5.6** Typical gas chromatogram. A mixture of *A,* 1-propanol; *B,* 1-butanol; *C,* 1-pentanol

The temperature during a gas chromatographic process is critical. To set the temperature range of the instrument, the boiling range of the sample must be known. No tars or inorganic material should ever be injected into the chromatograph. The temperature of the injection port should be higher than the boiling point of the highest boiling component, so that the mixture will vaporize as soon as it is introduced. The temperature of the column is lower than that of the injection port, but high enough to prevent condensation of the components. The temperature of the detector should be much higher than the boiling point range of the mixture, so that the sample clears it before a new sample is injected into the chromatograph. The current of the inert gas passing through the system should be sufficient to transport the components, as well as to cool the detector.

The time elapsed between the injection of the sample and the elution of a component is called the component's **retention time, $t_R$**. Each compound has a characteristic retention time, which is a function of the nature of the stationary phase, the column length, the column temperature, and the inert gas pressure (flow). Retention times may be measured in minutes or, because the recorder paper moves at a constant rate, in centimeters.

In this experiment, a homologous series of alcohols (1-propanol, 1-butanol, and 1-pentanol) will be gas chromatographed, and the retention time of each gas measured. 1-Propanol will be the first to be detected (lowest retention time) while 1-pentanol will be the last. A unique feature of gas chromatography is that it allows the relative concentrations of the components to be measured. The peak area of each component in the gas chromatogram is proportional to its concentration in the original mixture. One way to measure the peak area is to consider each peak as a triangle and multiply the height by half the base length (or width at half height).

**Suggested References**

Christian, G. D., *Analytical Chemistry.* 151–64. New York: John Wiley, 1977.

"Chromatography." In *McGraw-Hill Encyclopedia of Science and Technology,* 7th ed., vol. 3, 594–602. New York: McGraw-Hill, 1992.

Harris, D. C., *Quantitative Chemical Analysis.* 567–82. San Francisco: W. H. Freeman, 1982.

Flaschka, H. A., A. J. Barnard, Jr., and P. E. Sturrock, *Quantitative Analytical Chemistry.* 530–41. Boston: Willard Grant, 1980.

McNair, H. M., and E. J. Bonelli, *Basic Gas Chromatography.* Palto Alto, CA: Varian, 1968.

Skoog, D. A. and J. J. Leary. *Principles of Instrumental Analysis.* 4th ed., 605–23. Fort Worth: Saunders, 1992.

**Experimental**

1. Set the gas chromatograph conditions as suggested by the instructor. The set of conditions differs from column to column, so that the proper settings for the gas chromatograph should be decided by the instructor. Normally an injection port temperature of 150°C and a column temperature of 50–60°C should provide a good separation.

2. Inject the unknown three-to-five carbon alcohol mixture, and record the time of injection.

3. Determine the retention times in centimeters and write down these numbers. Also measure the retention time in minutes by dividing the distance between each peak by the rate of the recorder, which is expressed in cm/min.

4. Identify each alcohol in the chromatogram by comparing each peak with the chromatograms of the individual compounds run by your instructor under the same conditions. Calculate the percentage composition of the alcohol mixture by measuring the area under each peak.

**Questions**    1.  What will the effect on the chromatogram be if:
    a.  there is no carrier gas flow?
    b.  the temperature of the injection port is too low?
    c.  the temperature of the column is too high?
    d.  the temperature of the detector is too low?
    e.  the boiling point range of the components is not known?
2.  The column in a gas chromatograph is polar, and a mixture of a polar compound and a nonpolar compound is injected. Which of the two compounds will have the higher retention time? Why?
3.  Compare column, thin-layer, and gas chromatography in terms of:
    a.  mobile phase
    b.  stationary phase.

# chapter 6

# Extraction and Recrystallization

## 6.1   DISTRIBUTION COEFFICIENT

**Extraction** with a solvent is the separation technique most frequently employed to isolate one or more components of a mixture. In most cases, an organic solvent is used to remove an organic solute from an aqueous solution or suspension. The mutually immiscible solvents are shaken together in a separatory funnel to allow the solute to distribute itself in both solvents according to its solubility in each. The solute will partition itself according to the ratio

$$K_p = \frac{[\text{solute}]_A}{[\text{solute}]_B}$$

where $K_p$ is the **distribution coefficient,** $[\text{solute}]_A$ is the concentration of solute (in moles/L) in solvent $A$ (usually the organic solvent) and $[\text{solute}]_B$ is the concentration of solute (in moles/L) in solvent $B$ (usually water).

For example, 6.67 g of phenol will dissolve in 100 mL of water, while 8.33 g will dissolve in 100 mL of benzene. From this, the distribution coefficient is calculated to be

$$K_p = \frac{(8.33 \text{ g}/100 \text{ mL})_{benzene}}{(6.67 \text{ g}/100 \text{ mL})_{water}} = 1.25$$

Note that in this example, g/100 mL has replaced the units of moles/L that appeared in the original definition. This is because units cancel out in such a ratio.

Now assume that we have a solution composed of 1 g of phenol in 50 mL of water, and we extract that solution with 50 mL of benzene. The phenol will distribute itself between the water and the benzene in the following way

$$K_p = \frac{[\text{x g}/50 \text{ mL}]_{benzene}}{[(1-\text{x})/50 \text{ mL}]_{water}} = 1.25$$

where

$x = $ mass (in g) of phenol dissolved in the benzene layer.

Solving for $x$:

$$x = 0.56 \text{ g}$$

In other words, 0.56 g of phenol will be extracted by the benzene, and 0.44 g will remain in the water layer.

In the following experiment, the value of $K_p$ for the distribution of propanoic acid between water and ethyl acetate will be determined first. Then toluene will replace ethyl acetate, and the value of $K_p$ for propanoic acid in water and toluene will be calculated. Both determinations will be done by titration with a standard sodium hydroxide solution.

To calculate the weight of propanoic acid present in either the water or the organic layer, recall that at the end point of a titration

$$(\# \text{ of equivalents})_{acid} = (\# \text{ of equivalents})_{base}$$

or

$$\frac{(\text{weight in g})_{acid}}{(\text{Eq. wt.})_{acid}} = (\text{Volume})_{base} \times (\text{Normality})_{base}$$

where Eq. wt. = equivalent weight of acid in g, and $(\text{Volume})_{base}$ = the volume in liters of base needed to neutralize the acid. Thus,

$$(\text{weight in g})_{acid} = (\text{Volume} \times \text{Normality})_{base} \times (\text{Eq.wt.})_{acid}$$

**Suggested References**

"Extraction." In *McGraw-Hill Encyclopedia of Science and Technology,* 7th ed., vol. 6, 572–74. New York: McGraw-Hill, 1992.

Flaschka, H. A., A. J. Bernard, Jr., and P. E. Sturrock, *Quantitative Analytical Chemistry.* 2nd ed., 512–18. Boston: Willard Grant, 1980.

Ghayourmanesh, S., "Solvent Extraction." In *Magill's Survey of Science: Applied Science,* 2391–96. Pasadena, CA: Salem Press, 1993.

Harris, D. C., *Quantitative Chemical Analysis,* 533–51. San Francisco: W. H. Freeman, 1982.

Ramette, R. W., *Chemical Equilibrium and Analysis.* 571–85. Reading, MA: Addison-Wesley, 1981.

Svoronos, P., "Solvation and Precipitation." In *Magill's Survey of Science: Physical Science.* 2290–96. Pasadena, CA: Salem Press, 1992.

 **SAFETY TIPS**

✔ The vapors of propanoic acid can be irritating. High concentrations of toluene are harmful (recommended level [RL] = 375 mg/m$^3$), and therefore, the lab should be well ventilated.

✔ Ethyl acetate is highly flammable; open flames should be avoided.

**Experimental**

**Figure 6.1** Proper handling of the separatory funnel in the open extraction process

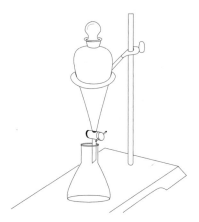

**Figure 6.2** A separatory funnel properly supported, with an Erlenmeyer flask placed beneath the exit tube

1. Dissolve 4.8 mL ($d = 1.04$ g/mL, 5.0 g) of propanoic acid in enough water to make a 100 mL solution. Mix well.
2. Using a pipette, transfer 30 mL of the acid solution to a conical flask. Add 2 drops of phenolphthalein and titrate with a standardized $0.5N$ sodium hydroxide solution. Record your results on the data sheet (item (a), first titration).
3. Again using a pipette, transfer a second 30 mL portion of the acid solution to a separatory funnel. Add 30 mL of ethyl acetate and stopper the funnel. While holding the stopper in place with your finger, invert the funnel to mix the contents. Release the pressure which may have built up inside the funnel by opening the stopcock using your other hand (Figure 6.1). (*Do not point the funnel at yourself or at your neighbor!*) Repeat this operation several times to thoroughly mix the liquids. Carefully empty the (lower) water layer into a clean conical flask (Figure 6.2) and titrate as before. Record the volume of sodium hydroxide used as item (c) of the data sheet. Calculate the weight of acid left in the water layer, item (d). Calculate the weight of propanoic acid that was extracted into the ethyl acetate layer for item (e) by subtracting item (d) from item (b).
4. Transfer a third 30 mL portion of the acid solution to the separatory funnel. Add 10 mL of ethyl acetate and shake the mixture. When the mixture has separated into two distinct layers, collect in a beaker the water layer and discard the ethyl acetate layer. Return the aqueous layer to the separatory funnel and repeat the procedure using a second 10-mL sample of ethyl acetate. Separate as before. Finally, add a third 10-mL sample of ethyl acetate, and after shaking the mixture, separate the lower aqueous layer into another clean conical flask and titrate for item (f), third titration. Calculate the mass of acid in the water layer for item (g).
5. Repeat steps 1 through 4 and extract using toluene (instead of ethyl acetate) as the organic solvent.
6. Wash the burette after the last titration, *while still clamped*. Drain the burette, then slowly add about 10 mL of water and drain; then 10 mL of dilute hydrochloric acid and drain; finally 50 mL of distilled water and drain. *Do not place the burette under the faucet in order to wash it.*

**Questions**

1. Based on your data, is propanoic acid more soluble in water or ethyl acetate? Explain. Do the same with toluene.
2. Which of the following organic solvents can be used in this experiment: ethanol, acetone, diethyl ether, methylene chloride, acetic acid, acetonitrile? Explain (*Hint:* Look up the water solubilities of each compound in a reference book [e.g., the *CRC Handbook of Chemistry and Physics* or this lab manual, Table 6.1] before answering this question).
3. Compared to propanoic acid, is hexanoic acid ($CH_3(CH_2)_4COOH$) more or less soluble in water? Explain.
4. Predict the solubility of propanoic acid in 5% NaOH solution. If propanoic acid were dissolved in ethyl acetate, and you wanted to remove *all* of it from the ethyl acetate layer by extraction, which would be a better extraction solvent, water or 5% sodium hydroxide? Explain.
5. Calculate $K_p$ using data items (d) and (e) for ethyl acetate. Suppose 1 g of propanoic acid were dissolved in 50 mL of ethyl acetate and the solution extracted with 50 mL of water. How much propanoic acid would be left in the ethyl acetate layer?

6. In step 4 of this experiment, the propanoic acid was removed from the water layer by three successive extractions with a total of 30 mL of ethyl acetate. Was more or less acid removed from the water layer by three 10-mL successive extractions or by one single extraction using a 30-mL sample of ethyl acetate? Compare data items (d) and (g). Do the same for toluene by comparing items (dd) and (gg).

7. Review the introduction of this experiment to answer this question: Assume you have a solution consisting of 1 g of phenol dissolved in 50 mL of water, and you extract it with two successive 25-mL portions of benzene. Calculate the mass of acid that will be removed from the water by each extraction. Is the total amount extracted more or less than what would be removed by one extraction with 50 mL of benzene?

# DATA SHEET: DISTRIBUTION COEFFICIENT: PAGE 1
## Show Your Calculations

### Ethyl Acetate

a. Volume of base required for first titration _____ mL

b. Mass of propanoic acid in 30 mL sample _____ g
   Calculate by using the formula:
   $\text{Mass} = (N \times V)_{base} \times (\text{Eq.Wt.})_{acid}$
   (Note: $V_{base}$ should be in L)

c. Volume of base required for second titration _____ mL

d. Mass of acid left over in water layer _____ g
   Use the same formula as in (b)

e. Mass of acid in ethyl acetate layer after a single 30-mL extraction _____ g
   Item (b) – item (d)

f. Volume of base required for third titration _____ mL

g. Mass of acid in water layer _____ g
   Use the same formula as in (b)

h. Mass of acid in ethyl acetate layer after three 10-mL extractions _____ g
   Item (b) – item (g)

i. Distribution coefficient of propanoic acid in ethyl acetate _____
   Use items (d) and (e)

# DATA SHEET: DISTRIBUTION COEFFICIENT: PAGE 2
## Show Your Calculations

## Toluene

aa. Volume of base required for first titration                    _____ mL
bb. Mass of propanoic acid in 30 mL sample                         _____ g


cc. Volume of base required for second titration                   _____ mL
dd. Mass of acid left over in water layer                          _____ g


ee. Mass of acid in toluene layer after a single 30-mL extraction  _____ g


ff. Volume of base required for third titration                    _____ mL
gg. Mass of acid in water layer                                    _____ g


hh. Mass of acid in toluene layer after three 10-mL extractions    _____ g


ii. Distribution coefficient of propanoic acid in toluene          _____

## 6.2  SEPARATION BY EXTRACTION

In this experiment, an organic solvent (dichloromethane) is used to dissolve a carboxylic acid (benzoic acid) and a neutral organic compound (*p*-dichlorobenzene), and the two solutes are then separated by extraction. This is accomplished by converting the carboxylic acid to its salt by reacting it with a base (sodium hydroxide):

$$\text{COOH} + \text{NaOH} \longrightarrow \text{COO}^-\text{Na}^+ + \text{H}_2\text{O}$$

| benzoic acid | sodium hydroxide | sodium benzoate | water |

In this neutralization reaction, a covalent organic compound (benzoic acid) that is soluble in organic solvents but rather insoluble in water is converted to an ionic compound (sodium benzoate) that is water soluble. Extraction with aqueous sodium hydroxide thus separates the two compounds in two immiscible layers: water (the *aqueous* layer), in which benzoic acid in its salt form is dissolved, and dichloromethane (the *organic* layer), in which *p*-dichlorobenzene is dissolved. Once the two layers are separated, the aqueous layer can be acidified to regenerate benzoic acid, which can then be isolated by filtration.

$$\text{COO}^-\text{Na}^+ + \text{HCl} \longrightarrow \text{COOH} + \text{NaCl}$$

| sodium benzoate | hydrochloric acid | benzoic acid | sodium chloride |

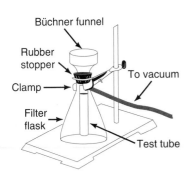

**Figure 6.3** Vacuum filtration (suction) assembly. The filter flask is clamped and a test tube is inserted to collect the filtrate

*p*-Dichlorobenzene can be recovered by evaporating the organic solvent.

**Vacuum** filtration is used extensively in organic experiments because it is much faster than **gravity** filtration (simple filtration through filter paper and a funnel) and it results in a drier product. Figure 6.3 shows the vacuum filtration setup, which is employed in the isolation of benzoic acid after the acidification process. It consists of a filter flask, a Buchner funnel, and a thick-walled tube connected to a vacuum aspirator. A good aspirator should be able to provide a reduced pressure of 30 to 35 mm Hg. Because of the low pressure, the filtered solutions should be *cold* to prevent the solvent from evaporating. The filter flask is secured with a clamp (preferably, a three-prong, with two of the prongs on the outlet side of the filter flask, one above and one below the outlet).

The filtrate (the filtered solution) is collected in a large test tube inside the filter flask, which will serve as the trap. The presence of a trap is necessary to avoid contamination by water that might back up into the vacuum line because of pressure changes. This assembly eliminates the necessity of an external trap which often results in a less efficient vacuum suction because of multiple connections.

**Suggested References**    Day, R. A. and A. L. Underwood. *Quantitative Analysis.* 5th ed., 506–20. Englewood Cliffs, NJ: Prentice Hall, 1991.

"Extraction." In *McGraw-Hill Encyclopedia of Science and Technology*, 7th ed., vol. 6, 572–74. New York: McGraw-Hill, 1992.

Flaschka, H. A., A. J. Barnard, Jr., and P. E. Sturrock, *Quantitative Analytical Chemistry.* 2nd ed., 512–18. Boston: Willard Grant, 1980.

Ghayourmanesh, S., "Solvent Extraction." In *Magill's Survey of Science: Applied Science,* 2391–96. Pasadena, CA: Salem Press, 1993.

Harris, D. C., *Quantitative Chemical Analysis.* 533–51. San Francisco: W. H. Freeman, 1982.

Ramette, R. W., *Chemical Equilibrium and Analysis.* 571–85. Reading, MA: Addison-Wesley, 1981.

Svoronos, P., "Solvation and Precipitation." In *Magill's Survey of Science: Physical Science,* 2290–96. Pasadena, CA: Salem Press, 1992.

---

 **SAFETY TIPS**

✔ Methylene chloride is harmful when ingested or when its vapors are inhaled (RL = 350 mg/m$^3$). Therefore, it should be used with adequate ventilation.

✔ Both concentrated sodium hydroxide and hydrochloric acid are extremely corrosive and should be handled with care.

---

**Experimental**

1. Place 2 g of a 1:1 mixture of benzoic acid and *p*-dichlorobenzene, and 30 mL of methylene chloride, in a stoppered separatory funnel; shake to dissolve (see Figure 6.1). After shaking briefly, invert the funnel and open the stopcock cautiously to release the pressure.

2. Close the stopcock, shake again, and release the pressure as before. Repeat until no more gas is released.

3. Clamp the funnel upright at the joint or suspend it through a ring (see Figure 6.2), and add 10 mL of 5% sodium hydroxide through a short stem funnel. It is also a good idea to place a beaker under the separatory funnel while it is suspended, in case the stopcock leaks.

4. When the two layers are clearly separated, decide which is the aqueous layer. This can be accomplished by adding a drop of water to the separatory funnel to see in which layer it disappears. That layer is the water layer. An alternative method is to put a drop of the fluid from each layer in a separate test tube containing water, and then check the miscibilities. A third method is to check the pH of each layer. This is accomplished be dipping a clean stirring rod on each layer and then touching a red litmus (or alkacid) paper that has been placed on a watch glass.

5. Remove the stopper and separate the two layers into two different beakers. Acidify the water layer (which one is it?) by adding concentrated hydrochloric acid dropwise with stirring until the solution is distinctly acidic to litmus. To test for acidity, place a blue litmus paper on a clean watch glass, then touch it with a clean stirring rod that has been previously dipped into the solution. Once the pH is below 7, the white crystals of benzoic acid will start forming. Add more concentrated hydrochloric acid (no more than 2–3 mL) until the pH = 1–2. You may use the indicator paper for this.

6. Cool the solution in an ice/water bath and isolate the crystals of benzoic acid by vacuum filtration (see Figure 6.3).

7. Check again to ensure that the filtrate (otherwise known as **mother liquor**) is acidic to litmus. If it is not, add more hydrochloric acid until it is definitely acidic.

8. Drain the organic (methylene chloride) layer into a 50-mL round-bottomed flask, and remove the solvent by simple distillation using a

hot water bath. *Do not distill to dryness!* Place the distilled methylene chloride in a bottle labeled "Waste Methylene Chloride."

9. Let the white crystals of *p*-dichlorobenzene stand until they are completely dry.

10. Weigh both the benzoic acid and the dichlorobenzene and determine their melting points. The literature values are 122°C for benzoic acid and 53°C for *p*-dichlorobenzene. Calculate the percent mass recovery of each.

**Questions**

1. Explain the purpose of the following in the above experiment and give pertinent equations when applicable:
   a. adding sodium hydroxide
   b. adding hydrochloric acid
   c. releasing the pressure in the separatory funnel while extracting
   d. using vacuum filtration instead of gravity filtration to isolate the benzoic acid
   e. placing a test tube inside the filter flask during the vacuum filtration process.

2. *p*-Nitroaniline is a basic compound. Draw a flow-chart in which you demonstrate the way to separate the components of a *p*-nitroaniline/ *p*-dichlorobenzene mixture.

## 6.3   RECRYSTALLIZATION OF A SOLID

The purification of a solid is achieved through **recrystallization.** In this process, the mixture that contains the compound of interest is dissolved in the *minimum* amount of *hot* solvent. An Erlenmeyer flask is particularly well adapted for this because it can be held at its neck using a clamp while it is heated over a steam bath. In addition, solvent loss by evaporation is minimized. Beakers should *never* be used for recrystallization because rapid evaporation of solvent allows the formation of crystals that are contaminated with impurities above the solvent line (i.e., along the walls of the beaker).

The proper choice of solvent is of great importance for the process. The ideal solvent should

- be nontoxic
- be inexpensive
- be inert toward the solute
- dissolve the solute at its boiling point, but not at low temperatures (0–25°C)
- be capable of dissolving impurities entirely or not at all *both* at its boiling point and at room temperature
- be volatile so that it can be readily removed from the crystals, but not too volatile, so that the difference between solubility at room temperature and at the boiling temperature is pronounced
- have a boiling point that is lower than the melting point of the solute.

After the compound is dissolved, the solution is filtered *while still hot* to keep the solute dissolved while the suspended, insoluble solids are removed by gravity filtration. The funnel should be stemless to avoid possible crystallization of the compound of interest in the stem as the solvent cools. If the solution is undesirably contaminated with colored impurities, a pinch of decolorizing carbon can be added before the hot filtration step. Decolorizing carbon, with its large surface area, has an affinity for such colored impurities and aids in their removal by filtration.

The hot solution should then be allowed to cool slowly, which allows the crystals to grow at a moderate pace. Immersing the solution in ice while it is still hot will lead to entrapment of the otherwise soluble impurities in the crystals of the compound of interest or to the formation of oils.

Table 6.1 lists some common recrystallization solvents and their boiling points, dielectric constants, flammability, toxicity, and water solubility.

After the solution is allowed to cool to room temperature, the flask is immersed in ice to ensure complete precipitation. The cold solution is then vacuum filtered (see Figure 6.3) to obtain the pure product.

The **second crop** is the batch of crystals obtained from the mother liquor (the filtered solution obtained in the last step). These crystals can be obtained by boiling down the mother liquor and further cooling. Often **seeding** is necessary: A few crystals of the previously isolated product are placed in the cool solution to promote crystallization. Because of the smaller volume of the solution, the otherwise soluble impurities might be entrapped, which may lead to the isolation of a second crop of somewhat lower purity. Nevertheless, its weight and melting point should be determined after isolation via vacuum filtration. In calculating the total percent yield, the partial yields of the two crops should be combined. The steps in the recrystallization process are outlined in Figure 6.4.

**Table 6.1** Common recrystallization solvents

| Solvent | BP[a] | e[b,c] | Flammability[d] | Toxicity[d] | Water Solubility[e] | Good for | Second Solvent in Mixture[f] | Comments |
|---|---|---|---|---|---|---|---|---|
| Water | 100 | 78.5[c] | 0 | 0 | — | Amides, salts, some carboxylic acids | Methanol, ethanol, acetone, dioxane, acetonitrile | Difficult to remove from crystals |
| Acetic acid | 118 | 6.15[b] | 1 | 2 | ∞ | Amides, some carboxylic acids, some sulfoxides | Water | Difficult to remove from crystals |
| Acetonitrile | 81.6 | 37.5[b] | 3 | 3 | vs | Some carboxylic acids, hydroquinones | Water, ether, benzene | |
| Methanol | 64.5 | 32.63[c] | 3 | 1 | ∞ | Nitro-compounds, esters, bromo-compounds, some sulfoxides, sulfones and sulfilimines, anilines | Water, ether, benzene | |
| Ethanol | 78.3 | 24.30[c] | 3 | 0 | ∞ | Same as methanol | Water, ethyl acetate, hydrocarbons, methylene chloride | |
| Acetone | 56 | 20.7[c] | 3 | 1 | ∞ | Nitro compounds, osazones | Water, ether, hydrocarbons | |
| Methyl cellosolve | 124 | | 2 | 2 | i | Carbohydrates | Water, ether, benzene | |
| Pyridine | 116 | 123[c] | 3 | 3 | s | Quinones, thiazoles, oxazoles | Water, methanol | Difficult to remove from crystals |
| Methyl acetate | 57 | 6.68[b] | 4 | 2 | vs | Esters, carbonyl compounds, sulfide derivatives, carbinols | Water, ether | |
| Ethyl acetate | 77.1 | 6.02[c] | 3 | 1 | i | Same as methyl acetate | Water, ether, chloroform, methylene chloride | |
| Methylene chloride (dichloromethane) | 40 | 9.08[b] | 0 | 2 | i | Low-melting compounds | Ethanol, hydrocarbons | Easily removed |
| Ether (diethyl ether) | 34.5 | 4.34[b] | 4 | 2 | i | Low-melting compounds | Acetone, acetonitrile, methanol, ethanol, acetate esters | Easily removed; can create peroxides |

**Table 6.1**    Common recrystallization solvents, cont'd.

| Solvent | BP[a] | e[b,c] | Flammability[d] | Toxicity[d] | Water Solubility[e] | Good For | Second Solvent in Mixture[f] | Comments |
|---|---|---|---|---|---|---|---|---|
| Chloroform | 61.7 | 4.81[b] | 0 | 4 | $i$ | Polar compounds | Ethanol, acetate esters, hydrocarbons | Easily removed; suspected carcinogen |
| Dioxane | 102 | 2.21[c] | 3 | 2 | $\infty$ | Amides | Water, hydrocarbons, benzene | Can form complexes with ethers |
| Carbon tetrachloride | 76.5 | 2.24[b] | 0 | 4 | $i$ | Acid chlorides, anhydrides | Ether, benzene, hydrocarbons | Can react with strong organic bases; suspected carcinogen |
| Toluene | 110.6 | 2.38[c] | 3 | 2 | $i$ | Aromatics, hydrocarbons | Ether, ethyl acetate, hydrocarbons | A little difficult to remove from crystals |
| Benzene | 80.1 | 2.28[b] | 3 | 4 | $i$ | Aromatics, hydrocarbons, molecular complexes, sulfides, ethers | Ether, ethyl acetate, hydrocarbons | Suspected carcinogen |
| Ligroin (naphtha solvent) | 90–110 | — | 3 | 1 | $i$ | Hydrocarbons, aromatic heterocycles | Ethyl acetate, benzene, methylene chloride | |
| Petroleum ether (ACS) | 35–60 | — | 4 | 1 | $i$ | Hydrocarbons less polar than ethanol | Any solvent | Easy to separate |
| n-Pentane | 36.1 | 1.84[b] | 4 | 1 | $i$ | Hydrocarbons less polar than ethanol | Any solvent | Easy to separate |
| n-Hexane | 69 | 1.89[b] | 4 | 1 | $i$ | Hydrocarbons less polar than ethanol | Any solvent | |
| Cyclohexane | 80.7 | 2.02[b] | 4 | 1 | $i$ | Hydrocarbons less polar than ethanol | Any solvent | |
| n-Heptane | 98.4 | — | 4 | 1 | $i$ | Hydrocarbons less polar than ethanol | Any solvent | |

[a]Normal boiling point (°C).
[b]Dielectric constant (20°C).
[c]Dielectric constant (25°C).
[d]Scale varies from 4 (highly flammable, highly toxic) to 0 (not flammable, not toxic).
[e]∞: infinitely soluble; vs: very soluble; s: soluble; i: insoluble.
[f]Second solvent used to facilitate dissolving the crystals in a solvent mixture.

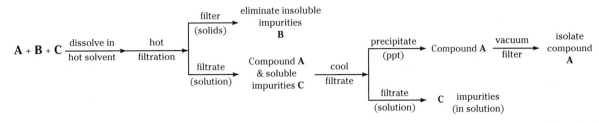

**Figure 6.4** Flow chart for recrystallization of compound **A** (soluble in hot, insoluble in cold solvent) from impurities **B** (insoluble both in hot and cold solvent) and impurities **C** (soluble both in hot and cold solvent)

In this experiment, acetanilide will be recrystallized from water, which serves as the solvent. There are two impurities: charcoal (which is insoluble in both hot and cold water) and sodium chloride (which is soluble in both hot and cold water). The charcoal will be removed during the hot filtration step, while acetanilide, the compound of interest, will be isolated by vacuum filtration of the cold solution, leaving sodium chloride in the filtrate. The experiment can also be performed using benzoic acid instead of acetanilide.

**Suggested References**
Bruno, T. J., and P. D. N. Svoronos, *CRC Handbook of Basic Tables for Chemical Analysis.* 472–74. Boca Raton, FL: CRC, 1989.
"Crystallization." In *McGraw-Hill Encyclopedia of Science and Technology,* 7th ed., vol. 4, 604–5. New York: McGraw-Hill, 1992.
Flaschka, H. A., A. J. Barnard, Jr., and P. E. Sturrock, *Quantitative Analytical Chemistry.* 2nd ed., 512–18. Boston: Willard Grant, 1980.
Harris, D. C., *Quantitative Chemical Analysis.* 533–51. San Francisco: W. H. Freeman, 1982.
Ramette, R. W., *Chemical Equilibrium and Analysis.* 571–85. Reading, MA: Addison-Wesley, 1981.
Svoronos, P., "Solvation and Precipitation." In *Magill's Survey of Science: Physical Science,* 2290–96. Pasadena, CA: Salem Press, 1992.

**Experimental**

1. Place 2 g of a mixture of acetanilide (80%), sodium chloride (19%) and charcoal (1%) in a 50-mL Erlenmeyer flask. Add a small amount (about 25 mL) of water, and boil over a hot plate.
2. While the solution in step 1 is heating, place 5 mL of water in a second Erlenmeyer flask. Secure a stemless funnel on a ring *just above* the flask. Place a piece of fluted paper (Figure 6.5) on the funnel and start heating the flask over the same hot plate (Figure 6.6).
3. Using a clamp as a handle for the flask containing the hot acetanilide solution, pour the boiling solution from step 1 through the funnel of the assembly of step 2. Rinse the Erlenmeyer flask with about 2 to 3 mL of water, heat to boiling, and pour through the filter. To avoid evaporation and cooling of the solution, cover the stemless funnel with a watch glass during filtration.
4. Remove the Erlenmeyer flask from the hot plate and allow the filtered solution to cool slowly. *To prevent entrapment of impurities, do not chill rapidly.* After the solution cools to room temperature, chill the flask in an ice/water bath for about 5 minutes and filter via vacuum filtration (see Figure 6.3).
5. Rinse the flask with a few (2–3) mL of *ice cold* water and pour over the crystals. Allow the crystals to dry while the water in the vacuum aspirator is running. Determine the MP (literature value 114°C) and percent mass recovery of acetanilide.

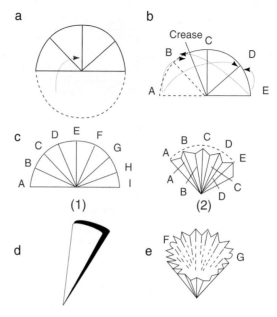

**Figure 6.5**   Preparing a fluted filter paper. *a,* The circle of filter paper is folded in half. The semicircle is folded half, then each edge is brought into the center fold and a crease is made. *b,* Corner *A* is brought to point *D,* and a crease is made. Corner *E* is brought to point *B,* and a crease is made. Corner *A* is brought to point *B,* and corner *E* is brought to point *D* and a crease is made each time. *c,* When the folding is complete, the filter paper should have seven folds as shown in (*1*). Point *B* is then brought to corner A so that the resulting fold lies in the direction opposite to the folds made in the preceding steps, as shown in (*2*). *d,* The filter paper as it appears when completely folded. *e,* The fluted paper open with the two inward folds at *F* and *G* and ready to use

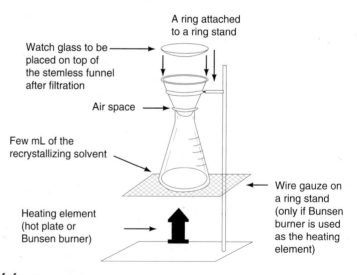

**Figure 6.6**   Funnel/Erlenmeyer flask setup for hot filtration

**Questions**

1. Explain the purpose of the following in this experiment:
   a. using hot filtration
   b. not cooling (chilling) the hot filtered solution immediately
   c. washing the crystals in the final step with ice cold water
   d. using gravity filtration (not vacuum filtration) in the hot filtration step
   e. using a stemless funnel in the hot filtration step.
2. What are the requirements for a good recrystallizing solvent?
3. Using the data of Table 6.1, compare the following pairs of liquids as recrystallization solvents (*Note:* Review the conditions for a good recrystallizing solvent):
   a. water vs. acetone
   b. methanol vs. toluene
   c. benzene vs. cyclohexane
   d. chloroform vs. ethyl acetate.

## 6.4   MICRORECRYSTALLIZATION OF A SOLID

The purification of a solid from a mixture of impurities is generally achieved through the process of recrystallization. The physical process involves dissolving the mixture in the minimum amount of hot solvent and filtering the solution to separate the insoluble impurities. When the filtrate cools, the solid to be purified crystallizes and can be isolated by filtration while the rest of the impurities remain dissolved in solution.

The proper choice of solvent is an important part in recrystallization (see the introduction of "Recrystallization of a Solid," Experiment 6.3).

The solution is filtered while still hot to keep the solute dissolved while the suspended, insoluble solids are being removed. Following this step, the crystals grow as the solution's temperature is slowly decreased to prevent entrapment of any impurities.

The procedure for filtration and microrecrystallization is applicable in cases where small quantities of a solid are prepared as derivatives of an unknown compound (such as the 2,4-dinitrophenylhydrazine derivative of an aldehyde or a ketone) or in testing new reactions by running them with small quantities.

The only special equipment necessary for microrecrystallization is a **capillary filter stick** (referred to as a **U-shaped capillary**), which is shown in Figure 6.7. This is a thick glass capillary, with an internal diameter of 2 mm, that is bent twice at right angles. The two parallel arms have different lengths, the longer one is about 10 mm long, and it ends with a large opening. The short arm is equipped with a rubber stopper (#1 to match the regular 16-by-160-mm laboratory test tube). This stick will be used in

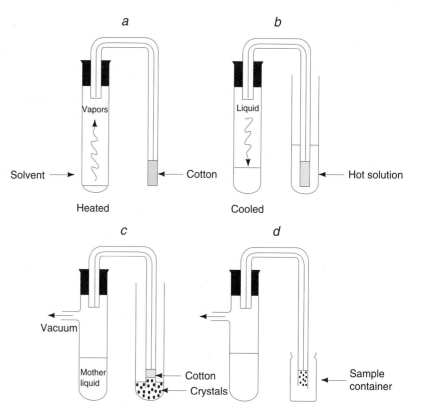

**Figure 6.7** Steps in the microrecrystallization process. *a,* Preparation of suction test tube. *b,* Hot filtration. *c,* Collection of crystals on the cotton after cooling the hot filtrate. *d,* Drying and transferring of crystals to a sample container

the experiment below for two operations: hot filtration of the solution to remove the insoluble impurities before the crystallization step, and separation of the crystals from the filtrate (mother liquor) after the crystallization has been completed.

The filtration is done in a special way that makes the operation particularly easy, safe, and, more importantly, quantitative. Instead of using the water aspirator, suction is accomplished by a vacuum resulting from condensation of hot vapors of a boiling liquid in a closed system. The filter stick is connected by a rubber stopper to a test tube. The test tube contains a small volume of the solvent used in the recrystallization (in this experiment, water), and the open end of the filter stick is packed with cotton. The test tube is heated until the boiling solvent starts dripping from the cotton packing. In this process, all the air previously present in the test tube is replaced by the vapors of the liquid. At the same time, the filter stick is warmed by the hot vapors passing through its entire length. This is required to prevent crystallization of the hot solution in the capillary during filtration. If the long stem of the filter stick is then immersed into the solution to be filtered (hot or cold), a slight decrease in temperature of the solvent vapors spontaneously pulls the solution through the filter. Deliberate cooling of the suction test tube facilitates the filtration process, which is usually complete in a few seconds. Volumes of less than 1 to 25 mL (or even larger volumes when conical flasks of adequate size are used) can be handled by this technique.

The crystals can be separated from the mother liquor by inverse filtration. The same filter stick can be used; fit it to a test tube equipped with a side arm. Suction can be applied by a water aspirator, which is suitable for this operation since the solvent is now cold. The enlarged end of the longer arm is packed with a fresh piece of cotton and placed in the test tube containing the precipitated product. The main portion of the solution above the crystals is removed first. Finally, after most of the crystals are collected in the filter stick, fresh cold solvent is added to the test tube. This helps trap the rest of the crystals, and it washes away traces of the mother liquor. At this point the test tube is replaced by a small sample container (vial or bottle). Air drying should be adequate, but suction may be continued to speed up the process. The crystals should be dried while the filter stick is in the sample container; the small quantity and light mass may cause the crystals to loosen and fall from the filter stick. The rest of the recrystallized solid can be released by gently tapping the filter stick or by using a microspatula without scraping the cotton fibers.

The advantages that the U-shaped capillary hot-filtration method offers are numerous. First, the conventional water-aspirator method causes evaporation of solvent, especially if the solvent has a low boiling point (methylene chloride, ether, petroleum ether), and consequent premature precipitation of the solid and possible entrapment of impurities will occur. Moreover, the filter-stick method allows easy quantitative separation of the insoluble impurities in the cotton-covered end of the capillary. Hot filtration with a stemless funnel and filter paper may lead to evaporation of the solvent and consequent precipitation on the filter paper.

In this experiment, benzoic acid will be recrystallized from a mixture that also contains sodium chloride and charcoal. Benzoic acid is soluble in hot water (the solvent) but insoluble in cold. Sodium chloride is soluble in both hot and cold water, while charcoal is insoluble in both hot and cold water and will be trapped in the cotton during the hot filtration step.

**Suggested References**    Horak, V., "Modification of the Svoboda Filtration Technique." *Chem. Listy.* 54 (1960): 723.

Ma, T.S., and V. Horak, *Microscale Manipulations in Chemistry.* New York: John Wiley, 1976.

Horak, V., and D.R. Crist, "Small Scale Organic Techniques: Filtration and Crystallization." *J. Chem. Educ.* 52 (1975): 664.

Svoronos, P., "Solvation and Precipitation." In *Magill's Survey of Science: Physical Science,* 2290–96. Pasadena, CA: Salem Press, 1992.

---

**Experimental**

1. Dissolve 0.5 g of a mixture of benzoic acid (75%), sodium chloride (24%), and charcoal (1%) in about 10 mL of water placed in a 25-mL Erlenmeyer flask or a large test tube, and heat over a hot plate or a water bath.

2. Place 2–3 mL of water in another test tube and secure the filter stick equipped with a rubber stopper (see Figure 6.7a). Pack the enlarged end of the filter stick with cotton. Heat the test tube or flask directly on a hot plate until practically all the water has dripped off the cotton plug.

3. At this point, place the test tube in a beaker of cool water. Insert the filter stick in the hot solution of the mixture (see Figure 6.7b) and allow the suction that is created by the cooling of the capillary-carrying test tube to filter the boiling solution. Disconnect the U-shaped capillary, and set the solution aside to cool to room temperature.

4. While waiting for the completion of crystallization, remove the cotton from the filter stick and clean the filter stick by washing it with acetone.

5. Pack the open end of the long arm with a fresh small piece of cotton, and connect the filter stick to the suction test tube (see Figure 6.7c) with its side arm attached to the vacuum line. Insert the long arm of the filter stick into the test tube containing the precipitated benzoic acid. Draw off the upper layer of the cold solution first. Lower the filter stick and continue the suction until the crystals are collected on the filter stick.

6. Rinse the walls of the test tube with about 10 drops of ice cold water and collect the crystals as quantitatively as possible. Remove the long arm of the filter stick from the test tube and insert it in a dry sample container. Continue suction for another 3–4 minutes (Figure 6.7d). Tap the stem to shake the crystals into the sample container, and use a microspatula to complete the process. Do not scrape the cotton.

7. Air dry the crystals for a few minutes, and then determine the melting point. Calculate the percent mass recovery.

**Questions**

1. In the above experiment what is the purpose of the following:
   a. using cotton?
   b. the rubber stopper on which the filter stick is attached?
   c. hot filtration?
   d. not cooling the hot filtered solution immediately?
   e. washing the crystals in the final step with *ice cold* water

2. What are the requirements for a good recrystallizing solvent?

3. What are the impurities in this experiment? How are they separated? Explain.

4. What will be the consequences if:
   a. hot filtration is performed before the dripping of the solvent occurs?
   b. not enough cotton is placed at the enlarged end of the capillary just before the hot filtration is performed?
   c. too much cotton is placed at the enlarged end of the capillary before the hot filtration is performed?

## 6.5   ISOLATION OF CAFFEINE FROM TEA

Many useful organic compounds are present in, and can be extracted and purified from, naturally occurring sources. One of the most widely used drugs in the United States is caffeine, an alkaloid (that is, a basic nitrogen-containing organic compound) present in coffee, tea, cola, chocolate and many non-prescription drugs. The purpose of this experiment is to demonstrate the isolation of a natural product from a biological source using extraction techniques, and to demonstrate purification by sublimation.

Caffeine is soluble in water because it has several polar and basic functional groups.

Caffeine
(Soluble in organic solvents)

Protonated form of caffeine
(Soluble in water)

This property also makes it insoluble in aqueous base. Thus, by adding a weak base (e.g., sodium or calcium carbonate) to an aqueous solution of tea extract, one can *decrease* its solubility in water and increase its solubility in a less polar organic solvent (such as $CH_2Cl_2$), into which the caffeine can easily be extracted, using a separatory funnel. Any neutral compounds will also be extracted into the organic phase, but fortunately, there are very few of these present in tea. Any acidic compounds (such as tannic acid, a major component of tea) will be deprotonated and will remain in the aqueous phase.

After evaporation of the organic phase, the crude caffeine can be purified by **sublimation**. In this process, a solid is heated while under partial vacuum so that it goes directly from the solid phase to the vapor phase. The vapors are then re-condensed on a cooled surface. Volatile impurities (such as excess solvent) are sucked out by the vacuum, while nonvolatile impurities, such as many salts, do not sublime and are left behind.

**Suggested References**

"Extraction." In *McGraw-Hill Encyclopedia of Science and Technology,* 7th ed., vol. 6, 572–74. New York: McGraw-Hill, 1992.

Flaschka, H. A., A. J. Bernard, Jr., and P. E. Sturrock, *Quantitative Analytical Chemistry.* 2nd ed., 512–18. Boston: Willard Grant, 1980.

Ghayourmanesh, S., "Solvent Extraction." In *Magill's Survey of Science: Applied Science,* 2391–96. Pasadena, CA; Salem Press, 1993.

Harris, D. C., *Quantitative Chemical Analysis.* 533–51. San Francisco: W. H. Freeman, 1982.

Kihlman, B. A., *Caffeine and Chromosomes.* Amsterdam: Elsevier, 1977.

Ramette, R. W., *Chemical Equilibrium and Analysis.* 571–85. Reading, MA: Addison-Wesley, 1981.

Ritchie, J. M., "Central Nervous System Stimulants. II: The Xanthines." In L. S. Goodman and A. Gilman, *The Pharmacological Basis of Therapeutics,* 7th ed. New York: Macmillian, 1985.

Svoronos, P., "Solvation and Precipitation." In *Magill's Survey of Science: Physical Science,* 2290–96. Pasadena, CA: Salem Press, 1992.

**Safety Tips**

✔  Methylene chloride is harmful when ingested or when its vapors are inhaled (RL = 350 mg/m³). Therefore, it should be used with adequate ventilation.

✔  No flammable solvents should be present while using a Bunsen burner.

✔  Avoid exposure to UV light; turn off the lamp when finished (step 18).

**Experimental**

1. In a 400 mL beaker, place 10 tea bags and approximately 100 mL of water. Bring to boil on the hotplate, and continue to boil for about 15 minutes.

2. Carefully remove the beaker from the hotplate, remove and discard the teabags, and dissolve 15 g of $Na_2CO_3$ in the tea solution by stirring.

3. Cool the tea solution in an ice-water bath, then transfer it into a separatory funnel, using a funnel.

4. Add 20 mL of methylene chloride, again via a funnel, and shake the mixture *very gently*. Do not shake too vigorously, or you will get an emulsion (i.e., a mixture consisting of droplets of one phase suspended in the other). If this happens, let it stand for about 15 minutes to allow separation of the layers.

5. Use the "water drop" technique (Experiment 6.2, step 4) to decide which layer is which, and then separate the layers.

6. Extract the aqueous layer with another 20 mL portion of methylene chloride.

7. Combine the organic extracts in a 125 mL conical flask, and dry them with about 1 g of anhydrous $MgSO_4$.

8. Allow the solution to stand for about 10 minutes, swirling it occasionally to complete the drying.

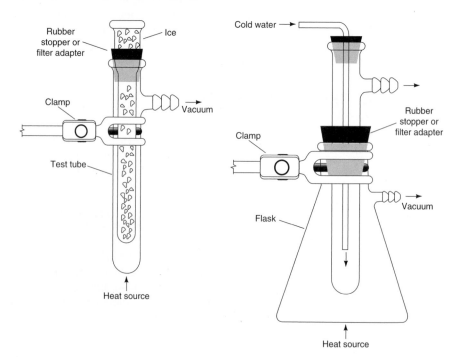

**Figure 6.8**   Assemblies for sublimation

9. Keep the aqueous phase in a beaker—do not discard it until you are sure you do not need it anymore!

10. Gravity-filter the methylene chloride solution into a small, pre-weighed beaker. Add one boiling chip, and carefully evaporate the solvent to dryness in a hot water bath or a large beaker of water on a hotplate in the hood.

11. Remove the boiling chip, re-weigh the beaker, and calculate the yield of crude caffeine.

12. The crude caffeine can be further purified by sublimation, using one of the assemblies pictured in the diagram. Scrape the caffeine from the beaker, and place it in the bottom of your side-arm test tube or filtration flask. (*Note:* check the apparatus carefully to be sure there are no cracks in it.)

13. Place the test tube condenser (securely seated in a rubber stopper) in the top of the test tube or filtration flask, and clamp the apparatus to the rack.

14. Attach the cooling-water tubing to the condenser (with the "in" line on the top inlet and the "out" line on the side inlet) and start a slow flow of water, or fill the test tube condenser with ice.

15. Connect the vacuum line to the side-arm of your apparatus, and turn on the vacuum.

16. After making sure that there are no flammable solvents open anywhere in the lab, light a Bunsen burner and gently heat the bottom of the sublimation apparatus until the caffeine begins to sublime, depositing a white crystalline solid on the condenser. When there appears to be no more sublimation, shut off the Bunsen burner and vacuum, and *carefully* remove the condenser from the test tube. If you are using a water aspirator, be sure to disconnect the vacuum line *before* turning off the water or you will suck water back into your apparatus.

17. Scrape your purified caffeine onto a piece of pre-weighed weighing paper, re-weigh it, record the mass, and calculate and record your yield. Place your product in a vial or test tube.

18. Dissolve a few milligrams of your caffeine in about 5 drops of acetone, and spot the solution on a silica gel-coated TLC slide. Develop the slide in 20:1 ($CH_2Cl_2/CH_3OH$) (chloroform-methanol, available in the hood) and visualize with UV light. Mark the spot with a pencil, and calculate the $R_f$ value. Be sure to label the TLC slide clearly; save it so you can tape it to your lab report before submitting the report.

**Questions**

1. Tea contains approximately 2% caffeine by weight. Assuming that you started with 25 g of tea leaves, calculate your percent recovery.

2. What information did you obtain from taking a TLC of your sublimed caffeine?

3. Suppose you forgot to add sodium carbonate to your tea solution prior to extraction. What effect would this have on your yield, and why?

4. Draw a phase diagram for an unspecified substance (that is, draw a plot with pressure on the *y*-axis, temperature on the *x*-axis, and solid lines denoting the regions of stability for solid, liquid and gas phases). Show points that indicate the boiling point and melting point of the substance at a given pressure ($P_1$), and the sublimation point at another pressure ($P_2$). What property of caffeine makes it possible to sublime it?

5. What might have happened to your caffeine if you had tried to sublime it at atmospheric pressure?

# chapter

# 7

# Equilibrium Constant

During the course of a chemical reaction, the concentrations of starting materials and products change with time until the equilibrium state is reached. A condition of **equilibrium** is reached in a system when two opposing changes occur simultaneously at the same rate. At any given temperature, the equilibrium state can be described by an equilibrium constant, $K_{eq}$. For the general equation

$$aA + bB \rightleftharpoons cC + dD$$

the equilibrium constant $K_{eq}$ is equal to

$$K_{eq} = \frac{[C]^c [D]^d}{[A]^a [B]^b}$$

where the molar concentration of each substance is raised to a power equal to the coefficient in the balanced chemical equation for that substance.

The value of $K_{eq}$ is constant for a specific reaction at a specific temperature. It is related to the standard Gibbs free energy ($\Delta G°$) of the reaction by the expression

$$\Delta G° = -RT \ln K_{eq}$$

where $\Delta G°$ = Gibbs free energy (cal or J), $T$ is the temperature (°K), $R$ is the gas constant (1.98 cal/mole K, or 8.31 J/mole K).

Thus, $K_{eq}$ values greater than 1 indicate a favorable free energy change and a reaction in which the position of equilibrium lies in the direction of the products. $K_{eq}$ values less than 1 indicate that the starting materials are more thermodynamically favored than the products, and that the position of equilibrium lies in the direction of the starting materials.

In this experiment, the equilibrium constant for an esterification reaction will be measured. **Fischer esterification** is an acid-catalyzed reaction between a carboxylic acid and an alcohol to form an ester and water. As with most esterifications, the formation of products is favored. Yields of ester at equilibrium are typically about 50–60%. The uncatalyzed reaction is very slow, but in the presence of sulfuric acid catalyst, the reaction reaches equilibrium in about one hour. The reaction described in this experiment is between acetic acid and *n*-propyl alcohol to form *n*-propyl acetate and water.

$$\text{CH}_3\text{COOH} + \text{CH}_3\text{CH}_2\text{CH}_2\text{OH} \overset{\text{H}^+}{\rightleftharpoons} \text{CH}_3\text{COOCH}_2\text{CH}_2\text{CH}_3 + \text{H}_2\text{O}$$

acetic acid    *n*-propyl alcohol                 *n*-propyl acetate    water

(a carboxylic    (an alcohol)                    (an ester)

    acid)

The experiment can also be performed using other reagents. A list of carboxylic acids and alcohols that give satisfactory results is given in Table 7.1.

**Table 7.1**   Densities of selected carboxylic acids and alcohols

| Compound | Formula | Density* |
|---|---|---|
| *Alcohols* | | |
| Methyl | $\text{CH}_4\text{O}$ | 0.79 |
| Ethyl | $\text{C}_2\text{H}_6\text{O}$ | 0.79 |
| *n*-Propyl | $\text{C}_3\text{H}_8\text{O}$ | 0.80 |
| Isopropyl | $\text{C}_3\text{H}_8\text{O}$ | 0.79 |
| *n*-Butyl | $\text{C}_4\text{H}_{10}\text{O}$ | 0.81 |
| | | |
| *Carboxylic acids* | | |
| Acetic | $\text{C}_2\text{H}_4\text{O}_2$ | 1.05 |
| Propionic | $\text{C}_3\text{H}_6\text{O}_2$ | 1.04 |
| Butyric | $\text{C}_4\text{H}_8\text{O}_2$ | 0.97 |
| Isobutyric | $\text{C}_4\text{H}_8\text{O}_2$ | 0.95 |
| Valeric | $\text{C}_5\text{H}_{10}\text{O}_2$ | 0.94 |

*In g/mL

The equilibrium constant value will be calculated from the expression

$$K_{\text{eq}} = \frac{[\text{ester}][\text{water}]}{[\text{acid}][\text{alcohol}]}$$

The concentration of the acid, and indirectly the concentrations of the other components, will be determined titrimetrically. The initial concentration of the acid (and the alcohol, since an equimolar quantity will be used) will be proportional to the concentration of the sodium hydroxide consumed in the first titration and can also be represented by that quantity. The molar concentration of the acid at equilibrium will be represented by the moles of base used for the second titration after it is corrected for the catalytic amount of sulfuric acid that was added. The difference between the two titrations will represent the amount of acid consumed, which is also equivalent to the amount of ester formed (which is equivalent to the water formed) since a 1:1 relative ratio is given by the equation. The equilibrium constant can then be calculated solely from the titration data.

**Suggested References**    Carey, F. A., *Organic Chemistry*. 3rd ed., 619-20, 784–88. New York: McGraw-Hill, 1996.

McMurry, J., *Organic Chemistry*. 4th ed., 157–61, 812–15. Pacific Grove, CA: Brooks/Cole, 1996.

Morrison, R. T., and R. N. Boyd, *Organic Chemistry.* 6th ed., 730, 737–38. Englewood Cliffs, NJ.: Prentice-Hall, 1992.

Sarlo, E., P. Svoronos, and P. Kulas, "Equilibrium Constants of Esterification Reactions." *J. Chem. Educ.* 67 (1990): 796.

Solomons, T. W. G., *Organic Chemistry.* 6th ed., 812–15. New York: John Wiley, 1992.

Wade, L. G., *Organic Chemistry.* 3rd ed., 962–64, 1003–05. Englewood Cliffs, NJ: Prentice-Hall, 1995.

Vollhardt, K. P. C., and N. E. Schore, *Organic Chemistry.* 2nd ed., 739–741. New York: W. H. Freeman, 1994.

**Experimental**

*Note:* This experiment can be performed with a series of simple carboxylic acids and alcohols using equimolar quantities of each reactant. The quantities may be measured volumetrically. Table 7.1 lists the densities and formulas of several simple carboxylic acids and alcohols. The following procedure uses 0.25 moles each of acetic acid and *n*-propyl alcohol.

1. Place 14.3 mL (15.0 g, 0.25 mole) of acetic acid and 18.8 mL (15.0 g, 0.25 mole) of *n*-propyl alcohol in a 100-mL round bottomed flask, add a few boiling chips and swirl the flask.
2. Transfer 1.0 mL of the mixture with a volumetric pipette to a 125-mL conical flask, add 20 mL of water, 2 drops of phenolphthalein, and titrate with a standardized 0.5$N$ sodium hydroxide solution. Record the volume used in the data section.
3. Add 8 drops of concentrated $H_2SO_4$ to the remaining alcohol-acid mixture in the round bottomed flask, and reflux for about 1 1/2 hours (Figure 7.1).
4. Remove 1.0 mL of the mixture using the *same clear* volumetric pipette and titrate as before. Record the volume of the second titration on the data sheet. Assume that 0.5 mL of sodium hydroxide were used to neutralize the catalytic amount of sulfuric acid present. This amount should be subtracted from the volume of the second titration.
5. Calculate the equilibrium constant for the reaction.
6. Wash the burette after the last titration, *while still clamped*. Drain the burette, then slowly add about 10 mL of water and drain; then 10 mL of dilute hydrochloric acid and drain; finally 50 mL of distilled water and drain. *Do not place the burette under the faucet in order to wash it.*

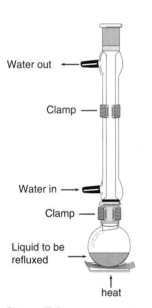

Water out →

Clamp —

Water in →

Clamp —

Liquid to be refluxed →

↑ heat

**Figure 7.1**   Setup for reflux

**Questions**

1. What would be the effect on the amount of ester formed if:
   a. an 80% 1-propanol solution in water was used instead of anhydrous 1-propanol?
   b. a 1.0$N$ NaOH instead of a 0.5$N$ NaOH solution was used for titration?
2. Calculate the value of $\Delta G°$ from your calculated $K_{eq}$. Assume the reaction temperature was 100°C.
3. Calculate the value for $\Delta G°$ (at 27°C) for:
   a. a reaction whose equilibrium constant is 10.
   b. a reaction whose equilibrium constant is 0.1.
      Given that R = 1.98 cal/K.mole
4. Suppose that the initial amount of acetic acid was doubled and the concentration of *n*-propyl alcohol remained the same in this experiment.
   a. How would these conditions affect the amount of ester formed?
   b. How would the added acid affect the value of $K_{eq}$? Explain.

# DATA SHEET: EQUILIBRIUM CONSTANT

1. Volume of _____ (alcohol)                    _____ mL

2. Volume of _____ (carboxylic acid)            _____ mL

3. Volume of base, first titration ($V_1$)               _____ mL

4. Volume of base, second titration ($V_2$)              _____ mL

5. Volume of base required for $H_2SO_4$                  __0.5__ mL

6. Corrected volume for the second titration
   (#4) − (#5) = $V_3$                                   _____ mL

## Calculation of Equilibrium Constant

$$K_{eq} = \frac{(V_1 - V_3)^2}{(V_3)^2}$$

# chapter

8

# The Sodium Fusion Test

To identify an unknown organic substance, it is often necessary to verify the presence or absence of elements other than the usual carbon, hydrogen, and oxygen. Usually these other elements are nitrogen, sulfur, and/or halogen(s). The preliminary qualitative tests that are usually performed depend on first transforming the covalently bound elements into simple ions. This is conveniently done by fusing the organic compound with sodium metal.

A preliminary ignition of a 0.1–0.2 g sample of the unknown on the tip of a spatula held over a free flame can offer some information about the make-up of the compound. A sooty flame suggests unsaturation (possibly aromatic), while a small residue left on the spatula could indicate the presence of metal ion. Condensed water vapor confirms the presence of hydrogen.

The detection of sulfur, nitrogen, and halogen(s) depends on the fact that fusion with sodium metal converts these elements into easily identifiable ions.

$$(C, H, O, N, S, X) \xrightarrow[\text{fusion}]{\text{Na}} NaCN + Na_2S + NaX + H_2O$$

For safety reasons the sodium fusion test should be conducted with dry compounds. The test is **not** recommended for alcohols and organic acids (because of their vigorous reaction with sodium metal) or for low molecular weight ethers (because of their flammability). Several polyhalogenated (e.g., carbon tetrachloride, chloroform) liquids also react vigorously.

Sulfur is identified as a black precipitate of lead (II) sulfide after addition of lead (II) acetate to the acidified test solution. Sodium nitroprusside can also be used in the detection of sulfur; it forms a red complex with the sulfide anion.

Halogens are identified by treating the acidified test solution with silver nitrate. When sulfur or nitrogen is present, the solution must be boiled to remove the very small amounts of $H_2S$ or HCN gas that would form in the acidified test solution and interfere with the halogen analysis.

Nitrogen is identified as a blue complex of Prussian blue, $KFe[Fe(CN)_6]$, after successive treatment of the test solution with ferrous ammonium sulfate, potassium fluoride (to stabilize the complex), sulfuric acid, and ferric chloride. Urea and its derivatives rarely give the Prussian blue complex, probably because urea's nitrogen is not converted into sodium cyanide during the fusion with sodium. The presence of sulfur can interfere with the nitrogen analysis, especially in the case of nitro compounds.

**Suggested References**

Bruno, T.J., and P.D.N. Svoronos, *CRC Handbook of Basic Tables for Chemical Analysis.* 416–17. Boca Raton, FL: CRC Press, 1989.

Morrison, R. T., and R. N. Boyd, *Organic Chemistry.* 6th ed., 72. Englewood Cliffs, NJ: Prentice-Hall, 1992.

"Sodium." In *Reagents for Organic Synthesis,* vol. 1, 1022-23 by L. F. Fieser and M. Fieser. New York: John Wiley, 1967.

Tucker, S. H., "A Lost Centenary: Lassaigne's Test for Nitrogen: The Identification of Nitrogen, Sulfur and Halogens in Organic Compounds." *J. Chem. Educ.* 22 (1945): 212.

**Experimental**

Perform each of the following tests on a known sample until you obtain satisfactory test results from each. Perform the tests on an unknown and report your results on the data sheet. The unknowns can be selected from the compounds listed in Table 8.1.

**Table 8.1**   Suggested compounds for the sodium fusion test

| Compound | Positive Test for |
|---|---|
| Benzamide | N |
| Diphenyl sulfone | S |
| *p*-Nitrotoluene | N |
| Sulfanilamide | S, N |
| *p*-Chloroacetanilide | N, Cl |
| *p*-Nitroacetanilide | N |
| Acetamide | N |
| *o*-Iodobenzamide | N, I |
| *p*-Chlorobenzamide | N, Cl |
| *p*-Nitroanisole | N |
| *p*-Bromoacetanilide | Br, N |
| *p*-Bromobenzenesulfonamide | Br, N, S |
| *o*-Chlorobenzaldehyde | Cl |
| *p*-Toluenesulfonamide | N, S |
| *p*-Nitrobenzenesulfonamide | N, S |
| 2, 4, 6-Tribromoaniline | Br, N |

## 8.1   SODIUM FUSION

*Caution:* **Sodium metal is dangerous and should be treated with respect. Never allow it to come into contact with water. Handle sodium with forceps, cut with a sharp knife, and dry with a piece of filter paper. Scraps of sodium metal should be destroyed by treating them with a large volume of ethanol under a fume hood.** *Safety goggles must always be worn.*

1. Place a piece of sodium metal (the size of a small pea) in a clean, *dry* 3-inch test tube. Hold the test tube with a test tube holder and make sure it is not pointed at anyone.

2. Heat the lower part of the test tube with a hot flame until the sodium metal melts and its vapors rise in the tube.
3. Remove the test tube from the flame and quickly (but carefully) allow 5–10 drops of the unknown (or a pinch if it is a solid) to fall directly into the sodium vapor without touching the side of the tube.
4. Heat the bottom of the test tube for two minutes until it is red hot.
5. Plunge the red hot tube into 5–10 mL of distilled water in a 25-mL beaker. Have your wire gauze ready to cover the beaker in case of a small fire. The bottom of the test tube will disintegrate, and its contents should dissolve in water. This solution is the test solution.
6. Transfer the broken glass and any undissolved solids to a mortar and crush the mixture with a pestle. Return the contents of the mortar to the test solution in the small beaker. Rinse the mortar with 4–5 mL of water and add this to the beaker. Boil the contents of the beaker for a few minutes to extract any sodium salts from the glass. Filter into a clean test tube or into a small conical flask. The filtrate should be alkaline (test with red litmus paper).

## 8.2   TEST FOR SULFUR

1. To 10 drops of the filtered test solution add 2 drops of a freshly prepared 1% sodium nitroprusside solution. A deep red-violet color indicates the presence of sulfur.
2. To 10 drops of the test solution, add 2 drops of dilute acetic acid, followed by 5 drops of a 5% lead (II) acetate solution. A black precipitate indicates the presence of sulfur.

## 8.3   TEST FOR HALOGENS

Acidify 3 mL of the test solution with dilute nitric acid. If no sulfur or nitrogen is present, add 4–5 drops of 5% silver nitrate solution. A precipitate indicates the presence of halogens. If sulfur or nitrogen is present, boil the acidified 3 mL of the test solution down to about 1 mL. Add 4–5 drops of 5% silver nitrate solution. A precipitate indicates the presence of halogen. If the precipitate is

a. white and soluble in concentrated ammonium hydroxide, it is silver chloride.
b. pale yellow and sparingly soluble in concentrated ammonium hydroxide, it is silver bromide.
c. bright yellow and insoluble in concentrated ammonium hydroxide, it is silver iodide.

## 8.4   TEST FOR NITROGEN

Add 5 drops of a freshly prepared saturated solution of ferrous ammonium sulfate and 5 drops of a 10% potassium fluoride solution to 3 mL of the filtrate. Boil the mixture for about 30 seconds, allow the suspension of iron hydroxides to cool, and then add 2 drops of a 5% ferric chloride solution. Finally, add sufficient dilute sulfuric acid to dissolve the iron hydroxides and make the solution acidic to litmus. If nitrogen (which has been converted to sodium cyanide upon fusion with sodium metal) is present, a brilliant blue solution or suspension of Prussian blue, $KFe[Fe(CN)_6]$, will appear. Formation of a greenish-blue color suggests that nitrogen is present, but that fusion was incomplete. In this case, the entire procedure should be repeated.

**Questions**      1. Provide equations for the reactions of the following in the sodium fusion test solution:
   a. chloride
   b. sulfide
   c. cyanide ions.

2. What is the purpose of:
   a. grinding the disintegrated test tube in a mortar before filtering the solution?
   b. boiling down the solution to approximately one third of its original volume before adding silver nitrate, when testing for halogens?
   c. acidifying the solution before testing for halogens?

# DATA SHEET: SODIUM FUSION TEST

*Test for Sulfur:*       Positive _____       Negative _____

*Test for Halogen:*      Positive _____       Negative _____

*Test for Nitrogen:*     Positive _____       Negative _____

# chapter 9

# Spectroscopy

After a compound has been synthesized, isolated, and purified, its identity must be verified. This is accomplished by comparing its physical properties (melting point, boiling point, refractive index) with the literature values, as well as by performing tests for the presence of functional groups (e.g., decolorization of $Br_2/CCl_4$, which indicates the presence of double or triple bonds; formation of an orange 2,4-DNP precipitate, which indicates the presence of a carbonyl group; determination of the compound's solubility in dilute $NaHCO_3$, which indicates the presence of a carboxyl, —COOH, group). The large quantity of compound required for such qualitative tests is a disadvantage. **Spectroscopy** is of great help in verifying the identity of a small amount of a compound, especially in establishing its structure if it has not been previously prepared. Various spectroscopic procedures can enable the chemist to unambiguously and confidently ascertain the presence of functional groups and, eventually to postulate the compound's complete structure. In this chapter, three spectroscopic techniques are presented: **infrared (IR), proton magnetic resonance (PMR, and UV-visible)** spectroscopy. The task of spectroscopy is greatly simplified if supplementary data are given, such as elemental analysis and mass spectroscopy information. Only a small amount of material is required for a routine spectroscopic analysis, and this material can be easily recovered since the process is nondestructive. Solid, liquid, and gaseous substances can all be analyzed; the handling and processing of solids and liquids are discussed in this chapter.

## 9.1   INFRARED SPECTROSCOPY

Electromagnetic radiation can be described in terms of either **frequency** ($v$) or **wavelength** ($\lambda$). These two parameters are related to each other by the **velocity of light, c,** which is equal to $3 \times 10^{10}$ cm/sec, or approximately 186,000 miles/sec. The equation is

$$E = hv = hc/\lambda$$

where **E** is the energy of the electromagnetic radiation and $h$ is **Planck's constant,** $6.63 \times 10^{-34}$ joule.sec, or $6.63 \times 10^{-27}$ erg.sec.

The broad electromagnetic spectrum is subdivided into smaller regions of energy ranges; absorption of different energies produce different kinds of excitations in molecules. The absorption of ultraviolet radiation, for instance, causes electronic transitions. The absorption of infrared radiation is related to molecular vibrations, and provides energies that *stretch* bonds (vibrate a bond along the bond axis, resulting in higher energies) and **bend** bonds (vibrate a bond perpendicular to the bond axis, leading to lower energies).

The infrared spectrophotometer records the spectrum produced when a sample that is exposed to continuously changing wavelengths of infrared radiation absorbs specific frequencies corresponding to the energy of a particular molecular vibration. Modern instruments record the absorption of energy as a function of decreasing frequency (decreasing energy) along the x-axis from left to right as the **wave number,** a frequency-related measurement with units of cm$^{-1}$. **Wavelength** increases from left to right, and is usually recorded in **microns,** μ (1 micron = 1 micrometer = $1 \times 10^{-6}$ m). The y-axis measures the **percent transmittance.** The IR spectrum of polystyrene is seen in Figure 9.1. A wave number at which there is almost 100% transmittance (3500 cm$^{-1}$), shows virtually no absorbance, while a wave number at which there is very little transmittance (1601 cm$^{-1}$) indicates that the molecule is undergoing a transition that leads to a very significant peak in the spectrum. The intensities of the infrared absorption bands vary from strong to medium to weak, depending on the degree of change in **dipole moment** upon excitation.

Figure 9.2 is a schematic diagram of a simple spectrophotometer. Infrared energy is provided by a source rich in IR radiation, such as the Nernst glower or the Nichrome wire. The double-beam nature of dispersion instruments eliminates various interfering atmospheric absorptions

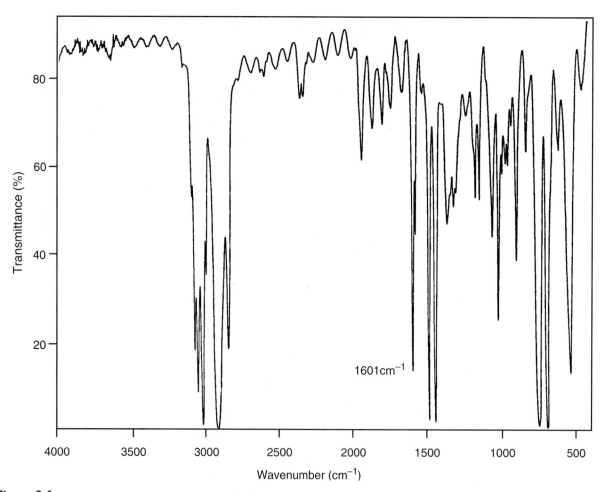

**Figure 9.1**  The infrared spectrum of polystyrene

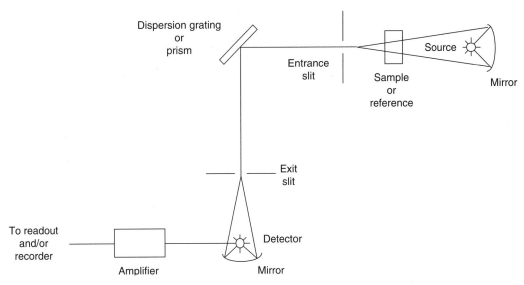

**Figure 9.2** General schematic diagram of a simple spectrophotometer

and energy variations from the recorded spectrum. The radiation is split into two parallel beams of equal intensity: one beam passes through the sample, and the other through the reference space (which is composed of air, if the sample is neat [undiluted]; or of pure solvent, if the sample is in solution). Upon leaving the sample and reference spaces, each beam is chopped by a rotating mirror and directed through the **dispersion grating** (monochromator or prism), which sorts out the individual frequencies of the radiation transmitted by the sample and the reference. The double-beam setup is adjusted so that when the beams that have passed through the sample and reference have the same intensity, no signal is observed. After passing through the dispersion grating, the light falls onto the **detector** (the thermocouple in the double-beam dispersion spectrophotometer), and is further amplified and recorded. Modern instruments are FT-IR (Fourier transform infrared) spectrophotometers that utilize computers.

Quantitative determinations in IR spectrophotometry use Beer's law in the same way visible-ultraviolet spectrophotometry does (see the section "UV-Visible Spectrophotometry," Experiment 9.3). However, various IR spectrophotometers currently in use differ considerably in dispersion, resolving power, and stray radiation—to a great degree because of the variation in the slit widths that determine the amount of infrared radiation passing through. As a result, the relative intensities of bands may sometimes appear to be dissimilar, and the differences much more pronounced from one instrument to another than they are in the visible-ultraviolet region. However, the positions of peaks (in wavelengths or wavenumbers) should *always* agree no matter what instrument is used.

Organic compounds in the IR region have complex spectra unlike inorganic compounds in the visible-UV region. Most organic compounds show several absorption peaks in the IR. However, traces of impurities usually show only the most intense peaks (such as the >C = O stretch in ethyl acetate or acetone, or the —OH stretch in ethanol or water). Qualitative infrared measurements to detect functional groups present in a compound are done with the following goals:

1. To establish the identity of a compound;
2. To verify the purity of a sample; and
3. To help identify any impurities that might be present in the sample.

**Table 9.1**   Characteristic infrared absorption bands

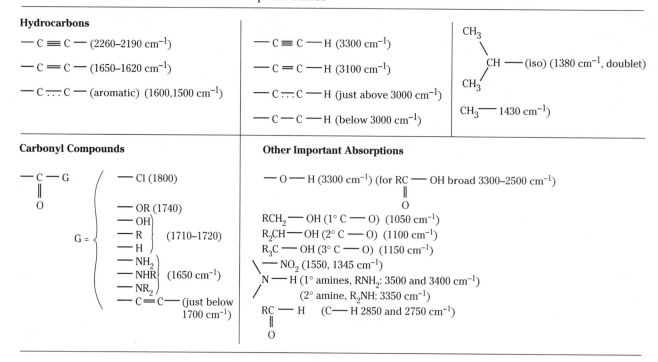

Table 9.1 lists the characteristic absorption bands of the different functional groups of organic families. It should be noted that **not all the absorption bands that appear in an infrared spectrum are useful for determining the compound's structural formula.** Many different fundamental absorptions occur in the spectrum—especially in the region below 1300 cm$^{-1}$. Table 9.2 gives the aromatic substitution patterns that help identify the positions of substituents in a benzene ring.

It is very difficult to establish the structure of a compound using only its infrared spectrum. For some simple compounds, however, the molecular formula and the IR spectrum may be enough, although usually more information is required, such as NMR and mass spectral data.

In this section, three experiments will be performed to demonstrate the use of the IR spectrophotometer:

1.  The identification of plastic films or wrappers
2.  The depolymerization-repolymerization of Plexiglass
3.  The identification of an unknown compound.

The instructor will show the students how to run the IR spectrophotometer. Each instrument has its own design, and the student should be aware of the individual steps involved in obtaining a spectrum.

### Handling the Liquid Sample

A substance in the liquid state can be examined neat or in solution. Neat samples will be used to simplify the procedure and the data interpretation. The liquid-film technique is most commonly used; it requires a demountable cell. Traditionally, a cell consists of two sodium chloride windows fitted into a metal holder. The cell can either be held in place by a metal plate that is then bolted to the base of the holder (Figure 9.3a), or it can be inserted into a metal cylinder that is attached to a rectangular base (Figure 9.3b). In the latter configuration, a friction-fit metal retaining collar is slipped over the O-ring around the outside of the cylinder (Figure 9.3b).

**Table 9.2** Infrared patterns of substituted aromatic compounds

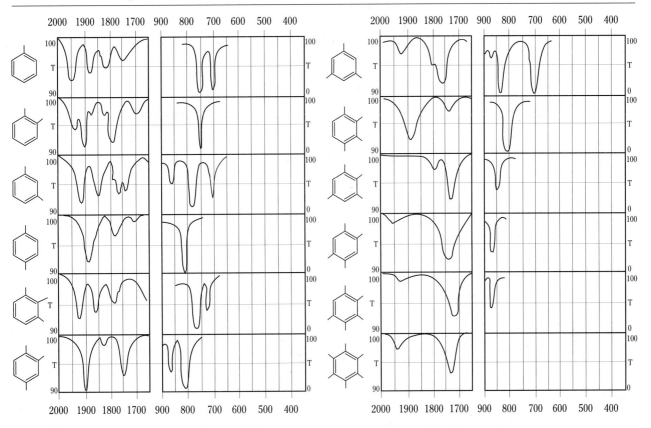

Sodium chloride plates are increasingly being replaced by silver chloride plates which employ a plastic holder that is much easier to handle (Figure 9.4). There are several advantages to silver chloride plates. First, they are much smaller, thinner, and cheaper than sodium chloride plates. Second, they are relatively inert towards practically all organic compounds, *except the amines.* Third, they are not as susceptible to water. Sodium chloride plates are extremely susceptible to moisture, whether from the air or from the sample itself. In the distillation of some compounds (e.g., cyclohexene, t-butyl chloride, cyclohexanone) traces of water cannot be excluded; this water is capable of damaging (dissolving) the sodium chloride plate. The only precaution associated with silver chloride plates is to avoid prolonged exposure to sunlight. After the IR spectrum is run, the plates are washed with a volatile organic solvent (preferably methylene chloride or acetone) and dried with tissues. The plates should then be placed in their black cloth sleeves to avoid further exposure to light. If sodium chloride plates are used, they should be stored in a dessicator.

## Obtaining the Infrared Spectrum of a Liquid Sample

1. Carefully remove the silver chloride windows from the black cloth sleeve in the storage container, and place them on a clean tissue or Kimwipe (Figure 9.4).
2. Use a disposable pipette to deposit one small drop of the liquid sample on one of the windows.
3. Place the other window on top of the window on which the sample was deposited.

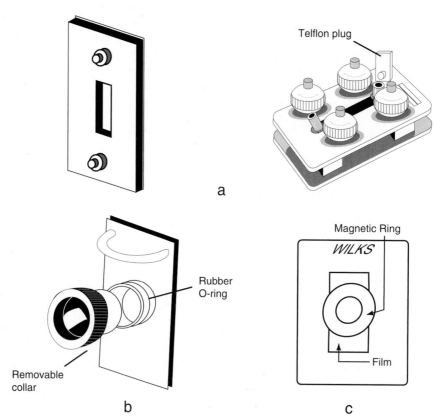

**Figure 9.3**  Infrared spectrophotometry cells: *a,* demountable cell with rectangular salt plate window; *b,* cell with circular salt plate window; *c,* magnetic film holder

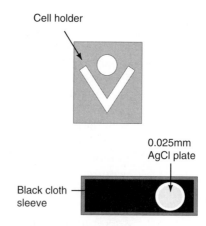

**Figure 9.4**  The silver chloride and cell holder plate

4. Carefully place both windows in the minicell body that comes with the AgCl plates and replace the top making sure that the unit is properly sealed.
5. Place the minicell body on the cell holder and measure the spectrum.
6. Remove the minicell body from the spectrophotometer as soon as the spectrum has been obtained.
7. Disassemble the minicell body, and wash the silver chloride plates quickly with a few milliliters of a volatile solvent (preferably acetone or methylene chloride) in the hood, using a disposable pipette. Wipe the cells with clean Kimwipes or tissue.
8. If the plastic minicell body has been contaminated by the compound whose spectrum was obtained, clean it with a few drops of dry acetone and wipe out with clean kimwipes or tissue.

9. Place the silver chloride windows in the black cloth sleeve. Place the cells and the minicell body in the storage container.

10. If sodium chloride plates were used instead of silver chloride windows, care should be taken to avoid their exposure to moisture. The cleaning solvent should also be dry.

### Handling the Solid Sample

There are two ways to prepare solids. In the **mull method,** a suspension of the finely powdered sample in mineral oil is smeared between two blank plates. In the **KBr disc,** or **pellet method,** a solid "solution" of the compound in dry potassium bromide is prepared. In the latter method, any peaks that are due to the mineral oil used in the mull method will be absent. The KBr method will be used here because of the "clean" spectra obtained and because the only precaution necessary is to keep the potassium bromide dry in a dessicator or an oven.

### Obtaining the Spectrum of a Solid Sample

1. Carefully grind a 10:1 (or, at most, a 5:1) *dry* potassium bromide-sample mixture for about 3–4 minutes or until a homogeneous, fine powder is obtained using a mortar and pestle.

2. Place a polished bolt on one of the ends of the minipress (Figure 9.5a), and then secure the die on a vise, with the polished bolt facing down.

3. Pour the ground mixture inside the die (Figure 9.5b). Tap the minipress to make the KBr powder layer more uniformly thick. Keep in mind that eventually the pellet will have to be thin, so only enough mixture to just cover the inside part of the bolt should be added.

4. Place the second bolt in the die, and then turn it tight with a wrench. Be careful to avoid stripping the threads of the die. Allow 30 seconds for the pressed powder mixture to set (Figure 9.5c).

5. Loosen one of the bolts carefully while the minipress is still mounted on the vise. Release the die from the vise and remove the bottom polished bolt, being careful not to disturb the pressed pellet that is lodged inside the body of the minipress.

6. The pellet should be held between grooves of the die and should be transparent. If too much powder was originally added, the pellet will not be transparent. If insufficient mixture was originally added, the pellet will not remain on the grooves of the die. If the mixture is wet,

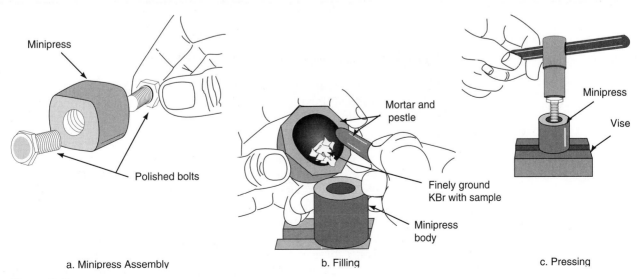

a. Minipress Assembly                    b. Filling                    c. Pressing

**Figure 9.5**   Steps for the preparation of the KBr pellet

the pellet will collapse upon removal of the bolts. In either case, the procedure must be repeated.

7. Place the body of the minipress on the special cell holder (Figure 9.4) facing the sample beam of the IR spectrophotometer, and run the spectrum.

8. Clean the KBr die with warm water until it is free of the solid. Rinse with acetone to remove the water, and dry in the oven. Place the individual parts in their respective containers.

*Note:* To obtain a good KBr pellet, you need experience. You might have to repeat the procedure more than once, or at least until you get it right. ***Patience is a virtue very much associated with scientists!***

**Suggested References**

Bruno, T. J., and P. Svoronos, *Handbook of Basic Tables for Chemical Analysis.* Boca Raton, FL: CRC Press, 1989.

Carey, F. A., *Organic Chemistry.* 3rd ed., 542–47. New York: McGraw-Hill, 1996.

Conley, R. T., *Infrared Spectroscopy.* Boston: Allyn and Bacon, 1972.

McMurry, J., *Organic Chemistry.* 4th ed., 433–46. Pacific Grove, CA: Brooks/Cole, 1996.

Morrison, R. T., and R. N. Boyd, *Organic Chemistry.* 6th ed., 590–96. Englewood Cliffs, NJ: Prentice-Hall, 1992.

Nakanishi, K., and P. H. Solomon, *Infrared Absorption Spectroscopy.* 2nd ed. San Francisco: Holden Day, 1977.

Silverstein, R. M., C. Bassler, and T. C. Morrill, *Spectrometric Identification of Organic Compounds.* New York: John Wiley, 1991.

Solomons, T. W. G., *Organic Chemistry.* 6th ed., 541–48. New York: John Wiley, 1996.

Vollhardt, K. P. C., and N. E. Schore, *Organic Chemistry.* 2nd ed., 387–92. New York: W. H. Freeman, 1994.

Wade, L. G., *Organic Chemistry.* 3rd ed., 477–505. Englewood Cliffs, NJ: Prentice-Hall, 1995.

Wilks, Jr., P. A., "Infrared Equipment for Teaching." *J. Chem. Educ.* 46 (January 1969): A9.

## Comparison of Plastic Films or Wrappers

1. The IR spectra of several common polymer films will be provided by the instructor. The student should compare the spectra for similarities and differences to identify his or her unknown. Although the films look the same, they can be easily distinguished by their IR spectra.

2. Place your (unknown) film in a magnetic film holder (Figure 9.3c) and obtain its IR spectrum. Compare your unknown with the spectra provided by the instructor. The spectra of Reynolds Oven Cooking Bag (Figure 9.6), Hefty Baggies (Figure 9.7), and polystyrene (Figure 9.1) are given. Which one is an ethylene glycol polymer?

## Depolymerization-Repolymerization of Plexiglass

Plexiglass is a polymer of methyl methacrylate, from which it is formed with a free-radical polymerizing agent such as benzoyl peroxide. **Depolymerization** (reformation of the monomer) can be achieved by heating Plexiglass.

methyl methacrylate                                        Plexiglass

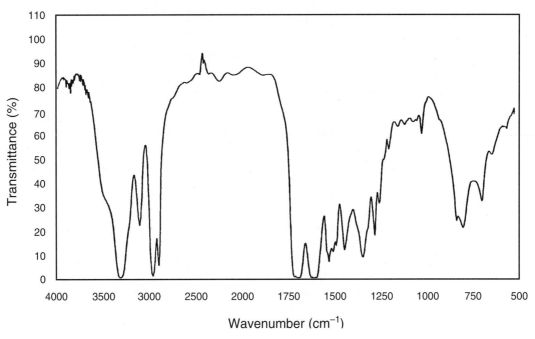

**Figure 9.6**   Infrared spectrum of Reynolds Oven Cooking Bag

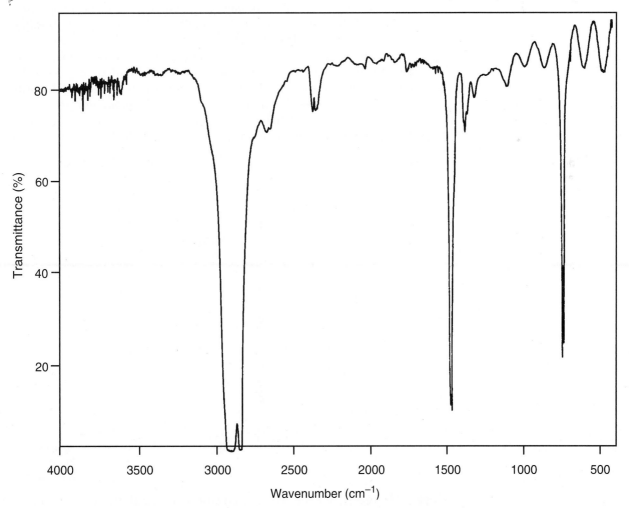

**Figure 9.7**   Infrared spectrum of Hefty Baggies

**Experimental**

1. Place 3–4 small pieces (about 1–2 g each) of Plexiglass (or Lucite) in a test tube, and attach a cork stopper containing a bent (90° to 120° angle) 6–8 mm glass tube 8 inches long. Clamp the test tube onto a ring stand as seen in Figure 9.8.

2. Place another clean test tube at the long end of the glass tube to receive the distilled methyl methacrylate.

3. Heat the polymer-containing test tube gently with a flame until no solid remains. Keep moving the flame along the test tube to prevent the vapor from condensing, but keep it away from the cork stopper and the open receiver tube. Collect the distilled monomer in an ice bath or a cold-water bath.

4. Use a dropper to transfer 1–2 drops of the distillate between two AgCl plates, and obtain the IR spectrum of pure methyl methacrylate.

5. Add about 10 mg (a few grains) of benzoyl peroxide to the remaining contents of the receiver tube, as in Figure 9.8. ***Caution: Use a wooden (never a metal!) spatula to transfer benzoyl peroxide. Although the compound is stable and has a long shelf life, it can cause sparks when in contact with metals.*** Loosely stopper the receiver tube, and heat in a bath of boiling water. After 20–30 minutes, the mixture will become viscous and will eventually turn into a clear (or slightly yellow) solid.

6. Use your spatula to transfer a small piece of the polymer into a tiny test tube. Add a few drops of methylene chloride to make a saturated solution of the polymerized methyl methacrylate. Using a disposable pipette or a dropper, transfer a few drops of the solution to the cleaned silver chloride plates and allow the methylene chloride to evaporate, thus forming a thin film of the polymer. Run the IR spectrum of the polymer and compare it with the spectrum obtained in step 4. Identify the main differences. Are there any traces of the monomer present?

7. If Plexiglass is not available, one can perform the same experiment using styrene as the monomer. Polystyrene can be obtained in a

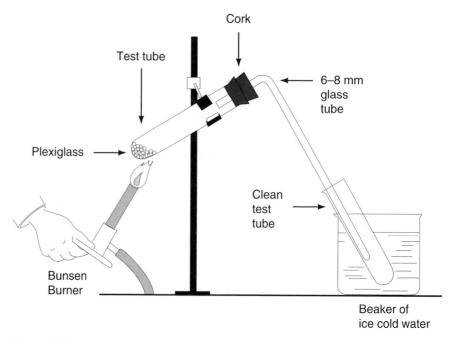

**Figure 9.8**   Setup for the depolymerization of Plexiglass

procedure similar to that outlined in step 5. In this case methylene chloride (step 6) may be replaced with toluene, which, unfortunately, may take longer to evaporate.

### Running an unknown sample

Run the IR spectrum of an unknown obtained from your instructor. The unknown is one of the compounds listed in Table 9.3. Check the literature for physical constants to determine whether each compound in Table 9.3 is a solid or liquid at room temperature. For each compound, determine the functional group(s) present, and predict the major absorption bands. Compare the physical state and IR spectrum of your unknown with the data you entered in Table 9.3. Which is the most likely compound for your unknown?

**Table 9.3** IR unknown samples selected from this list

| Name | Formula | Liquid/ Solid | Expected IR Peaks |
|---|---|---|---|
| *t*-Butyl alcohol | $(CH_3)_3C-OH$ | Liquid | |
| Ethyl acetate | | | |
| *m*-Xylene | | | |
| Isobutyl alcohol | | | |
| *sec*-Butyl alcohol | | | |
| Benzonitrile | | | |
| Benzoic acid | | | |
| Methyl salicylate | | | |
| Ethyl acetoacetate | | | |
| Benzophenone | | | |
| Acetanilide | | | |
| Diphenyl acetylene | | | |
| *p*-Dichlorobenzene | | | |
| *n*-Butyl alcohol | | | |
| Phenylacetylene | | | |
| Acetamide | | | |
| Phenylacetic acid | | | |
| Cyclohexene | | | |
| Isopropyl ether | | | |
| *o*-Xylene | | | |
| *p*-Nitrobenzaldehyde | | | |
| *n*-Butyl bromide | | | |
| Benzalacetophenone | | | |
| Maleic anhydride | | | |
| *p*-Xylene | | | |

**Questions**    1. Identify the compounds that display IR spectra with the following peaks:

*A* **($C_2H_3O_2Cl$):** 3200–2500 (broad), 1720, 1410 cm⁻¹.

*B* **($C_8H_8O$):** 3030, 2820, 2760, 1715, 1605, 1595, 1495, 1410, 750, 695 cm⁻¹.

*C* **($C_8H_7N$):** 3080, 3050, 3020, 2950, 2220, 1600, 1495, 1410, 745, 695 cm⁻¹.

*D* **($C_8H_8O$):** 3020, 2970, 1695, 1600, 1480, 1435, 760, 690 cm⁻¹.

2. Identify the compounds whose IR spectra are seen below:

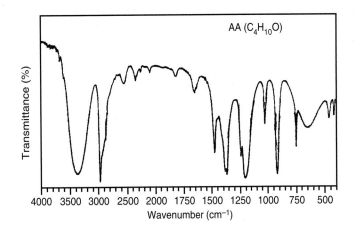

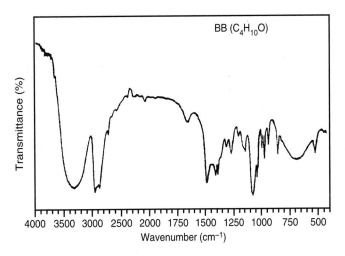

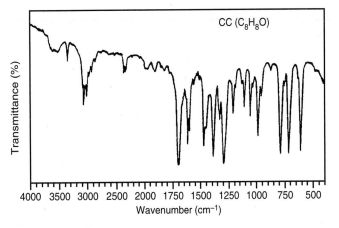

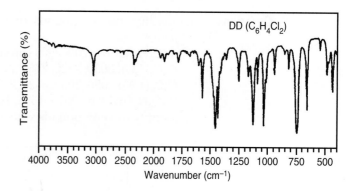

3. Match the following compounds with the IR spectra seen below:
   a. cyclohexene
   b. ethyl acetoacetate
   c. cyclohexanol
   d. cyclohexanone
   e. 2-pentanol
   f. benzonitrile
   g. methyl benzoate
   h. 2-methyl-3-buten-2-ol.

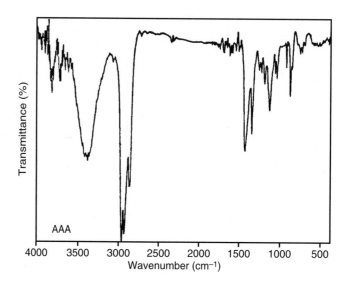

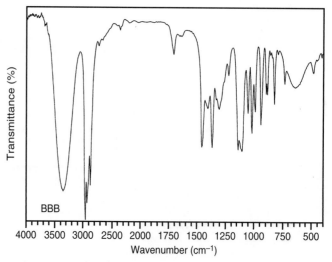

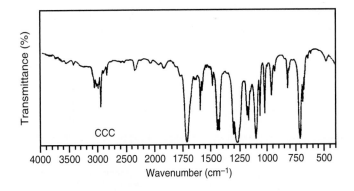

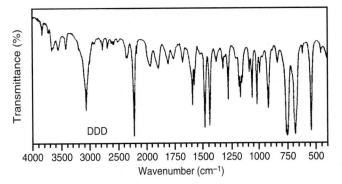

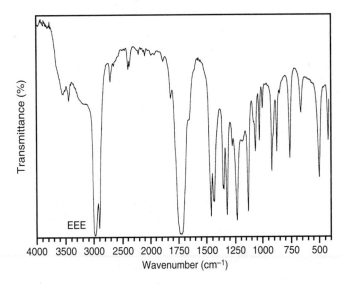

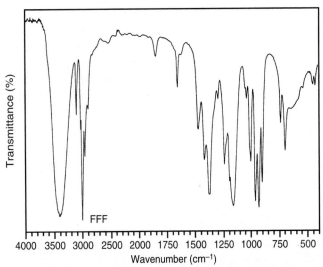

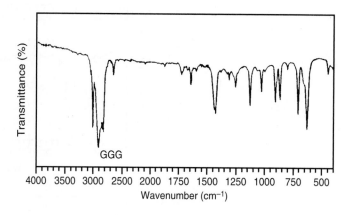

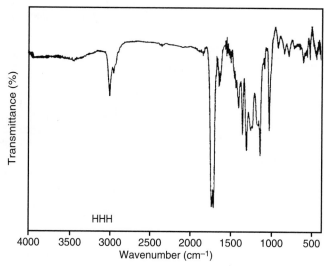

4. Identify an IR peak that would help distinguish between the following pairs of compounds:
   a. cyclohexene and cyclohexane
   b. 1-pentyne and 2-pentyne
   c. 1-pentene and 1-pentyne
   d. 1-butanol and 2-butanol
   e. 1-butanol and diethyl ether
   f. o-xylene and p-xylene
   g. acetic acid and methyl acetate
   h. 2-pentene and 2-pentyne
   i. 1-pentyne and 1-pentanol
   j. acetone and propionaldehyde
   k. phenol and cyclohexanol
   l. 1-butyne and propionitrile
   m. 3-buten-2-one and 2-butanone
   n. ethylbenzene and m-xylene.

## 9.2  NUCLEAR MAGNETIC RESONANCE

The progress in **nuclear magnetic resonance** (NMR) spectrometry has caused a revolution in the practice of organic chemistry. Almost every article published in an organic chemical journal today involves an NMR spectrometry technique at one point or another to solve a problem under consideration. The commercialization and easy access of the instrument enables many undergraduate institutions to use NMR spectrometry in the identification of an isolated product.

To increase the applicability of an NMR experiment, the procedure described here can be used in two ways: in departments where a spectrometer is available for undergraduate use, the products' ratio in the acid-catalyzed dehydration of pinacol can be experimentally determined. Or, in departments where a spectrometer is not available, the experiment can be used as a pencil-and-paper exercise using the data obtained from the spectrum of Figure 9.10.

### Theory

Nuclear magnetic resonance spectrometry involves transitions between nuclear spin states with an accompanying absorption of energy. The nuclei of several isotopes of elements commonly found in organic compounds (such as $^1H$, $^2H$, $^{13}C$, $^{19}F$, and $^{31}P$) have a non-zero magnetic moment, and therefore, like electrons, these nuclei are associated with a **spin.** Because of this spin, these nuclei can provide usable NMR signals. The spin of these excited nuclei generates a magnetic moment along the axis of the spin, and as a result, they act as small magnets. Thus, when a proton ($^1H$) is located in an external magnetic field, $H_o$, its magnetic moment can be aligned with or against the external field, and it will assume $(2I + 1)$ spin orientations (**spin states**), $I$ being the **nuclear spin quantum number** of the nucleus. For example, for $I = 1/2$, *two* spin states of $-1/2$ and $+1/2$ are possible; for $I = 1$, *three* spin states of $-1$, $0$, and $+1$ are possible. The spin quantum number $I$ of any nucleus is related to its mass and atomic number, as presented in Table 9.4.

**Table 9.4**  Spin quantum number, ($I$), atomic number, and mass number of various nuclei

| Atomic Number | Mass Number | I | Examples |
|---|---|---|---|
| Even or odd | Odd | ½, ³⁄₂, ⁵⁄₂ | $^1H$, $^{13}C$, $^{31}P$ |
| Even | Even | 0 | $^{12}C$, $^{16}O$, $^{32}S$ |
| Odd | Even | 1, 2, 3 | $^2H$, $^{14}N$ |

In the absence of a magnetic field at room temperature, the energies of the spin state of a nucleus are degenerate (i.e., equivalent). As a magnetic field is imposed on the nucleus, the spin states lose their degeneracy and become separated by an energy difference ($\Delta E$), that is directly proportional to the strength of the applied field. Because various nuclei differ in their values of magnetic moment and spin quantum number, they undergo nuclear spin transitions at different frequencies in the same applied field. Table 9.5 lists some common nuclei and their associated properties.

**Table 9.5**   Resonance frequency and related nuclear properties*

| Nucleus | Relative Natural Abundance (%) | Spin Quantum Number | Magnetic Moment** | Resonance Frequency (MHz) |
|---------|-------------------------------|---------------------|-------------------|---------------------------|
| $^1$H   | 99.985  | 1/2 | 2.79628  | 42.5759 |
| $^2$H   | 0.015   | 1   | 0.857387 | 6.53566 |
| $^{13}$C | 1.108  | 1/2 | 0.702199 | 10.7054 |
| $^{14}$N | 99.63  | 1   | 0.40347  | 3.0756  |
| $^{15}$N | 0.37   | 1/2 | −0.28298 | 4.3142  |
| $^{17}$O | 0.037  | 5/2 | −1.8930  | 5.772   |
| $^{19}$F | 100    | 1/2 | 2.62727  | 40.0541 |
| $^{31}$P | 100    | 1/2 | 1.1305   | 17.235  |
| $^{33}$S | 0.76   | 3/2 | 0.64257  | 3.2654  |

*$H_o$ = 10,000 Gauss.
**Multiples of the nuclear magneton, eh/4πMc.

Certain nuclei ($^2$H, $^{13}$C, $^{15}$N, $^{17}$O) are present in low abundance, and as a result their sensitivities to NMR spectrometric techniques are extremely low. In the past, to record the NMR spectra of compounds using these low-abundance nuclei, it was necessary to increase the atom concentration of low-abundance nuclei by chemical synthesis. Recently, fast Fourier analysis transform techniques have enabled quick NMR spectra recordings for low-abundance nuclei. In this section only proton ($^1$H) spectra will be considered, in an area of study called **proton magnetic resonance (PMR).**

The recording of an NMR spectrum of a sufficient quantity of sample with a high-abundance nucleus (such as $^1$H) requires a single pass (sweep) through the appropriate electromagnetic region in a process called **continuous-wave spectroscopy.** The absorption of energy results in the excitation of the nucleus to a higher spin-energy state. When **saturation** is achieved, equal populations of ground-state and excited-state have been attained. In principle, the technique involves exposing the sample to a strong, homogeneous magnetic field, and then passing steadily changing frequencies of radio waves through the sample. However, in practice it is more convenient to maintain a constant radio-wave frequency while varying the strength of the magnetic field. The proton absorbs the radio energy and moves from its ground state to its excited state (i.e., it "flips") at energies proportional to particular magnitudes of the magnetic field. The absorption of the radio energy is easily detected, and a signal is electronically transmitted to a recorder. The spectrum is a plot of the intensity of absorption of the radio-frequency signal ($y$ axis) as a function of increasing magnetic field strength ($x$ axis). The magnetic field strength, **delta,** (δ), is expressed in parts per million (ppm).

Protons in different electron environments absorb at different field strengths because they "feel" a magnetic field strength that is not exactly the same as the *applied* field strength. The *effective* field strength at each proton (the field "felt" by the proton) depends greatly on factors that affect the proton's magnetic environment—the *electron density* around the proton. The electron density creates a small magnetic field that is either opposed to or aligned with the applied magnetic field.

If the electron density around a particular proton is low, that proton is **deshielded** from the applied magnetic field, which results in higher **chemical shift** (δ) values. High electron density creates a **shielding** effect that leads to lower δ values. Table 9.6 gives a summary of the most important $^1$H NMR chemical shifts observed in organic families. The chemical shifts of various hydrogens are designated with circled, Ⓗ, hydrogens.

**Table 9.6** Important ¹H NMR chemical shifts

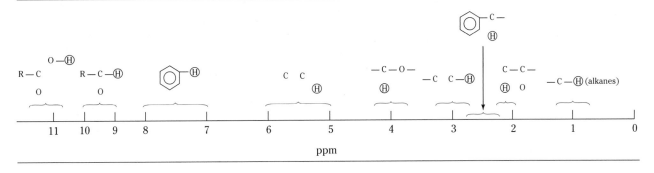

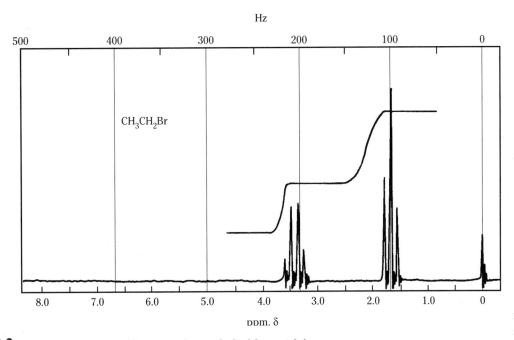

**Figure 9.9** PMR spectrum of bromoethane (ethyl bromide)

As in other forms of spectroscopy, the intensity of the absorption is proportional to the number of protons involved in the absorption. Consequently, the spectrum presents not only the number of different environments in which the protons are located, but also the relative numbers of protons in each environment.

The PMR spectrum of ethyl bromide displays two sets of nonequivalent hydrogens (Figure 9.9): a quartet (four peaks) at approximately 3.5 ppm, and a triplet (three peaks) at approximately 1.7 ppm. These **multiplet absorptions** occur because neighboring nuclei with non-zero magnetic moments have a regular, predictable effect on the protons actually measured. For every ***n* neighboring protons** (protons that are attached to the carbon(s) directly bonded to the carbon on which the protons of interest are attached) there are ***n* + 1** lines observed. This phenomenon is known as **spin-spin splitting.** The two methylene (—CH₂) hydrogens of ethyl bromide are split by the three neighboring protons of the methyl (—CH₃) group to yield a quartet (3 + 1 peaks). On the other hand, the three methyl hydrogens are split by the two hydrogens of the adjacent methylene group to yield a triplet (2 + 1 peaks). The area under each signal is proportional to the number of protons generating that peak. It is the ***area*** under each signal, and not simply the *height* of the peak, that is important. Spectrometers are equipped with an electronic **integrator** that shows a peak area as a stepped curve. The height of each step is proportional to the number of protons responsible for that step. To find the relative number of each type of proton in a spectrum, each step can be measured with a ruler, and the relative heights of the steps quantitatively compared. Thus, in the spectrum of bromoethane (see Figure 9.9), the triplet at 1.7 ppm has an **integral height** of 2.1 cm, while the quartet at 3.5 ppm has an **integral height** of 1.4 cm—a **relative ratio** of 3:2.

Most PMR spectra are taken on liquid samples. It is common to use solutions (usually 5%–20%), although a pure liquid can sometimes be used in some simple instruments. The solvent has to be either **deuterated** (e.g., $CDCl_3$ instead of "regular" chloroform) or **nonprotonated** (e.g., carbon tetrachloride). Normally a 0.5 mL sample of a nonviscous solution gives the best results.

In this experiment, PMR spectroscopy will be used to determine the relative percentages of pinacolone and dimethylbutadiene in the distillate from the reaction of pinacol (2,3-dimethyl-2,3-butanediol) with an acid. The ratio of the products varies with the acid used. Concentrated hydrobromic acid results in more butadiene (about 16%) than does concentrated sulfuric acid (about 6%). The relative amounts of dimethylbutadiene and pinacolone formed in each case can be determined from the PMR spectra by measuring the integral peak heights and proceeding as described in the Experimental section. If a PMR spectrometer is unavailable, the student can calculate the product ratio from the spectra presented in Figure 9.10.

## Suggested References

Carey, F. A., *Organic Chemistry.* 3rd ed., 512–37. New York: McGraw-Hill, 1996.

McMurry, J., *Organic Chemistry.* 4th ed., 454-77. Pacific Grove, CA: Brooks/Cole, 1996.

Morrison, R. T., and R. N. Boyd, *Organic Chemistry.* 6th ed., 600–29. Englewood Cliffs, NJ: Prentice-Hall, 1992.

Solomons, T. W. G., *Organic Chemistry.* 6th ed., 548–80. New York: John Wiley, 1996.

Vollhardt, K. P. C. and N. E. Schore, *Organic Chemistry.* 2nd ed., 324–56. New York: W. H. Freeman, 1994.

Wade, L. G., *Organic Chemistry.* 3rd ed., 525–92. Englewood Cliffs, NJ: Prentice-Hall, 1995.

## Experimental

1. Half the class should perform the "Pinacol Rearrangement" experiment in Experiment 17.4 using 20 mL 20% sulfuric acid; the other half should use 20 mL 20% hydrobromic acid. Collect the distillate at a temperature between 70°C and 100°C (in ice).

2. Run a PMR spectrum of the distillate, and calculate the relative percentage of pinacolone and dimethylbutadiene using the following procedure:

   $a$ = Relative integration of $CH_3$ singlet (at 1.9 ppm) in dimethylbutadiene

   = _____

   $b$ = Relative integration of *t*-butyl singlet (at 1.2 ppm) in pinacolone

   = _____

   c = Relative percentage of pinacolone in distillate =

$$\frac{(b/9) \times 100}{(b/9) + (a/6)} = \underline{\quad\quad} \%$$

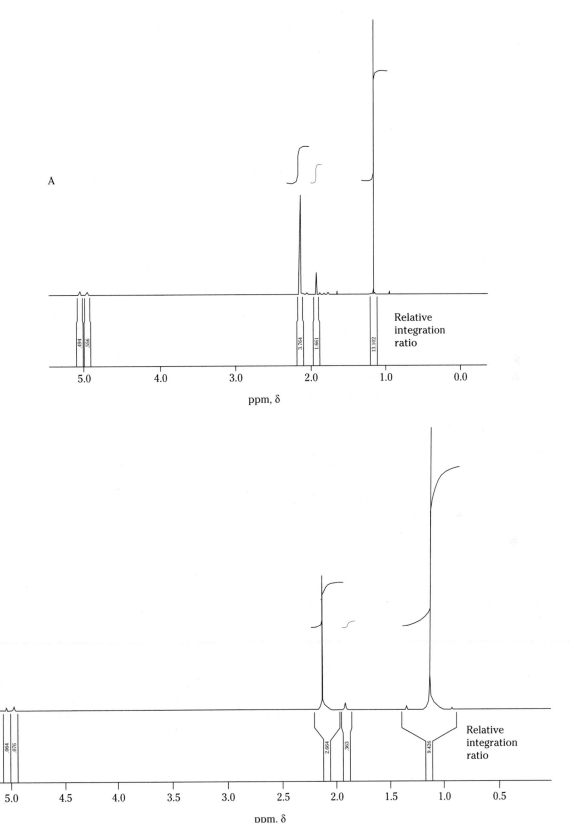

**Figure 9.10** The pinacol-pinacolene rearrangement, *A*, using HBr. *B*, using $H_2SO_4$

**Questions**   1. Give structures for the compounds below that show one PMR signal each.
 a. $C_2H_6O$
 b. $C_2H_4O$
 c. $C_4H_9Cl$
 d. $C_3H_6$
 e. $C_4H_6O_3$
 f. $C_3H_6O$

2. Give structures for the following compounds that show two PMR singlets each.
 a. $C_2H_4O_2$
 b. $C_8H_{10}$
 c. $C_5H_{11}Cl$
 d. $C_3H_6O_2$
 e. $C_7H_{15}Cl$
 f. $C_3H_5Cl_3$

3. How would you distinguish between the following compounds using PMR spectroscopy?
 a. methyl benzoate and phenylacetic acid
 b. phenylacetic acid and *p*-methylbenzoic acid
 c. *t*-butyl alcohol and *sec*-butyl alcohol
 d. *o*-xylene and *p*-xylene
 e. methyl propionate and ethyl acetate

4. Draw approximate PMR spectra expected for the following:
 a. methyl acetoacetate
 b. 1,4-dimethoxybenzene
 c. 3-deuteropropanal
 d. $CH_3COCH_2CH_2COCH_3$.

5. Select the solvents that can be used in PMR spectroscopy which do not interfere with the spectrum:
 a. carbon tetrachloride
 b. chloroform
 c. benzene-$d_6$
 d. hexachloroacetone
 e. acetonitrile
 f. acetone
 g. methylene chloride
 h. DMF (dimethylformamide)
 i. $D_2O$.

6. The PMR spectrum of dimethylformamide (DMF) shows three singlets. Explain.

7. When HBr is used as a catalyst in the pinacol rearrangement, more dimethylbutadiene is isolated than when $H_2SO_4$ is used. Why?

8. In the calculation of the dimethylbutadiene/pinacolone product ratio, the integral peak heights were divided by 6 and by 9, respectively. Explain the reason for this.

9. Identify the compounds whose PMR spectra are seen below:

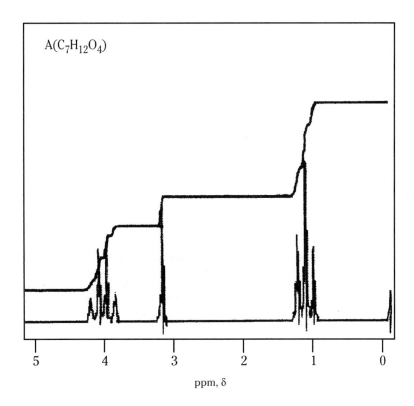

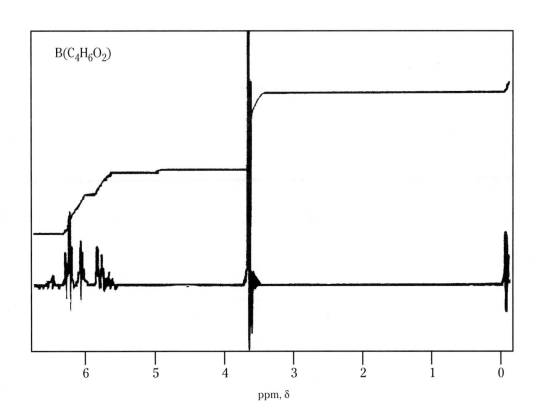

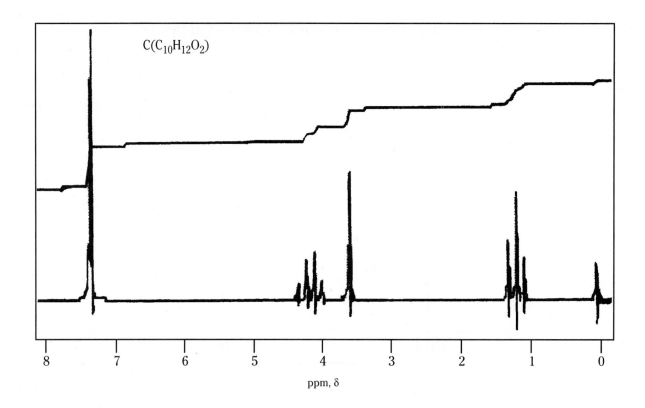

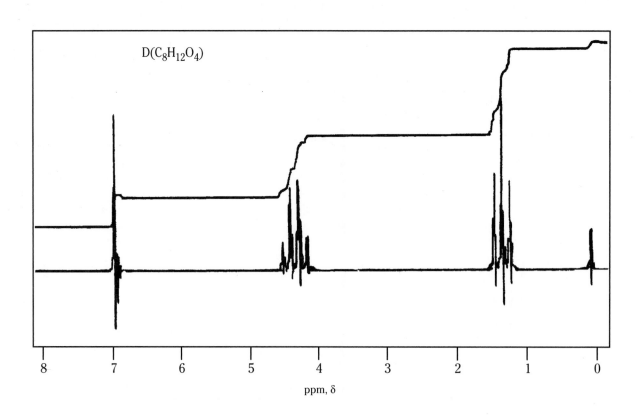

## 9.3 UV-VISIBLE SPECTROPHOTOMETRY

Many substances absorb photons of light in the visible and ultraviolet regions of the electromagnetic spectrum. The energy of the absorbed photons promotes the electrons of molecules and ions to higher energy levels. For the promotion to occur from a lower energy level ($E_1$) to a higher energy level ($E_2$), the wavelength of the light absorbed has to correspond exactly to the difference ($\Delta E$) between the electron energy levels:

$$\Delta E = E_2 - E_1 = hc/\lambda$$

where $\lambda$ is the wavelength of light absorbed; $c$ is the speed of light, $3 \times 10^{10}$ cm/sec; and $h$ is Planck's constant.

Visible light is that small portion of the electromagnetic spectrum with wavelengths in the range of 400–800 nm. The human eye responds to this region, which is perceived as a rainbow of colors from violet (shorter wavelengths) to red (longer wavelengths). In other words, the eye reacts to different wavelengths of light through the rainbow from violet to red, when it is subjected to the different colors.

When placed in a spectrophotometer the total number of photons absorbed by a given solution is directly proportional to the following:

1. the path length, $b$, that the incident light beam travels
2. the concentration, $c$ (in moles/liter), of the absorbing species in the solution and
3. the molar extinction coefficient, $\varepsilon$, which is a constant characteristic of the absorbing species and the specific wavelength.

At very low concentrations, $A$, the absorbance, can be expressed by the **Beer-Lambert law:**

$$A = \varepsilon bc.$$

It is common practice in spectrophotometric measurements to keep the optical path constant. This is accomplished by using a cell, or **cuvette,** of the same size for all solutions. As a result, absorbance and concentration are the only variables in a series of spectrophotometric measurements. By varying the concentration and then measuring the corresponding absorbance, one can get a straight line plot of the absorbance ($y$ axis) vs. the concentration ($x$ axis). The slope of the straight line is equal to the molar extinction coefficient ($\varepsilon$) if the path length ($b$) is equal to 1 cm. Although the Beer-Lambert law applies only to very dilute solutions, it affords very accurate measurements. Measurements are usually taken at the wavelength at which maximum absorption occurs. This wavelength is called the **optimum wavelength** and is indicated as $\lambda_{max}$.

In this experiment, the distribution coefficient $K_d$ will be determined spectrophotometrically. The **distribution coefficient** is defined as the concentration of a solute in an organic solvent (in g/mL) divided by the concentration of the solute in aqueous solution (in g/mL) (see chapter 6, pp. 59–62). The solute in this experiment is a dye. The absorbance of an aqueous solution of the dye will be determined before extraction ($A_1$) and after extraction ($A_2$) with an organic solvent, and the distribution coefficient will be calculated from the equation

$$K_d = \frac{[\text{Solute}]_{\text{in organic layer}}}{[\text{Solute}]_{\text{in aqueous layer}}} = \frac{A_1 - A_2}{A_2}$$

An organic solvent appropriate for the extraction process should be immiscible with water, nontoxic, and rather polar, so that the dye (which is usually a water-soluble organic salt) can be extracted to an appreciable degree. These conditions eliminate a great number of water-soluble polar solvents such as methanol, ethanol, 2-propanol, dimethyl sulfoxide, acetonitrile, acetone, and tetrahydrofuran, as well as chloroform, because of its toxicity. The solvents used in this experiment are 1-pentanol and ethyl acetoacetate.

**Suggested References**  Berberan-Santos, M. N., "Beer's Law Revisited." *J. Chem Educ.* 67 (1990): 757.

Carey, F. A., *Organic Chemistry.* 3rd ed., 547-49. New York: McGraw-Hill, 1996.

Christian, G., *Analytical Chemistry,* ch. 7. New York: John Wiley, 1977.

Conn, Eric E., and P. K. Stumpf, *Outlines of Biochemistry.* 4th ed., Appendix 2. New York: John Wiley, 1976.

Day, R. A., Jr., and A. L. Underwood, *Quantitative Analysis.* 5th ed., 506–20. Englewood Cliffs, NJ: Prentice-Hall, 1991.

"Extraction." In *McGraw-Hill Encyclopedia of Science and Technology.* 7th ed., vol. 6, 572–74. New York: McGraw-Hill, 1992.

Flaschka, H. A., A. J. Barnard, and P. E. Sturrock, *Quantitative Analytical Chemistry.* ch. 44. Boston: Willard Grant, 1980.

Ghayourmanesh, S., "Solvent Extraction." In *Magill's Survey of Science: Applied Science.* 2391–96. Pasadena, CA: Salem Press, 1993.

Lott, P. F., "Recent Instrumentation for UV-Visible Spectrophotometry." *J. Chem. Ed.* 45 (February–April 1968): A89, A169, A273.

McMurry, J., *Organic Chemistry.* 4th ed., 519–25. Pacific Grove, CA: Brooks/Cole, 1996.

Morrison, R. T., and R. N. Boyd, *Organic Chemistry.* 6th ed., 597-99. Englewood Cliffs, NJ: Prentice-Hall, 1992.

Ramette, R. W., *Chemical Equilibrium and Analysis.* ch. 19. Reading, MA: Addison-Wesley, 1981.

Solomons, T. W. G., *Organic Chemistry.* 6th ed., 536–41. New York: John Wiley, 1996.

Wade, L. G., *Organic Chemistry.* 3rd ed., 525–68. Englewood Cliffs, NJ: Prentice-Hall, 1995.

**Experimental**

## A. Determination of the optimum wavelength

1. Fill the cuvette with the stock solution, and determine the absorbance for each of the wavelengths indicated in the data sheet (p. 117).
2. Using the graph paper provided (p.119) plot absorbance vs. wavelength. Select the wavelength that shows the highest absorbance value, and set the spectrophotometer at this wavelength. All future determinations should be performed at that selected optimum wavelength.

## B. Determination of the distribution coefficient

1. Record the absorbance of the stock solution at the optimum wavelength ($A_1$) in part B of the data sheet (item b). Wash the cuvette thoroughly.
2. Place 10 mL of the stock solution (approximately $6 \times 10^{-6}$ g/mL) and 10 mL of 1-pentanol in a separatory funnel. Mix thoroughly and let stand on a ring stand for a couple of minutes or until the two layers can be clearly seen.
3. Fill the clean cuvette with the extracted aqueous layer by draining the lower layer of the separatory funnel. (Remember to remove the glass stopper before draining!) Determine the absorbance ($A_2$) of this liquid and place in part B of the data sheet (item c).
4. Calculate the distribution coefficient by using the formula

$$K_d = \frac{A_1 - A_2}{A_2}$$

5. Repeat steps 1 through 4 using ethyl acetoacetate instead of 1-pentanol.

*Note:* This experiment may also be performed using a methylene blue stock solution ($3.8 \times 10^{-6}$ g/mL) and the same organic solvents.

**Questions**

1. Which organic solvents from the following list can be used in this experiment? Explain. (*Hint:* See Table 6.1 when answering this question.)
   a. ethanol
   b. acetone
   c. diethyl ether
   d. methylene chloride
   e. benzene
   f. toluene
   g. methanol
   h. carbon tetrachloride
   i. acetic acid
   j. 2-propanol.

2. An aqueous solution of a dye ($\varepsilon = 60,000$) has an absorbance of 0.8 ($A_1$) when a 1-cm cuvette is used. Calculate the concentration of the dye from the Beer-Lambert Law (molecular weight of dye = 374).

3. A 20-mL portion of the solution in question 2 is extracted with 20 mL of ethyl acetate. After extraction, the aqueous layer has an absorbance of 0.2 ($A_2$).
   a. Calculate the value of the distribution coefficient.
   b. Calculate the concentration of the dye (in g/mL) left in the aqueous layer.

4. A second 20-mL solution of the dye from question 2 is extracted with *two* successive 10-mL samples of ethyl acetate (for a total of 20 mL).
   a. Calculate the concentration of the dye (in g/mL) remaining in the aqueous layer after both extractions, and compare this value with your answer from question 3.
   b. Based on this comparison, is it more efficient to perform an extraction with one large sample of the solvent or with two (or more) smaller samples of solvent whose total volume equals the volume used for the one-extraction process?
   (*Hint:* The calculation is based on the definition of $K_d$. Let $x$ be the total number of grams of dye extracted by ethyl acetate (10 mL). The number of grams of dye remaining in the aqueous layer (20 mL) is then equal to: (the original amount) – $x$.

5. After checking the dielectric constant of each of the organic solvents in this experiment (Table 6.1), what can you say about the relation of the dielectric constant and the distribution coefficient?

## DATA SHEET: UV-VIS DETERMINATION OF DISTRIBUTION COEFFICIENT
## PART A

1. Name of dye: methyl violet.
2. Concentration of dye's stock solution: _____ g/mL.

Determination of the optimum wavelength

| Wavelength | Absorbance | Wavelength | Absorbance |
|---|---|---|---|
| 400 | _____ | 555 | _____ |
| 405 | _____ | 560 | _____ |
| 410 | _____ | 565 | _____ |
| 415 | _____ | 570 | _____ |
| 420 | _____ | 575 | _____ |
| 425 | _____ | 580 | _____ |
| 430 | _____ | 585 | _____ |
| 435 | _____ | 590 | _____ |
| 440 | _____ | 595 | _____ |
| 445 | _____ | 600 | _____ |
| 450 | _____ | 605 | _____ |
| 455 | _____ | 610 | _____ |
| 460 | _____ | 615 | _____ |
| 465 | _____ | 620 | _____ |
| 470 | _____ | 625 | _____ |
| 475 | _____ | 630 | _____ |
| 480 | _____ | 635 | _____ |
| 485 | _____ | 640 | _____ |
| 490 | _____ | 645 | _____ |
| 495 | _____ | 650 | _____ |
| 500 | _____ | 655 | _____ |
| 505 | _____ | 660 | _____ |
| 510 | _____ | 665 | _____ |
| 515 | _____ | 670 | _____ |
| 520 | _____ | 675 | _____ |
| 525 | _____ | 680 | _____ |
| 530 | _____ | | |
| 535 | _____ | | |
| 540 | _____ | | |
| 545 | _____ | | |
| 550 | _____ | | |

The optimum wavelength for the dye_____is

_____ nm

# PLOT OF ABSORBANCE VS. WAVELENGTH TO DETERMINE THE OPTIMUM WAVELENGTH IN THE SPECTROPHOTOMETRIC DETERMINATION OF THE DISTRIBUTION COEFFICIENT

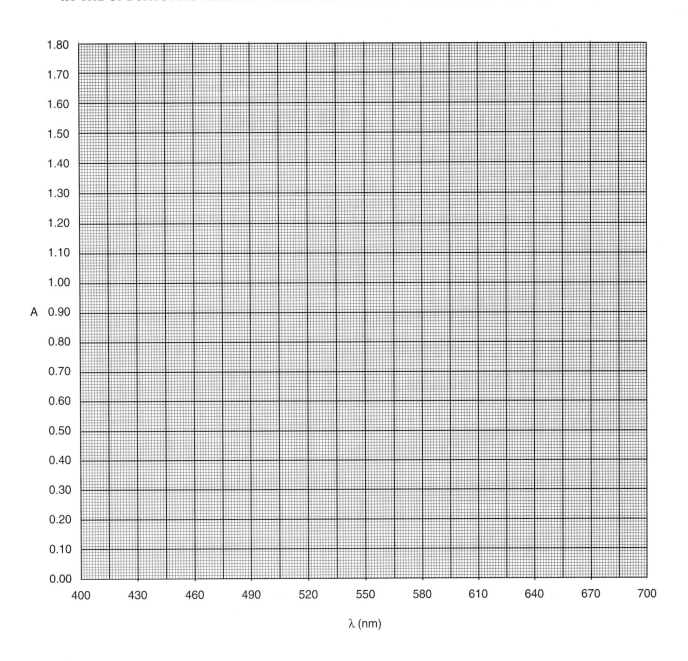

# DATA SHEET: SPECTROPHOTOMETRIC DETERMINATION OF DISTRIBUTION COEFFICIENT PART B

## Calculation of Distribution Coefficient

Dye: methyl violet        Optimum wavelength: _____

| | 1-Pentanol | Ethyl Acetoacetate |
|---|---|---|
| a. Concentration of stock solution | _____ g/ mL | _____ g/ mL |
| b. Absorbance of stock solution ($A_1$) | _____ | _____ |
| c. Absorbance of stock solution after extraction ($A_2$) | _____ | _____ |
| d. Distribution coefficient ($K_d$) = [$(b - c)/c$] | _____ | _____ |

# chapter
# 10

# Alkenes

## 10.1 SYNTHESIS OF CYCLOHEXENE

Alkenes can be conveniently synthesized from the corresponding alcohols by acid dehydration. The usual catalysts are concentrated sulfuric acid or concentrated phosphoric acid.

$$\underset{\text{alcohol}}{\overset{\displaystyle}{\text{C}-\text{C}}}\ \underset{\text{H}\quad\text{OH}}{\phantom{C}} \xrightarrow{\text{H}^+}\ \underset{\text{alkene}}{\text{C}=\text{C}}\ +\ \text{H}_2\text{O}$$

The following is an example of an elimination reaction in which a **leaving group** (X) departs *with* its bonding electrons, while another group (usually hydrogen) on an adjacent carbon is abstracted *without* its bonding electrons.

$$\text{C}-\text{C} \xrightarrow{\phantom{xxx}}\ \text{C}=\text{C}\ +\ \text{HX}$$

Such an elimination is generally classified according to the timing of the $\text{H}^+/\text{X}^-$ loss.

a. X might depart with its electrons, leaving behind a carbocation. In the second step, the carbocation loses a proton to form a double bond (**E1**, or first-order elimination).

$$\text{C}-\text{C} \xrightarrow{-\text{X}^-}\ \underset{\text{carbocation}}{\overset{+}{\text{C}-\text{C}}} \xrightarrow{-\text{H}^+}\ \text{C}=\text{C}$$

b. Both $\text{H}^+$ and $\text{X}^-$ might be lost simultaneously in a *one-step* (concerted) mechanism (**E2**, or second-order elimination). The $\text{H}^+$ is abstracted by a base :B.

$$\text{C}-\text{C} \xrightarrow{:\text{B}}\ \text{C}=\text{C}\ +\ \text{BH}^+\text{X}^-$$

**123**

c. The proton might be abstracted by the base prior to the elimination of X, thus creating a carbanion ($E1_{CB}$)

Carbanion

In this experiment, cyclohexanol will be dehydrated using a strong acid (concentrated phosphoric acid, $H_3PO_4$) to cyclohexene via an E1 mechanism. The reaction involves the initial protonation of the oxygen atom to form an oxonium ion, which then decomposes into water and the carbocation. This carbocation cannot rearrange to a more stable carbocation and yields cyclohexene upon elimination of an adjacent proton.

| cyclohexanol | oxoniumion | carbocation | cyclohexene |

The experiment calls for the azeotropic distillation of cylohexene as it is formed. This minimizes the possibility of the product's reaction with another carbocation and subsequent polymerization. The immediate removal of cyclohexene also drives the reaction to the right hand side, thus increasing the percent yield. The azeotropic mixture of water and cyclohexene is then washed (extracted) with a base (sodium hydroxide) to neutralize any traces of acid. Finally, the product is dried with *anhydrous* sodium or magnesium sulfate, and is eventually purified by simple distillation. The isolated cyclohexene is tested for the presence of a double bond using standard tests for unsaturation ($Br_2/H_2O$ and dilute $KMnO_4$ solution).

Use of sulfuric acid instead of phosphoric acid leads to a considerable amount of tarry products and obnoxious fumes but, nevertheless, yields cyclohexene in a reasonable amount.

---

⚠️ **SAFETY TIPS**

✔ Cyclohexanol (MP 23–25°C, BP 161°C) is a moderately toxic liquid, whose vapors may cause irritation of the eyes, nose and throat when inhaled in large quantities. Adequate ventilation is therefore recommended. Cyclohexene is a volatile (BP 83°C), flammable liquid, (flash point [closed cup] –11.6°C) and all flames should be extinguished during its preparation and distillation.

✔ To prevent peroxide formation, the reaction mixture must not be distilled to dryness (peroxides are potentially explosive compounds that are sensitive to shock, sparks, heat, and light). Any peroxide that would form would have a higher boiling point than cyclohexene, and would become concentrated in the distilling flask while cyclohexene and water are being distilled and removed from the system. Heating the distilling flask to dryness, therefore, results in heating an increasingly concentrated solution of peroxide, thus risking a possible explosion.

✔ Concentrated phosphoric acid is extremely corrosive and should be handled with care.

✔ The distilling flask should be washed with methylene chloride, water, and acetone immediately after distillation to remove the obnoxious-smelling, tarry residue.

✔ In the final distillation, cyclohexene is collected in a pre-weighed vial or test tube immersed in an ice bath to reduce evaporation.

| cyclohexene | + | KMnO$_4$ (cold, dilute) | → | *cis*-1,2-cyclohexane diol | + | MnO$_2$ |

cyclohexene    potassium permanganate (purple)    *cis*-1,2-cyclohexane diol    manganese dioxide (black ppt.)

cyclohexene    +    Br$_2$/H$_2$O    →    *trans*-1,2-dibromocyclohexane (colorless)

cyclohexene    bromine (orange solution)    *trans*-1,2-dibromocyclohexane (colorless)

**Suggested References**

Carey, F. A., *Organic Chemistry.* 3rd ed., 184–93. New York: McGraw-Hill, 1996.

Ege, Seyhan N., *Organic Chemistry: Structure and Reactivity.* 3rd ed., 297–98, 304–05. Lexington, MA: D. C. Heath, 1994.

Loudon, G. Marc, *Organic Chemistry.* 3rd ed., 443–46. Redwood City, CA: Benjamin/ Cummings, 1995.

March, J., *Advanced Organic Chemistry.* 4th ed., 1011. New York: John Wiley, 1992.

McMurry, J., *Organic Chemistry.* 4th ed., 648–51. Pacific Grove, CA: Brooks/Cole, 1996.

Morrison, R. T., and R. N. Boyd, *Organic Chemistry.* 6th ed., 310–15. Englewood Cliffs, NJ: Prentice-Hall, 1992.

Patnaik, P., *A Comprehensive Guide to the Hazardous Properties of Chemical Substances,* 81–423. New York: Van Nostrand Reinhold, 1992.

Solomons, T. W. G., *Organic Chemistry.* 6th ed., 308–15. New York: John Wiley, 1996.

Vollhardt, K. P. C., and N. E. Schore, *Organic Chemistry.* 2nd ed., 400–03. New York: W. H. Freeman, 1994.

Wade, L. G., *Organic Chemistry.* 3rd ed., 450–52. Englewood Cliffs, NJ: Prentice-Hall, 1995.

**Experimental**

1. Set up a fractional distillation apparatus (see p. 32), using a 100-mL round-bottomed flask as the distilling flask and a 25-mL round-bottomed flask as the receiving flask. Do not pack the fractionating column. Keep the receiving flask in an ice-water bath to reduce evaporation.

2. Place 10.4 mL (d = 0.96 g/mL; 10.0 g) of cyclohexanol and 3.0 mL of concentrated phosphoric acid in the distilling flask (via a funnel), and add a few boiling chips. Start heating the reaction flask and collect approximately 6–8 mL of the cyclohexene/water distillate in a round bottom flask that has been immersed in an ice-water bath.

3. Use a funnel to transfer the distillate to a separatory funnel and wash it with 10 mL of a 10% sodium hydroxide solution.

4. Transfer the (upper) organic layer to a 50 mL *dry* conical flask and add 0.5–1.0 g of *anhydrous* sodium or magnesium sulfate that has been weighed *immediately after* the transfer of the organic layer. Cover with a watch glass and let stand for about 5 minutes.

5. Gravity filter the solution into a *dry* 50-mL round-bottomed flask, and set up for a simple distillation using *dry glassware*. Remember that the connecting adaptor, adaptor outlet and condenser (used in step 1) might contain droplets of water that have to be removed (by washing with acetone) in order to obtain *dry* cyclohexene. The dark residue in the distilling flask can be removed using a few milliliters of methylene chloride and a test tube brush.

6. Collect the fraction that boils in the range of 80°–85°C in a pre-weighed vial or test tube that is immersed in an ice-water bath. Calculate the percent yield.

7. Perform the following tests for unsaturation:
   a. Add 5–10 drops of the distilled cyclohexene to 0.5 mL of a 1% potassium permanganate solution in a test tube.
   b. Add 5–10 drops of the distillate to 0.5 mL of 1% bromine-water solution in a test tube.

8. Observe and record the results of these two tests, then compare them with the results obtained when cyclohexane is used instead of cyclohexene.

**Questions**

1. Explain the purpose of the following in the above experiment:
   a. phosphoric acid
   b. fractional distillation
   c. sodium hydroxide extraction
   d. *anhydrous* sodium or magnesium sulfate
   e. ice-cooled receiving flask
   f. covering the flask with a watch glass in step 4.

2. Write balanced equations for the unsaturation tests of cyclohexene performed in step 7 of this experiment. Use the half-reaction $MnO_4^- \longrightarrow MnO_2$ for (a).

3. Give all possible alkenes formed in the dehydration of the following alcohols:
   a. 1-methylcyclohexanol
   b. 2-methylcyclohexanol
   c. 1-hydroxymethylcyclopentane
   d. 3,3-dimethyl-1-butanol.

4. What alcohol would be most appropriate for the synthesis of 1-methylcyclohexene? Write an equation.

## 10.2 SYNTHESIS OF 1,2-CYCLOHEXANEDIOL FROM CYCLOHEXENE: A SYN ADDITION

Addition to a carbon-carbon double bond can be either **syn** or **anti.** Anti additions include bromination (via $Br_2/CCl_4$) and diol formation via epoxidation ($RCO_3H$) and subsequent hydrolysis ($OH^-/H_2O$). This experiment is an example of a syn addition.

cyclohexene                                          1,2-cyclohexanediol

This reaction must be performed at a low temperature to prevent further oxidation of the diol to the adipate salt.

1,2-cyclohexanediol                              potassium adipate

A "mixed" solvent is used (*t*-butyl alcohol/water), to bring the organic reagent (cyclohexene) and inorganic reagents ($MnO_4^-/OH^-$) into intimate contact. Cyclohexene is dissolved in *t*-butyl alcohol, mixed with the sodium hydroxide solution (in water), and chilled before the addition of potassium permanganate. Manganese dioxide ($MnO_2$), a black precipitate, is gelatinous and cannot be easily separated with filter paper (which normally clogs). For this reason, a filter aid (Celite) is used. After filtering, the filter cake is washed thoroughly with water/*t*-butyl alcohol to wash out any trapped 1,2-cyclohexanediol. Reduction of the solvent volume is recommended, so that the maximum yield of product can be obtained. For safety reasons, this process is performed by a simple distillation, which takes about two hours. This is the principal reason why the experiment requires two laboratory sessions to complete. Solid potassium carbonate is added to convert any adipic acid that has formed as a side product into its water-soluble salt.

$$HOOC(CH_2)_4COOH + K_2CO_3 \longrightarrow K^+ \, {}^-OOC(CH_2)_4COO^-K^+ + H_2O + CO_2$$
adipic acid                                          potassium adipate

Hot filtration with ethyl acetate will separate the diol from the water-soluble potassium adipate.

---

### ⚠ SAFETY TIPS

✔ Cyclohexene is a volatile (BP 83°C), flammable liquid; therefore, flames should be extinguished in the laboratory during this experiment.

✔ *t*-Butyl alcohol (MP 25°C, BP 83°C) and ethyl acetate (BP 77°C) are flammable liquids; therefore, the volume reduction in steps 7 and 9 should be performed by a simple distillation. *t*-Butyl alcohol may irritate both skin and eyes upon contact.

> ✔ Sodium hydroxide is caustic and should be handled with care.
> ✔ Potassium permanganate is a strong, corrosive oxidant. Its stains
> may be removed with aqueous thiosulfate solutions.

**Suggested References**

"Celite." In *Reagents for Organic Synthesis,* edited by L. F. Fieser and M. Fieser, vol. 1, 120. New York: John Wiley, 1967.

Ege, Seyhan N., *Organic Chemistry: Structure and Reactivity.* 3rd ed., 332–34. Lexington, MA: D. C. Heath, 1994.

Loudon, G. Marc, *Organic Chemistry.* 3rd ed., 192–194. Redwood City, CA: Benjamin/Cummings, 1995.

McMurry, J., *Organic Chemistry.* 4th ed., 241–42. Pacific Grove, CA: Brooks/Cole, 1996.

Morrison, R. T., and R. N. Boyd, *Organic Chemistry.* 6th ed., 357–58. Englewood Cliffs, NJ: Prentice-Hall, 1992.

Patnaik, P., *A Comprehensive Guide to the Hazardous Properties of Chemical Substances.* 423. New York: Van Nostrand Reinhold, 1992.

"Potassium Permanganate." In *Reagents for Organic Synthesis,* edited by L. F. Fieser and M. Fieser, vol. 1, 942–52. New York: John Wiley, 1967.

Solomons, T. W. G., *Organic Chemistry.* 6th ed., 346–48. New York: John Wiley, 1996.

Vollhardt, K. P. C., and N. E. Schore, *Organic Chemistry.* 2nd ed., 436–38. New York: W. H. Freeman, 1994.

Wade, L. G., *Organic Chemistry.* 3rd ed., 370–75. Englewood Cliffs, NJ: Prentice-Hall, 1995.

**Experimental**

1. Dissolve 3.2 g of potassium permanganate in 100 mL of warm water, then cool to about 15°C.

2. While the permanganate solution of step 1 is cooling, dissolve 3.1 mL (d = 0.81 g/mL; 2.5 g) of cyclohexene in approximately 80 mL of *t*-butyl alcohol in a 250-mL Erlenmeyer flask.

3. To the mixture of step 2, add a solution of 0.5 g of solid sodium hydroxide in 3 mL of water. Cool the mixture to 0–5°C by adding 50 g of crushed ice to the conical flask.

4. Place the flask used in step 3 in an ice/water bath. Transfer the cold mixture of step 1 to a separatory funnel, and add the potassium permanganate solution dropwise, stirring continuously with a stirring rod or by swirling. Do not allow the temperature to go above 10°C.

5. After the addition is complete, prepare a filter cake in a Büchner funnel as follows: Mix 1.0 to 1.5 g of a filter aid (Celite) and 5 mL of water in a test tube. Swirl the test tube, and then pour the suspension on the Büchner funnel fitted with a piece of filter paper (do not forget to place a test tube inside the filter flask to trap the filtrate!), and then turn on the vacuum *slowly.* If the vacuum is turned on rapidly, the filter cake may crack. When the entire solution has been filtered, discard the filtrate. The filter aid has now formed a quarter-inch layer on top of the filter paper that will trap the fine particles of manganese dioxide.

6. Heat the reaction mixture almost to boiling, and filter the hot mixture through the filter cake. Wash the filter cake with 40 mL of a 1:1 water/ *t*-butyl alcohol solution as follows: Add a small volume of the solution to the cake *with the suction off* (to allow it to soak into the solid), then turn on the vacuum to create suction. Repeat several times until all the rinsing solution has been used. Do not allow the cake to dry out and crack at any point. If this happens prepare another filter cake (as in step 5) and repeat the filtration. The filtrate must be colorless.

7. Distill the filtrate until approximately 50 mL remain in the flask. This distillation is a long procedure that can take up to two hours. Place the distillate in a bottle labeled "waste *t*-butyl alcohol/ethyl acetate." Transfer the residue to a 250-mL conical flask, and start adding solid potassium carbonate in portions of 1 g to the solution, swirling until the salt no longer dissolves. Normally, 15–20 g of potassium carbonate should be enough. Cover the flask until the next laboratory period.

8. By the second period, the diol should have solidified. Use vacuum filtration to separate it. Transfer the solid to a 125-mL conical flask, add 20 mL of ethyl acetate, heat to boiling and filter the hot solution using a fluted filter paper. Rinse the filter paper with 5 mL of hot ethyl acetate.

9. Add a few boiling chips to the filtrate and remove most of the ethyl acetate by distillation. Transfer the ethyl acetate into the bottle designated "waste *t*-butyl alcohol/ethyl acetate." The diol might oil out, but should crystallize upon cooling. Isolate the product by vacuum filtration.

10. Determine the MP (lit. 99–101°C) and calculate the % yield.

**Questions**

1. Write the balanced equation for the reaction of cyclohexene with $MnO_4^-$ to form 1,2-cyclohexanediol and $MnO_2$ in basic medium.

2. Write the balanced equation for the formation of the adipate anion in basic $MnO_4^-$ from:
   a. cyclohexene
   b. 1,2-cyclohexanediol.

3. What is the purpose of the following in the above experiment?
   a. *t*-butyl alcohol
   b. Celite
   c. potassium carbonate (two reasons)
   d. hot filtering the ethyl acetate solution

4. What are the products of the reaction of 3-methylcyclohexene with the following reagents? Show all expected stereochemistry.
   a. cold, dilute $KMnO_4$
   b. hot, basic $KMnO_4$
   c. cold $HCO_3H/OH^-$
   d. $Br_2/CCl_4$

## 10.3   SYNTHESIS OF α, β-DIBROMOBENZALACETOPHENONE

Benzalacetophenone (chalcone) is a *trans-*α, β-unsaturated ketone prepared from the mixed aldol condensation of benzaldehyde with acetophenone.

benzaldehyde          acetophenone                    *trans*-benzalacetophenone

Benzalacetophenone reacts *stereospecifically* with $Br_2$ in methylene chloride to yield a racemic dibromide.

benzalacetophenone                    α, β-dibromobenzalacetophenone
                                              (a racemic mixture)

The anti addition of $Br_2$ to the *trans* alkene yields one pair of enantiomers.

equivalent to                              equivalent to

(2S, 3R)                                      (2R, 3S)

erythro-2,3-dibromo-1,3-diphenyl-propanone

The addition of bromine to an α, β-unsaturated system is slower than bromine addition to a nonconjugated alkene (e.g., cyclohexene) because of the electropositive character of the β carbon, as seen in the following resonance structures:

**Suggested References**   Carey, F. A., *Organic Chemistry*. 3rd ed., 237–41. New York: McGraw-Hill, 1996.

Ege, Seyhan N., *Organic Chemistry: Structure and Reactivity*. 3rd ed., 318–24. Lexington, MA: D. C. Heath, 1994.

Furniss, B. S., A. J. Hannaford, P. W. G. Smith, and A. R. Tatchell, *Vogel's Textbook of Practical Organic Chemistry*. 5th ed., 422, 1034. London: Longman, 1989.

Loudon, G. Marc, *Organic Chemistry*. 3rd ed., 175–76, 313–17. Redwood City, CA: Benjamin/Cummings, 1995.

McMurry, J., *Organic Chemistry*. 4th ed., 222–26. Pacific Grove, CA: Brooks/Cole, 1996.

Morrison, R. T., and R. N. Boyd, *Organic Chemistry*. 6th ed., 339–42. Englewood Cliffs, NJ: Prentice-Hall, 1992.

Patnaik, P., *A Comprehensive Guide to the Hazardous Properties of Chemical Substances*. 367. New York: Van Nostrand Reinhold, 1992.

Solomons, T. W. G., *Organic Chemistry*. 6th ed., 339–44. New York: John Wiley, 1996.

Vollhardt, K. P. C., and N. E. Schore, *Organic Chemistry*. 2nd ed., 423–25. New York: W. H. Freeman, 1994.

Wade, L. G., *Organic Chemistry*. 3rd ed., 360–63. Englewood Cliffs, NJ: Prentice-Hall, 1995.

---

### ⚠️ SAFETY TIPS

✔ Bromine (BP 59°C) is extremely corrosive (a concentration of 50 ppm in air is highly irritating to humans) and must be handled with gloves, and *always* in the hood. The liquid produces painful skin burns upon contact that should *immediately* be treated with glycerol. If the vapors are inhaled, relief may be obtained by soaking a clean piece of cloth (such as a handkerchief) in ethanol and holding it near the nose. Spills of liquid bromine should be treated with sodium thiosulfate or lime water.

✔ Benzalacetophenone (chalcone) is a skin irritant and should be handled with gloves.

✔ Methylene chloride (BP 40°C) is harmful when ingested or when its vapors are inhaled (RL = 350 mg/m$^3$). It should, therefore, be handled with adequate ventilation. Its spills should be treated with a dispersing agent or sand which should be transported afterwards to a safe open area for atmospheric evaporation.

---

**Experimental**

1. Dissolve in a large test tube 1.0 g of dry benzalacetophenone in 5 mL of dry methylene chloride and cool in ice.
2. Add dropwise 5 mL of a 15% (w/w) solution of bromine in methylene chloride. Allow the mixture to stand for 10 minutes, and isolate the crystals by vacuum filtration.
3. Wash the crystals with 4–5 mL of ice cold ethanol and air dry. Determine the MP (lit. 150–160°C) and percent yield.
4. Perform the same experiment in a test tube, using cyclohexene instead of benzalacetophenone. Compare the relative rates of reaction, but do not attempt to isolate the product.

**Questions**

1. 2.1 g of cyclohexene is reacted with 6.0 g of bromine to yield 2.0 g 1,2-dilonemocyclohexane.
   a. What is the expected product?
   b. Which is the limiting reagent?
   c. What is the expected yield?
   d. What is the percent yield?

  e. Can 1,2,3-tribromocyclohexane or 3-bromocyclohexene be formed as a side product if these experimental conditions are used? Explain.

2. What are the bromination products of the following:
   a. cyclohexene
   b. 1-methylcyclohexene
   c. 3-methylcyclohexene
   d. 4-methylcyclohexene.

   Draw all stereoisomers of the products, and name them using the R/S designations.

## 10.4 OXIDATION OF CYCLOHEXENE TO ADIPIC ACID

Alkenes are susceptible to oxidation, which leads to cleavage of the carbon-carbon double bond.

Oxidation can be accomplished by means of oxidizing agents such as ozone, dichromate, and potassium permanganate. The latter yields the corresponding *cis*-1,2-diol under mild conditions (cold, dilute solution).

Stronger conditions cleave the carbon-carbon double bond. If the alkene is not tetrasubstituted, the aldehyde which is formed is further oxidized to the corresponding carboxylic acid.

In this experiment, cyclohexene will be treated with basic potassium permanganate, and adipic acid will be formed via ring opening. The product will be precipitated with hydrochloric acid.

Manganese dioxide ($MnO_2$) will be removed by means of a filter acid (Celite, see experiment 10.2 pp. 127–129). Decolorizing carbon may be used to remove any colored impurities before the acidification step.

**Suggested References**

Carey, F. A. *Organic Chemistry*. 3rd ed., 211–17. New York: McGraw-Hill, 1996

"Celite." In *Reagents for Organic Synthesis,* edited by L. F. Fieser and M. Fieser, vol. 1, 120. New York: John Wiley, 1967.

McMurry, J., *Organic Chemistry*. 4th ed., 238–41. Pacific Grove, CA: Brooks/Cole, 1996.

Morrison, R. T., and R. N. Boyd, *Organic Chemistry*. 6th ed., 357–60. Englewood Cliffs, NJ: Prentice-Hall, 1992.

Patnaik, P., *A Comprehensive Guide to the Hazardous Properties of Chemical Substances.* 423. New York: John Wiley, 1992.

"Potassium Permanganate." In *Reagents for Organic Synthesis* edited by L. F. Fieser and M. Fieser, vol. 1, 942–52. New York: John Wiley, 1967.

Solomons, T. W. G., *Organic Chemistry.* 6th ed., 348–49. New York: John Wiley, 1996.

Wade, L. G., *Organic Chemistry.* 3rd ed., 370–75. Englewood Cliffs, NJ: Prentice-Hall, 1995.

## ⚠ SAFETY TIPS

✔ Cyclohexene is an irritant and a flammable liquid. All flames should be extinguished in the lab during the first three steps of the experiment.

✔ Potassium permanganate is a strong, corrosive oxidant.

✔ Concentrated hydrochloric acid is extremely corrosive. It should be poured carefully and with adequate ventilation.

**Experimental**

1. Add 5.2 mL (d = 0.81 g/mL; 4.2 g) of cyclohexene to a solution of 15 g of potassium permanganate in 125 mL of water in a 250-mL conical flask.
2. Add 1 mL of 10% sodium hydroxide, and allow the temperature to rise to 45°C, but not higher. Use an ice-water bath to control the temperature. The contents of the flask have to be swirled continuously to allow proper mixing.
3. When the temperature no longer rises above 45°C when the flask is removed from the ice bath, allow the solution to stand for about 5 minutes at room temperature.
4. Warm the mixture in a steam bath for about 5 minutes. Be sure to swirl the contents while the solution is being heated.
5. Test for the presence of permanganate by placing a drop of the reaction mixture on a piece of filter paper. Any unchanged permanganate will appear as a purple ring around the spot of the brown manganese dioxide. If permanganate remains, add small portions of a saturated aqueous sodium thiosulfate to the mixture until the spot test is negative.
6. Vacuum-filter the suspension through a pad of 1–2 g of filter aid (see "Synthesis of 1,2-Cycloclohexanediol from Cyclohexene: A Syn Addition" p. 128, Experimental, step 5), and wash the filter cake thoroughly with about 40 mL of water.
7. Concentrate the solution to a volume of about 35 mL by heating it on a hot plate or burner.
8. If the concentrate is colored, add a pinch of decolorizing carbon, reheat, and gravity-filter it *while it is hot.*
9. Carefully add concentrated hydrochloric acid to the filtrate (use the hood!) until the solution tests acidic to litmus paper, (pH = 4), and then add an additional 5 mL of the acid.
10. Allow the solution to cool to room temperature, and isolate the precipitated adipic acid by vacuum filtration. Recrystallize from about 15–20 mL of hot water.
11. Determine the MP (lit. 152°C) and percent yield.
12. Use a few drops of hydrochloric acid to remove the last traces of manganese dioxide from the Büchner funnel.

**Questions**

1. Explain the purpose of the following in this experiment:
   a. filter aid
   b. decolorizing carbon
   c. hydrochloric acid (step 9).

2. What are the products of the hot alkaline $KMnO_4$ oxidation of the following compounds:
   a. 1-methylcyclohexene
   b. 3-methylcyclohexene
   c. 1,2-dimethylcyclohexene
   d. 1,3-dimethylcyclohexene
   e. 1,2-dimethylcyclopentene
   f. 1,3-cyclopentadiene.
3. What will be the products in question 2 if hot $KMnO_4$ is replaced with $O_3/Zn$, $H_2O$?
4. What will be the products in question 2 if cold, dilute $KMnO_4$ is used instead?
5. Balance reactions 2(a), 2(b), 4(a) and 4(b) using $MnO_4^- \longrightarrow MnO_2$ (in base) as the oxidizing agent.

## 10.5   CATALYTIC HYDROGENATION OF AN ALKENE

Catalytic hydrogenation is conceptually very simple—it is merely the addition of $H_2$ across the C=C bond of an alkene:

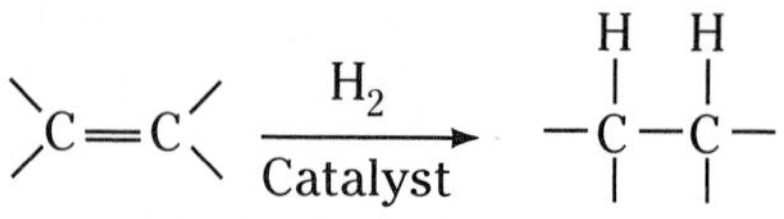

$$\text{>C=C<} \xrightarrow[\text{Catalyst}]{H_2} \quad \begin{array}{cc} H & H \\ | & | \\ -C-C- \\ | & | \end{array}$$

Several metals (including nickel, palladium, and platinum) catalyze this reaction. The mechanism is believed to involve the dissociation of hydrogen molecules into separate hydrogen atoms on the surface of a metal particle. The alkene then binds to the metal surface, breaking its π-bond. Hydrogen atoms then migrate to the bound carbon atoms and form C—H bonds. The driving force for this reaction is the formation of two strong C—H bonds (about 90–100 kcal/mole each) from a weak π-bond (about 70 kcal/mole) and an H—H bond (104 kcal/mole), which makes the reaction exothermic. Once the two new C—H bonds are formed, the molecule is no longer attracted to the metal surface and is released, freeing the catalyst surface for further reaction. The mechanism is depicted below:

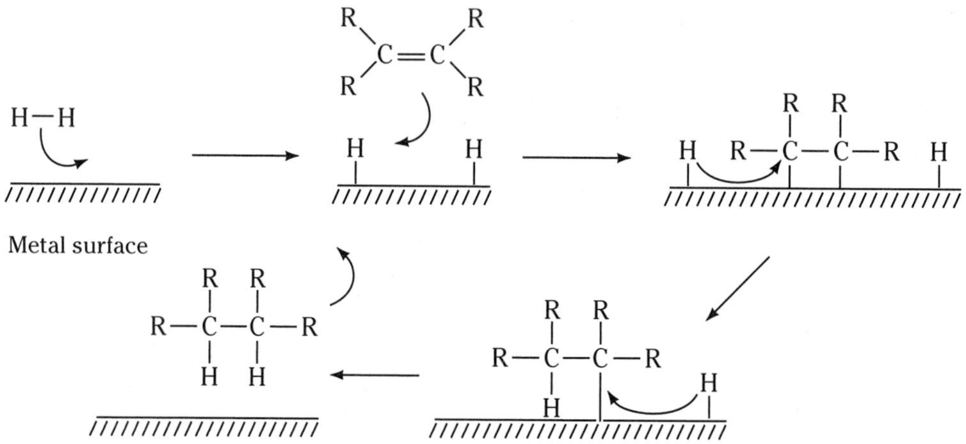

A convenient and commercially available hydrogenation catalyst is palladium metal deposited on finely divided carbon (i.e., Pd/C), which can be obtained as 5% or 10% Pd by weight. This catalyst can simply be weighed out and added to the reaction mixture as a solid reagent. The hydrogen gas required for the hydrogenation can be held over the solution using a balloon as a gas reservoir, and can be obtained either from a cylinder of compressed $H_2$, or generated *in situ* by reacting sodium borohydride with hydrochloric acid, according to the net reaction:

$$NaBH_4(aq) + HCl(aq) + 3H_2O \longrightarrow 4H_2(g) + B(OH)_3(aq) + NaCl(aq)$$

Since the palladium surface is a very active catalyst, the hydrogenation reaction will proceed to completion if we hold the hydrogen gas over the catalyst and substrate for only a few minutes. Remember that a catalyst is defined as a substance which (a) increases the rate of a reaction by lowering the energy of activation, but (b) is neither produced nor consumed in the reaction. Palladium metal meets both of these requirements.

In this experiment, *cis*-5-norbornene-*endo*-2,3-dicarboxylic acid will be hydrogenated to *cis*-norbornane-*endo*-2,3-dicarboxylic acid:

$$\xrightarrow[\textbf{PART 1}]{H_2O}$$

$$\xrightarrow[\substack{\text{EtOH-}H_2O \\ \textbf{PART 2}}]{H_2, Pd/C}$$

*cis*-5-norbornene-*endo*-
2,3-dicarboxylic anhydride
m.w. 164 g/mole
m.p. 165°C

*cis*-5-norbornene-*endo*-
2,3-dicarboxylic acid
m.w. 182 g/mole
m.p. 180–190°C (dec)

*cis*-5-norbornane-*endo*-
2,3-dicarboxylic acid
m.w. 184 g/mole
m.p. 170–175°C (dec)

The substrate is readily available via aqueous hydrolysis of the product formed in the Diels-Alder reaction between cyclopentadiene and maleic anhydride (Experiment 12). The hydrogenation reaction can be performed in aqueous ethanol. After the removal of the catalyst by filtration, and solvent evaporation, the product can be purified by recrystallization from water. A note on nomenclature: norbornane is the trivial name for bicyclo[2.2.1]heptane, *cis* indicates that both $CO_2H$ groups are on the same face of the molecule, and *endo* implies that they are on the face opposite the bridging $CH_2$ group.

**Suggested References**

Carey, F. A. *Organic Chemistry*. 3rd ed., 211–17. New York: McGraw-Hill, 1996.

Fessenden, R. J. and J. S. Fessenden, *Organic Chemistry*. 5th ed., 439–46. Pacific Grove, CA: Brooks/Cole, 1994.

Loudon, G. M., *Organic Chemistry*. 3rd ed., 161–62. Redwood City, CA: Benjamin/Cummings, 1995.

McMurry, J., *Organic Chemistry*. 4th ed., 238–41. Pacific Grove, Ca: Brooks/Cole, 1996.

Rylander, P. N., *Catalytic Hydrogenation in Organic Synthesis*. New York: Academic Press, 1979.

Rylander, P. N., *Hydrogenation Methods*. Orlando, FL: Academic Press, 1985.

Solomons, T. W. G. *Organic Chemistry*. 6th ed., 289–92. New York: John Wiley, 1996.

Vollhardt, K. P. C., and N. E. Schore, *Organic Chemistry*. 2nd ed., 414–17. New York: W. H. Freeman, 1994.

Wade, L. G. Jr., *Organic Chemistry*. 3rd ed., 357–58. Englewood Cliffs, NJ: Prentice-Hall, 1995.

⚠ **Safety Tips**

✔ The hydrogen gas used in this experiment is extremely flammable and explosive in air; no open flames should be used under any circumstance. Also, the hot plate from part 1 should be turned off and put away before the beginning of part 2.

✔ Sodium borohydride is corrosive; in case of skin contact, wash with dilute acetic acid and water.

✔ Ether is extremely flammable; avoid open flames. It also forms peroxides, which are extremely sensitive to shock, sparks, heat, and friction. Since the peroxides have a higher boiling point than ether, their residue remains in the distilling flask and can overheat and explode. Evaporation to dryness, therefore, must be avoided.

✔ The palladium catalyst collected by filtration in step 6 of part 2 should be kept moist at all times to prevent spontaneous combustion. Do not allow any Pd/C outside of the reaction flask during the hydrogenation.

**Experimental**

**Part I.** Hydrolysis of *cis*-5-norbornene-*endo*-2,3-dicarboxylic anhydride to *cis*-5-norbornene-*endo*-2,3-dicarboxylic acid (Note: This step may be omitted if the diacid is already available.)

1. Place 3.0 g of *cis*-5-norbornene-*endo*-2,3-dicarboxylic anhydride in a 100 mL beaker, add 40 mL of water, and bring to a boil on a hot plate. Heat until all of the anhydride has dissolved.
2. Remove the beaker from the heat source and allow it to stand undisturbed; crystallization should occur spontaneously. If not, scratch the sides of the beaker to induce crystallization.
3. After crystals begin to form and the beaker has cooled to room temperature, place it in an ice-water bath to complete crystallization.
4. Collect the crystals by suction filtration, dry them in an oven to constant weight, or allow them to air-dry overnight. Record the mass, calculate the percent yield, and measure the melting range. Note that the diacid does not really melt in the conventional sense but appears to undergo a chemical change over a wide temperature range. What do you think this chemical change is? (*Hint:* allow the capillary tube to cool and repeat the melting point determination.)

**Part II.** Catalytic hydrogenation of *cis*-5-norbornene-*endo*-2,3-dicarboxylic acid to *cis*-5-norbornane-*endo*-2,3-dicarboxylic acid

1. Place 1.0 g of the diacid (isolated in step 4 above) together with a Teflon-coated magnetic stir bar (if available) in a 25 mL Erlenmeyer flask with a side-arm.

### Method A: Hydrogen tank

2A. Dissolve the diacid in 15 mL of 95% ethanol; gentle heating with a hot plate may be required to complete the dissolution. Add 50 mg of 10% Pd/C; stir the suspension to mix well. Seal the flask securely with a rubber septum or cork.

3A. Your instructor will give you a hydrogen-filled gas reservoir device consisting of a balloon secured to a short piece of vacuum tubing with a rubber band (see Figure 10.1). The balloon will be clamped with a

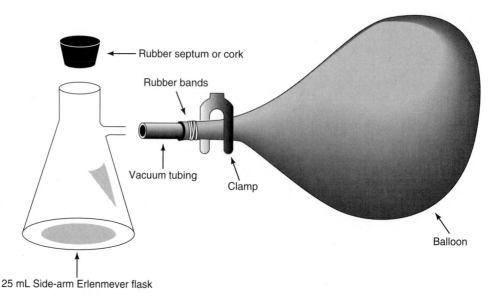

**Figure 10.1**   Hydrogenation apparatus

tubing clamp to prevent escape of the hydrogen. Attach the reservoir to your reaction flask by slipping the open end of the vacuum tubing over the side-arm and remove the clamp. This will allow hydrogen to come in contact with the catalyst suspension.

4A. Clamp the flask over a magnetic stirrer and stir the suspension vigorously for one hour. Continue with step 5.

## Method B: In situ generation of hydrogen gas

2B. Dissolve the diacid in a mixture of 5 mL of 1 M HCl and 10 mL of 95% ethanol. Add 50 mg of 10% Pd/C; stir the suspension to mix well. Attach the (empty) gas reservoir device to the reaction flask by slipping the open end of the vacuum tubing over the side-arm (see Figure 10.1).

3B. Weigh out 0.15 g of sodium borohydride. Transfer it in one portion to your reaction flask, and *RAPIDLY* seal the top of the flask with a rubber septum or cork. Hydrogen gas will be generated instantaneously by the reaction of sodium borohydride with HCl; it is *essential* that you minimize the amount of $H_2$ allowed to escape from the flask! The balloon will inflate partially, as it fills with the hydrogen gas.

4B. Clamp the flask over a magnetic stirrer and stir the suspension vigorously for one hour. Continue with step 5.

5. Set up a Büchner funnel for a suction filtation. Wet the filter paper, and under low vacuum, carefully pour a slurry of Celite (about 2 g) in water (about 15 mL) onto the center of the filter paper. After the water drains and the layer of Celite is firmly packed, disconnect the vacuum and remove the water from the filter flask.

6. After 1 hour (or after the hydrogen has been used up, as indicated by the "shrinking" of the balloon), remove the septum from your reaction flask in the hood. Slowly pour the suspension through the Celite on the Büchner funnel, being careful to avoid getting any charcoal through the filter. The filtrate should be clear. If not, repeat step 5 with a new Celite coating after you have properly disposed of the first Celite coating.

7. After all of the water has drained, rinse your reaction flask out with two or three small portions of ether (about 5 mL each), and pour each successively through the filter cake to wash through any product that may have become trapped on the catalyst.

8. Discontinue the vacuum upon completion of the filtration and discard the filter cake into the waste container labeled "Catalyst Waste."

9. Transfer the filtrate into a 100 mL beaker, add a boiling chip, and boil off the solvents on a hot water bath or steam bath in the hood. *Only* after the ether is evaporated should you boil off the water on a hot plate.

10. When the total volume of water remaining is less than 5 mL, remove the beaker from the hot plate, and allow it to cool; the product will spontaneously begin to crystallize. Cool the beaker in an ice-water bath to complete crystallization.

11. Collect the product by suction filtration, using a clean Büchner funnel. Rinse the collected crystals with a few mL of ice-water, and allow them to dry under suction for a few minutes. Transfer the crystals to a watch-glass, and dry them in an oven for 30 minutes (or at room temperature overnight).

12. Weigh the dried product and measure the melting (decomposition) range. Calculate the percent yield.

13. Test the product for unsaturation using 1% bromine water solution (see Experiment 10.1 step 7-b, p. 126). Repeat the test on a small sample of the starting alkene, and compare the results.

**Questions**    1. What is the main safety precaution to be observed during this experiment?

2. Describe your observations in the bromine-water tests of the alkene and the hydrogenation product (step 13). What can you conclude from these observations?

3. Write a balanced equation for the reaction of sodium borohydride and aqueous HCl to form hydrogen gas and boric acid, as described in Method B. Using the indicated amounts, how many millimoles of hydrogen gas are generated? How many millimoles of excess hydrogen are formed *beyond* the amount needed to hydrogenate the double bond of the alkene (i.e., how much excess $H_2$ is generated)?

4. Catalytic hydrogenation is widely used in the food industry to prepare "partially hydrogenated vegetable oils" from "polyunsaturated fats." Although less healthy, saturated fats (i.e., esters of long-chain carboxylic acids) are less susceptible to aerobic degradation. An example of this process is the hydrogenation of (Z)-9-octadecenoic acid (oleic acid) to octadecanoic acid (stearic acid). Write a balanced equation (showing correct chemical structures) that describes this reaction. (*Note:* You may need to use your lecture textbook.)

5. What is the major advantage in using palladium metal deposited on carbon, compared to simply adding a small amount of powdered palladium metal?

6. Show the product expected to be formed by addition of deuterium gas ($D_2$, where D = $^2$H) to norbornene (i.e., bicyclo[2.2.1]hept-2-ene) in the presence of Pd/C.

7. Fill in the missing products in the reactions below. Indicate any possible stereochemistry involved. Assume the conditions used in the above experiment are applied.
   a. 1-pentene + $H_2$/Pd $\longrightarrow$ ?
   b. *cis*-2-pentene + $H_2$/Pd $\longrightarrow$ ?
   c. *trans*-2-pentene + $H_2$/Pd $\longrightarrow$ ?
   d. 1-pentyne + $H_2$/Pd (1 equiv.) $\longrightarrow$ ? + $H_2$/Pd (2nd equiv.) $\longrightarrow$ ?
   e. 2-pentyne + $H_2$/Pd (1 equiv.) $\longrightarrow$ ? + $H_2$/Pd (2nd equiv.) $\longrightarrow$ ?
   f. 3-phenyl-1-propene + $H_2$/Pd (excess) $\xrightarrow[\text{temperature}]{\text{room}}$ ?

# chapter 11

# Synthesis and Reactions of Acetylene

**Acetylene** is the first member, and the only commercially important member, of the **alkyne** series. It is also a **terminal alkyne;** i.e., the triple bond comes at the end of a carbon chain. If the triple bond is located somewhere other than at the end of the carbon chain, the alkyne is called an **internal**, or a **non-terminal**, **alkyne.** Thus, 1-butyne, $CH_3CH_2C \equiv CH$, is a terminal alkyne; while 2-butyne, $CH_3—C{\equiv}C—CH_3$, is a non-terminal alkyne.

Acetylene is conveniently synthesized by the hydrolysis of calcium carbide—a reaction that can be viewed as one involving both a Brønsted acid ($H_2O$) and a base ($CaC_2$).

$$CaC_2 \ + \ 2H_2O \ \longrightarrow \ H—C{\equiv}C—H \ + \ Ca(OH)_2$$

| calcium carbide | water | | acetylene | calcium hydroxide |

Although pure acetylene is odorless, the acetylene synthesized by this hydrolysis method has a characteristic smell that is attributed to contamination by phosphorus, arsenic, and sulfur oxides.

In this experiment, the synthesis of acetylene will be followed by some characteristic reactions and comparisons with members of the other hydrocarbon families—the alkanes, alkenes, dienes and aromatic compounds—as well as with alkyl halides.

**Suggested References**

Carey, F. A., *Organic Chemistry.* 3rd ed., 349–72. New York: McGraw-Hill, 1996.

Ege, Seyhan N., *Organic Chemistry: Structure and Reactivity.* 3rd ed., 332–34. Lexington, MA: D. C. Heath, 1994.

Loudon, G. Marc, *Organic Chemistry.* 3rd ed., 653–65. Redwood City, CA: Benjamin/Cummings, 1995.

McMurry, J., *Organic Chemistry.* 4th ed., 264–88. Pacific Grove, CA: Brooks/Cole, 1996.

Morrison, R. T., and R. N. Boyd, *Organic Chemistry.* 6th ed., 425–42. Englewood Cliffs, NJ: Prentice-Hall, 1992.

Patnaik, P., *A Comprehensive Guide to the Hazardous Properties of Chemical Substances.* 414. New York: Van Nostrand Reinhold, 1992.

Solomons, T. W. G. *Organic Chemistry.* 6th ed., 351–54. New York: John Wiley, 1996.

Vollhardt, K. P. C., and N. E. Schore, *Organic Chemistry.* 2nd ed., 466–500. New York: W. H. Freeman, 1994.

Wade, L. G., Jr., *Organic Chemistry.* 3rd ed., 627–63. Englewood Cliffs, NJ: Prentice-Hall, 1995.

**SAFETY TIPS**

✔    Acetylene is a highly flammable gas that burns with a sooty flame. It also forms explosive mixtures with air (2–80% by volume). All flames should be extinguished in the lab during the experiment. Bromine is a corrosive liquid whose vapors are irritating to humans (maximum allowed concentration in air = 50 ppm).

**Experimental**

*Note:* It is suggested that a common setup be used for a class of 10 to 25 students. The quantities described below should be adequate for about 80 test tubes. The container holding the calcium carbide should be sealed immediately after use to protect its contents from moisture.

1. Place about 8 g of calcium carbide in the ***dry*** round-bottomed flask of the setup, as shown in Figure 11.1.
2. Add 20 mL of water *dropwise* through the funnel.
3. Collect the gas in inverted test tubes as it displaces the water. Discard the first two test tubes (they contain mostly air). Cork the test tubes as soon as the water has been displaced.
4. Use the corked test tubes in the experiments below.
5. Once the required acetylene is collected in the test tubes, attach a disposable pipette to the end of the tubing of the assembly. Place the pipette in a beaker containing a solution of 5% $KMnO_4$ to neutralize the excess acetylene, until no more bubbles are produced when water is added to the calcium carbide.
6. Clean up the assembly in the fume hood first with water and then with 6 M HCl.

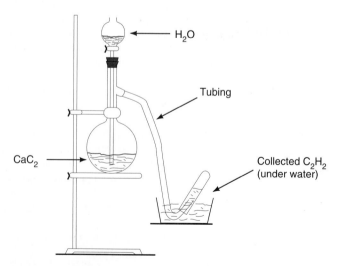

**Figure 11.1**   Setup for the synthesis of acetylene

## Qualitative tests for hydrocarbons and alkyl halides

Perform the following tests on separate acetylene test tubes and also on

a. cyclohexane
b. cyclohexene
c. diphenylacetylene
d. cyclopentadiene (freshly distilled) or dicyclopentadiene
e. toluene
f. *n*-butyl bromide
g. *t*-butyl bromide
h. benzyl bromide
i. bromobenzene.

Write your observations and equations on the data sheet ( p. 157) for any reactions that take place on the data sheet.

*Test 1:* Add 5 drops of 1% cold $KMnO_4$ to 5 drops of each compound in separate test tubes and mix well. Record your observations.

*Test 2:* Add 10 drops of a 1% solution of $Br_2/H_2O$ to 5 drops of each compound in separate test tubes and mix well. Record your observations.

*Test 3:* Add 5–10 drops of 5% ammoniacal $AgNO_3$ to 5 drops of each compound in separate test tubes and mix well. Record your observations. For compounds that do not react at room temperature, place the corked test tube in a water bath at 70°C for about 5 minutes with occasional shaking.

*Test 4:* Add 10 drops of concentrated $H_2SO_4$ to 5 drops of each compound in separate test tubes and mix well. Record your observations.

*Note:* Before you perform test 5, the instructor should make sure that all reagents are anhydrous.

*Test 5:* Add a tiny piece of sodium metal to 5 drops of each compound in separate test tubes. Record your results. For acetylene, use a test tube containing 2 mL of hexane in which acetylene has been bubbled for 2 or 3 minutes.

Obtain an unknown from your instructor and perform these tests on it. Is your unknown an alkane, an alkene, a terminal alkyne, a nonterminal alkyne, a diene, an aromatic compound, or an alkyl halide?

**Questions**

1. Give balanced equations for the reactions of acetylene with the following:
   a. $Br_2/CCl_4$(1 equivalent)
   b. $Br_2/CCl_4$(2 equivalents)
   c. $MnO_4^-$ (in acid to yield $Mn^{+2}$)
   d. $MnO_4^-$ (in base to yield $MnO_2$)
   e. ammoniacal $AgNO_3$.
2. Suggest a simple chemical test that would distinguish between the following compounds. State what you would do and what you would see.
   a. *n*-pentane and 1-pentene
   b. 1-pentene and 1-pentyne
   c. 1-pentyne and 2-pentyne
   d. 2-bromo-2-methylbutane and benzene
   e. 1-pentane and 2-bromo-2-methylbutane
   f. benzene and cyclopentadiene
   g. 2-bromo-2-methylpentane and 1-bromopentane
3. Calculate the amount of calcium carbide needed to produce 10.0 g of acetylene in a 50% yield.
4. Give a balanced equation for the complete combustion of propyne.

Name: _____

# DATA SHEET: QUALITATIVE TESTS FOR HYDROCARBONS AND ALKYL HALIDES

| Compound | Test 1 KMnO$_4$ | Test 2 Br$_2$/H$_2$O | Test 3 AgNO$_3$ | Test 4 H$_2$SO$_4$ | Test 5 Na |
|---|---|---|---|---|---|
| Cyclohexane | | | | | |
| Cyclohexene | | | | | |
| Acetylene | | | | | |
| Diphenylacetylene | | | | | |
| (Di)cyclopentadiene | | | | | |
| Toluene | | | | | |
| n-Butyl bromide | | | | | |
| t-Butyl bromide | | | | | |
| Benzyl bromide | | | | | |
| Bromobenzene | | | | | |
| Unknown # _____ | | | | | |

Based on the above observations, unknown # _____ is a(n) _____ (circle your answer[s])

alkane/alkene/terminal alkyne/non-terminal alkyne/diene/arene/1° alkyl halide/3° alkyl halide/benzyl halide/aryl halide

# 12

# Conjugated Dienes: The Diels-Alder Reaction

Hydrocarbons are classified as **saturated** (alkanes) and **unsaturated** (alkenes, alkynes, aromatics, polyenes). Compounds with two double bonds are called **dienes,** and they can be **conjugated** (i.e., have alternating double bonds) or **nonconjugated** (i.e., have non-alternating double bonds):

$$CH_2 = CH — CH = CH — CH_3 \qquad CH_2 = CH — CH_2 — CH = CH_2$$

1,3-pentadiene             1,4-pentadiene

(a conjugated diene)      (a nonconjugated diene)

A conjugated system can react with the double bond of an alkene (or the triple bond of an alkyne) in a concerted, one-step process to form a six-membered ring. The reaction, which has great synthetic application, was named as the **Diels-Alder** reaction after its two originators, Otto Paul Hermann Diels and Kurt Alder, the 1950 Chemistry Nobel Prize winners. The alkene is called a **dienophile** (i.e., diene-loving), and the product is often referred to as the **adduct**. The yield of the product is enhanced by the presence of electron-releasing groups (e.g., alkyl or alkoxy groups) on the diene and by electron-withdrawing groups (e.g., cyano groups, $— C \equiv N$, or carbonyl groups, $\overset{\diagdown}{\underset{\diagup}{C}} = O$) on the dienophile. Very often the yields are quantitative.

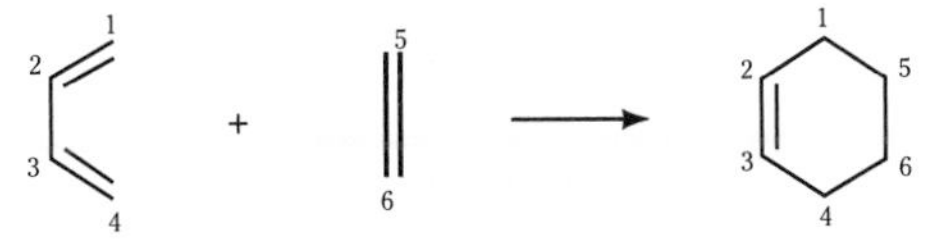

diene       dienophile      Diels-Alder adduct

In order for the closure to take place, the diene must be in the (s)-cis conformation. The ends of the diene in the (s)-trans arrangement are too far apart to react with the dienophile to form a cyclohexene ring. Because of this constraint, cyclic dienes which are entirely in the (s)-cis conformation are very reactive Diels-Alder dienes. In this experiment, 1,3-cyclopentadiene will be used. However, 1,3-cyclopentadiene is not commercially available because it rapidly undergoes Diels-Alder cycloaddition with itself to form a dimer.

cyclopentadiene        dicyclopentadiene
(dimer)

Because the Diels-Alder reaction is reversible when heat is applied, the dimer can be "cracked" by heat to regenerate the monomeric 1,3-cyclopentadiene, which is then collected by fractional distillation. Once regenerated, the monomer must be used immediately before it re-dimerizes.

The Diels-Alder reaction is a type of cycloaddition that involves four $\pi$ electrons of the diene and two $\pi$ electrons of the dienophile. For this reason it is classified as a [4 + 2] cycloaddition. As the diene and dienophile combine, three $\pi$ bonds (two in the diene and one in the dienophile) are broken, while two new (and stronger) $\sigma$ bonds and one $\pi$ bond are formed in the adduct. As a result, the reaction is exothermic. The reaction is also a stereospecific process in which the stereochemistry of the dienophile is retained.

In this experiment, monomeric 1,3-cyclopentadiene will be generated and then reacted with the very active dienophile maleic anhydride

cyclopentadiene     maleic anhydride     *endo*-s-norbornene-*endo*-2,3-dicarboxylic anhydride
(a Diels-Alder adduct)

**Suggested References**     Boger, D. L., and S. M., Weinreb, *Hetero-Diels-Alder Methodology in Organic Synthesis.* San Diego: Academic Press, 1987.

Carey, F. A., *Organic Chemistry.* 3rd ed., 397–402. New York: McGraw-Hill, 1996.

Ege, Seyhan N, *Organic Chemistry: Structure and Reactivity.* 3rd ed., 350–75. Lexington, MA: D. C. Heath, 1994.

"Cyclopentadiene." In *Reagents for Organic Synthesis,* edited by L. F. Fieser and M. Fieser, vol. 1, 415–24. New York: John Wiley, 1967.

"Diels-Alder Solvents." In *Reagents for Organic Synthesis,* edited by L. F. Fieser and M. Fieser, vol. 1, 236–43. New York: John Wiley, 1967.

Flett, L. H., and W. H. Gardner, *Maleic Anhydride Derivatives: Reactions of the Double Bond.* New York: John Wiley, 1952.

Fringuelli, F., and A. Taticchi, *Dienes in the Diels-Alder Reaction in Heterocyclic Systems.* New York: John Wiley, 1990.

Kagan, H. B., and O. Riant, "Catalytic Asymmetric Diels-Alder Reactions." *Chem. Reviews.* 92(1992): 1007–19.

Loudon, G. Marc, *Organic Chemistry.* 3rd ed., 690–98. Redwood City, CA: Benjamin/Cummings, 1995.

Mehta, G., "Molecular Design of Compounds via Intramolecular Diels-Alder Reactions." *J. Chem. Educ.* 53 (1976): 559.

McMurry, J., *Organic Chemistry.* 4th ed., 512–17. Pacific Grove, CA: Brooks/Cole, 1996.

Morrison, R. T., and R. N. Boyd, *Organic Chemistry.* 6th ed., 1013–18. Englewood Cliffs, NJ: Prentice-Hall, 1992.

Solomons, T. W. G., *Organic Chemistry.* 6th ed., 520–25. New York: John Wiley, 1996.

Taber, D. F., *Intramolecular Diels-Alder and Alder Ene Reactions*. Berlin: Springer-Verlag, 1984.

Vollhardt, K. P. C., and N. E. Schore, *Organic Chemistry*. 2nd ed., 518–25. New York: W. H. Freeman, 1994.

Wade, L. G., *Organic Chemistry*. 3rd ed., 683–91. Englewood Cliffs, NJ: Prentice-Hall, 1995.

---

 **SAFETY TIP**

✔ Cyclopentadiene (BP 42°C), petroleum ether and ethyl acetate (BP 77°C) are flammable liquids. All flames should be extinguished in the lab. Dicyclopentadiene has an obnoxious odor and should be handled with adequate ventilation. Maleic anhydride (MP 53°C) is a white crystalline powder, whose dust may irritate the skin and eyes upon contact. The addition of cyclopentadiene to the maleic anhydride solution is very exothermic and should be carried out in an ice/water bath.

---

**Experimental**

1. Place 10 mL of dicyclopentadiene in a 100-mL round-bottomed flask, and set up a fractional distillation apparatus. Do not pack the column.
2. Heat the dimer gently initially, and collect the fraction that distills between 40–45°C. Collect 3.0 mL (d = 0.95 g/mL, 2.9 g) of the monomer *in ice* to avoid evaporation and re-dimerization.
3. While distillation is in progress, place 3.0 g of powdered maleic anhydride in a 125-mL Erlenmeyer flask, and dissolve it in 10 mL ethyl acetate by heating the mixture on a steam bath. Cool the flask in an ice/water bath. Add 10 mL of petroleum ether, and then the 3 mL of freshly distilled cyclopentadiene *while the flask is still in ice*.
4. Swirl the solution until complete precipitation occurs.
5. Heat on a steam bath until the mixture has just dissolved, and then *allow the solution to cool slowly to room temperature*. Cool the solution in ice and isolate the crystals by vacuum filtration. Wash the crystals with 5 mL of ice-cold petroleum ether while the crystals are still on the Büchner funnel.
6. Determine the MP (lit. 156°) and percent yield.
7. Dissolve a few crystals of the product in 1–2 mL of ethyl acetate. Test with dilute $KMnO_4$ or $Br_2/H_2O$ solution for the presence of double bonds. Record your observations.

**Questions**

1. Give an explanation for the following:
   a. Cyclopentadiene dimerizes more readily than acyclic dienes.
   b. Cyclopentadiene reacts more rapidly with maleic anhydride than with another cyclopentadiene molecule.
   c. Commercial cyclopentadiene cannot be used as a conjugated diene in a Diels-Alder reaction.
   d. The distillate of the fractional distillation is collected in ice.
   e. Cyclopentadiene is added to a *cool* solution of maleic anhydride.

2. Give the products of the following Diels-Alder reactions. Identify the diene and dienophile.

a. 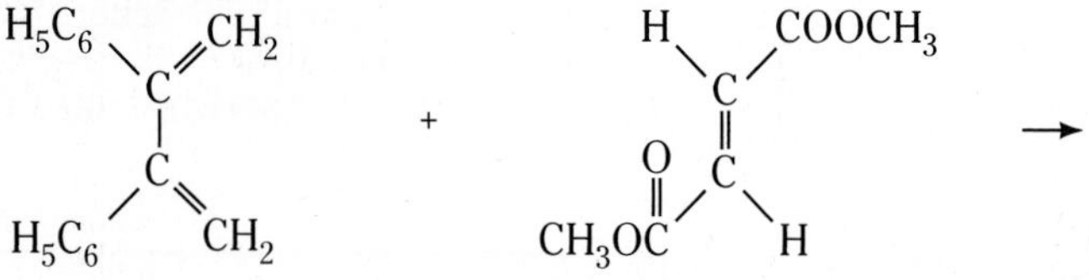 + →

b. 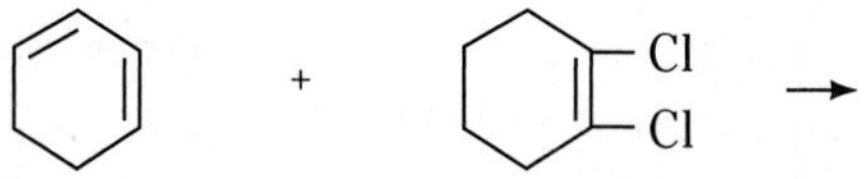 + →

c. 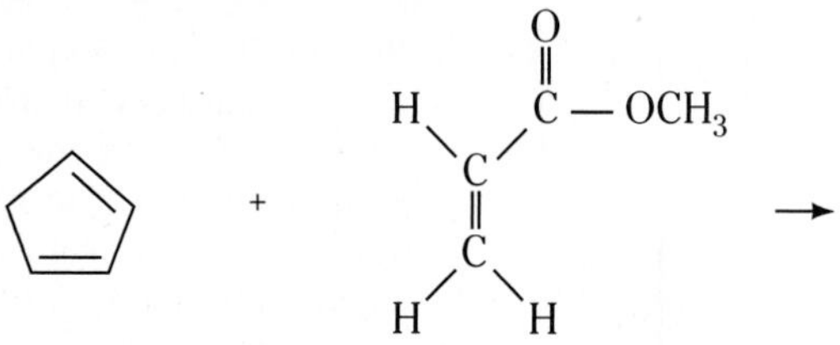 + →

d. 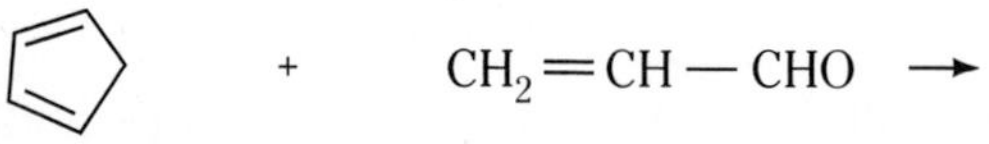 + $CH_2 = CH - CHO$ →

e. 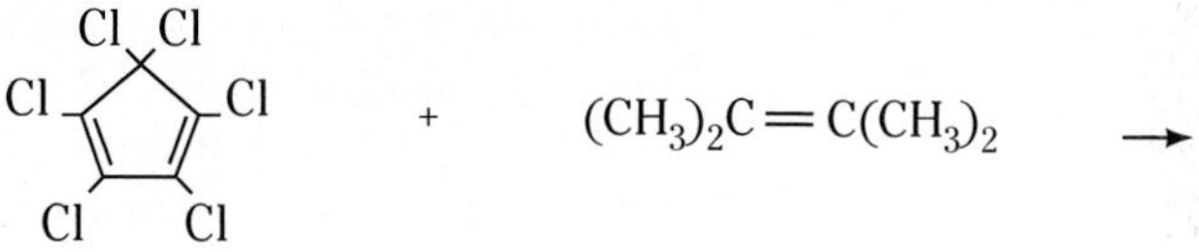 + $(CH_3)_2C = C(CH_3)_2$ →

f. 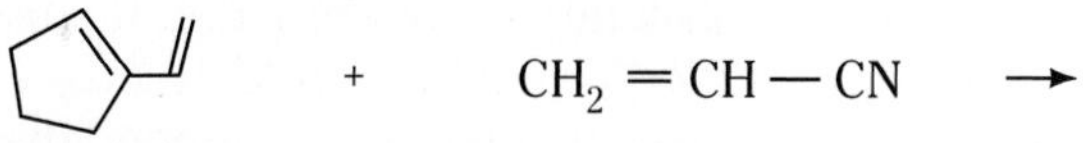 + $CH_2 = CH - CN$ →

3. Identify the diene and dienophile that would yield the following Diels-Alder adducts:

a.

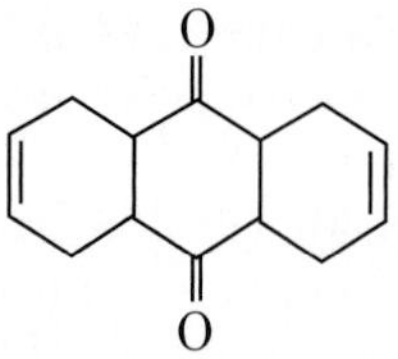

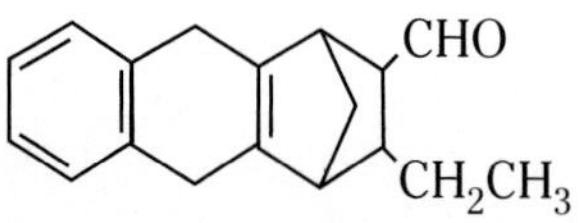

e.

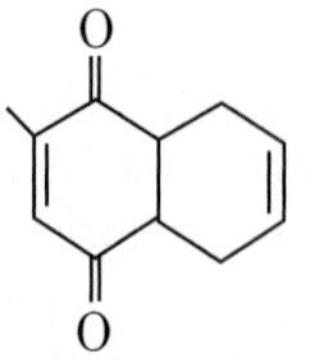

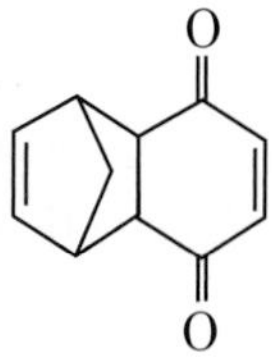

b.

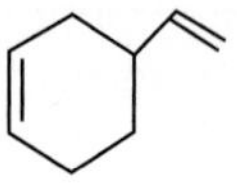

f.

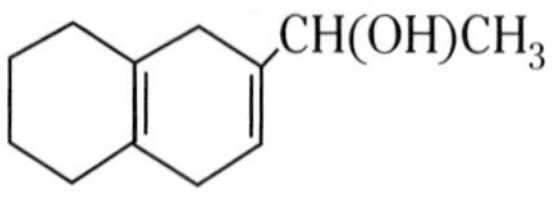

c.

g.

d.

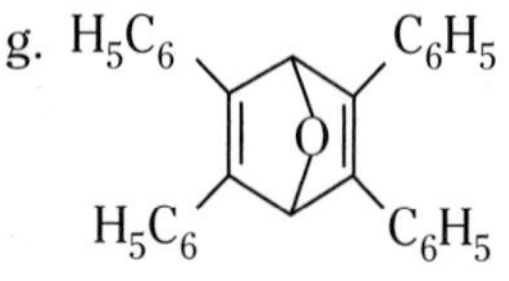

# chapter

# 13

# Alkyl Halides and Nucleophilic Aliphatic Substitution

## 13.1 SYNTHESIS OF *t*-BUTYL CHLORIDE

Alkyl halides can be prepared by reacting an alcohol with concentrated hydrogen halides:

$$R{-}OH + H{-}X \longrightarrow R{-}\overset{+}{O}H_2 \xrightarrow{-H_2\ddot{O}:} R^+ \xrightarrow{X^-} R{-}X + H_2O$$

where X = Cl, Br, I. The reaction of hydrochloric acid with 2° and 3° alcohols involves the formation of a carbocation, R$^+$. When possible, 2° carbocations rearrange to form the more stable 3° carbocations. On the other hand, 3° carbocations are relatively stable and seldom yield rearranged products. However, elimination side reactions (E1) in which an alkene is formed can be significant.

In this experiment, *t*-butyl alcohol is reacted with concentrated hydrochloric acid, and the product (*t*-butyl chloride) is removed as it is formed via an azeotropic distillation.

$$(CH_3)_3C{-}OH + HCl \longrightarrow (CH_3)_3C{-}Cl + H_2O$$

*t*-butyl alcohol  *t*-butyl chloride

The distillate is washed first with a sodium bicarbonate solution to neutralize any remaining traces of acid, and then with water to remove any basic bicarbonate that may regenerate the starting alcohol or form an alkene during the final distillation:

$$(CH_3)_3C{-}Cl \xrightarrow{HCO_3^-} (CH_3)_2C{=}CH_2 + H_2O + Cl^- + CO_2$$

**Suggested References**  Carey, F. A., *Organic Chemistry*. 3rd ed., 137–44. New York: McGraw-Hill, 1996.

Ege, Seyhan N., *Organic Chemistry: Structure and Reactivity*. 3rd ed., 452–56. Lexington, MA: D. C. Heath, 1994.

Loudon, G. Marc, *Organic Chemistry*. 3rd ed., 416–23. Redwood City, CA: Benjamin/Cummings, 1995.

March, J., *Advanced Organic Chemistry*. 4th ed., 298–99. New York: John Wiley, 1992.

McMurry, J., *Organic Chemistry*. 4th ed., 391–98. Pacific Grove, CA: Brooks/Cole, 1996.

Morrison, R. T., and R. N. Boyd, *Organic Chemistry*. 6th ed., 229–33. Englewood Cliffs, NJ: Prentice-Hall, 1992.

Patnaik, P., *A Comprehensive Guide to the Hazardous Properties of Chemical Substances.* 78. New York: Van Nostrand Reinhold, 1992.

Solomons, T. W. G., *Organic Chemistry.* 6th ed., 438–41. New York: John Wiley, 1996.

Vollhardt, K. P. C., and N. E. Schore, *Organic Chemistry.* 2nd ed., 518–25. New York: W. H. Freeman, 1994.

Wade, L. G., *Organic Chemistry.* 3rd ed., 443–47. Englewood Cliffs, NJ: Prentice-Hall, 1995.

 **SAFETY TIPS**

  *t*-Butyl alcohol is flammable (flash point 9°C) and all flames should be extinguished in the lab.

  Concentrated hydrochloric acid is corrosive; it should be poured with care and with adequate ventilation.

  Extraction of the distillate in step 2 should be done with frequent opening of the stopcock to vent the carbon dioxide that has formed and to reduce the built up pressure.

**Experimental**

1. Place 6.3 mL (d = 0.79g/mL; 5.0 g) of *t*-butyl alcohol and 15 mL of concentrated hydrochloric acid in a 50-mL round-bottomed flask, and perform a simple distillation.

2. Attach one end of a rubber hose at the vacuum adaptor outlet and the other end to a short stem funnel. Place the short stem funnel in a beaker containing a 10% $NaHCO_3$ solution. This will serve as a trap to neutralize the fumes of hydrogen chloride. Collect about 10 mL of the distillate in ice.

3. Transfer the distillate to a 125-mL separatory funnel, and extract it with 10 mL of a saturated sodium bicarbonate solution. Remember to vent any carbon dioxide that forms by opening the stopcock frequently.

4. Separate the aqueous bicarbonate layer from the organic layer and discard the aqueous layer. If you are unsure which is the aqueous layer, add a drop or two of water with a medicine dropper or a disposable pipette. If the drop of water falls through the upper layer into the bottom layer, the bottom layer is the aqueous layer. *Do not discard anything until you are sure that you are not throwing away your product.* Add 15 mL of water to the organic layer and extract again.

5. Pour the organic layer into a dry conical flask, add about 1 gram of anhydrous sodium or magnesium sulfate, and let stand covered with a watch glass for about 15 minutes.

6. Gravity filter into a 50-mL round-bottomed flask, and perform a simple distillation. Collect the fraction that distills between 48° and 51°C in a preweighed vial that is immersed in an ice/water bath. Calculate the percent yield.

7. Run a gas chromatograph of your product, and compare it with a gas chromatograph of a standard prepared by mixing equal quantities of *t*-butyl alcohol and water. Also run IR spectra for both *t*-butyl alcohol and the distillate.

8. Add 5 drops of the distillate to 1 mL of a 5% $AgNO_3$ solution in a test tube and shake the test tube. Do the same with 5 drops of *t*-butyl alcohol and compare. Explain your observations.

**Questions**

1. What is the purpose of the following in this experiment:
   a. concentrated (and not dilute) hydrochloric acid
   b. collecting the distillate in ice
   c. sodium bicarbonate extraction
   d. water extraction
   e. anhydrous sodium or magnesium sulfate.
2. Why do tertiary alcohols react faster with concentrated hydrochloric acid than do secondary or primary alcohols?
3. Why are rearrangements rare with tertiary alcohols but not with secondary or primary alcohols?
4. Compare the IR spectra of the reactant and final product of this experiment.
5. Would you expect the peak due to the $(CH_3)_3C$ — group in the $^1H$ NMR spectrum of the product to have a greatly different chemical shift from the corresponding chemical shift seen in the $^1H$ NMR spectrum of the reactant? Explain.

## 13.2   SOLVOLYSIS OF *t*-BUTYL CHLORIDE: A KINETIC STUDY

The rate at which a homogeneous chemical reaction takes place depends upon a number of factors, including the inherent properties of the reacting species, the concentration of the reactants, the particular solvent, the temperature, the presence of catalysts, and the existence of any surface effects.

The **order** of a reaction is defined as the sum of the powers to which the concentrations are raised in the experimentally determined rate equation.

The rate of decomposition of $N_2O_5$

$$2\,N_2O_5 \longrightarrow 4\,NO_2 + O_2$$

is found to obey the rate law

$$-d[N_2O_5]/dt = k_1\,[N_2O_5]$$

The order, therefore, is 1, and the reaction is said to be unimolecular.

The rate of decomposition of $NO_2$

$$2\,NO_2 \longrightarrow 2\,NO + O_2$$

is found to obey the rate law

$$-d[NO_2]/dt = k_1\,[NO_2]^2$$

The order, therefore, is 2, and the reaction is said to be bimolecular.

The rate of reaction of triethylamine with ethyl bromide

$$(C_2H_5)_3N: + C_2H_5Br \longrightarrow (C_2H_5)_4N^+Br^-$$

is found to obey the rate law

$$-d[(C_2H_5)_4N^+Br^-]/dt = k_1\,[C_2H_5Br][(C_2H_5)_3N]$$

and is first order with respect to triethylamine, first order with respect to ethyl bromide, and second order overall.

The **order** of a reaction may be zero or fractional. It is determined solely by the best fit of the experimental data, and is not necessarily related to the stoichiometry of the reaction. **Molecularity** of a reaction is defined as the number of molecules of reactants that are used to form the activated complex. Thus, molecularity must be a whole number, while the reaction order applies to the experimental rate equation and can be a fraction, or inverse number.

In this experiment, the solvolysis of *t*-butyl chloride will be studied. The term **solvolysis** means that the solvent is the nucleophile and is in large excess. The solvolysis of *t*-butyl chloride proceeds via an $S_N1$ mechanism—i.e., rate-determining loss of Cl⁻, followed by rapid attack of water, forming *t*-butyl alcohol and HCl. (Remember that the rate depends only on the concentrations of reactants in the *slowest* step—i.e., the rate determining step!)

$$(CH_3)_3C — Cl \xrightarrow{-Cl^-} (CH_3)_3\overset{+}{C} \longrightarrow (CH_3)_3C — \overset{+}{\underset{\ddot{}}{O}}H_2 \xrightarrow{-H^+} (CH_3)_3C — OH$$

t-butyl chloride     t-butyl     protonated     t-butyl alcohol
carbocation     alcohol

$$(CH_3)_2CH\ddot{O}H \longrightarrow (CH_3)_3C\overset{H}{\underset{+}{O}}CH(CH_3)_2 \xrightarrow{-H^+} (CH_3)_3C — OCH(CH_3)_2$$

isopropyl t-butyl ether

Thus, the rate law is expected to be first order:

$$\text{rate} = -d[RX]/dt = k[RX]$$

Rearranging and then integrating from $t = 0$ to $t$, gives

$$\int d[RX]/[RX] = -\int_0^t (k)dt$$

$$\ln[RX]_0 - \ln[RX] = rt \text{ or}$$

$$\ln([RX]_0/[RX]) = kt$$

where: $[RX]_0$ = initial concentration of t-BuCl and
$[RX]$ = concentration of t-BuCl at time $t$

A plot of the natural log of the ratio of the initial concentration of starting material to the concentration at time $t$, versus time $t$, gives a straight line with a slope equal to the rate constant.

In this experiment, one equivalent of t-BuCl yields one equivalent of HCl. We can determine the amount of HCl present (and thus the amount of t-BuCl that has reacted) at any time by titration of an aliquot of the reaction mixture with a standardized solution of NaOH. The initial concentration of t-BuCl is proportional to the amount of NaOH required to neutralize an aliquot at time $t$ = infinity (that is after essentially all of the t-BuCl has reacted). The remaining concentration of t-BuCl at any other time $t$ is equal to the initial amount present minus the amount that has reacted at time $t$. Thus:

$$[RX]_0 \text{ is proportional to (vol NaOH)}_\infty$$

and

$$[RX]_t \text{ is proportional to (vol NaOH)}_\infty \text{ minus (vol NaOH)}_t$$

Thus, the rate equation above becomes:

$$\ln \frac{(\text{vol NaOH})_\infty}{(\text{vol NaOH})_\infty - (\text{vol NaOH})_t} = kt$$

A plot of the left-hand side of the equation versus time will give a straight line with a slope equal to the first-order rate constant. [A] represents the volume of NaOH required for titration to neutrality at time $t$ (that is, (vol NaOH)$_t$), [A$_t$] is the difference between (vol NaOH)$_\infty$ and (vol NaOH)$_t$, and [A$_0$] is (vol NaOH)$_\infty$. To determine $k$, plot ln([A$_0$]/[A$_t$]) versus time and draw a best-fit line through your points using a ruler. Calculate the slope (as $\Delta x/\Delta y$), and report your rate constant, using the correct units. Calculate the half-life (i.e., the time required for half of the remaining $t$-BuCl to react) using the standard first-order relationship:

$$t_{1/2} = 0.693/k$$

The volume of NaOH used in each titration, therefore, can be used to represent the concentration of $t$-butyl chloride. At intervals a 10-mL sample of the reaction mixture will be withdrawn, quenched with isopropyl alcohol to stop the reaction, and titrated with NaOH. The rate of solvolysis, as measured by the quantity of the base consumed per unit time, is greatest in the initial stages of the reaction, and then steadily decreases.

**Suggested References**　　Allen, A., A. J. Haughey, Y. Hernandez, and S. Ireton, "A Study of Some 2-Chloro-2-methylpropane Kinetics Using a Computer Interface." *J. Chem. Educ.* 68 (1991): 609–11.

Carey, F. A. *Organic Chemistry.* 3rd ed., 323–27. New York: McGraw-Hill, 1996.

Ege, Seyhan N., *Organic Chemistry: Structure and Reactivity.* 3rd ed., 241–60. Lexington, MA: D. C. Heath, 1994.

Loudon, G. Marc, *Organic Chemistry.* 3rd ed., 417. Redwood City, CA: Benjamin/Cummings, 1995.

March, J., *Advanced Organic Chemistry.* 4th ed., 298–99. New York: John Wiley, 1992.

McMurry, J., *Organic Chemistry.* 4th ed., 386–91. Pacific Grove, CA: Brooks/Cole, 1996.

Morrison, R. T., and R. N. Boyd, *Organic Chemistry.* 6th ed., 258–61, 268–71, 188–91. Englewood Cliffs, NJ: Prentice-Hall, 1992.

Patnaik, P., *A Comprehensive Guide to the Hazardous Properties of Chemical Substances.* 73–4. New York: Van Nostrand Reinhold, 1992.

Solomons, T. W. G., *Organic Chemistry.* 6th ed., 247–48, 255. New York: John Wiley, 1996.

Vollhardt, K. P. C., and N. E. Schore, *Organic Chemistry.* 2nd ed., 178–97. New York: W. H. Freeman, 1994.

Wade, L. G., *Organic Chemistry.* 3rd ed., 256–70. Englewood Cliffs, NJ: Prentice-Hall, 1995.

 **SAFETY TIPS**

 Isopropyl alcohol is a flammable liquid even in 91% solutions in water. Its flash point (anhydrous grade) is 17.2°C and, therefore, all flames should be extinguished in the lab.

 $t$-Butyl chloride is an irritant and a flammable liquid.

**Experimental**

1. Place 100 mL of a 1:1 isopropyl alcohol-water mixture in an Erlenmeyer flask that is tightly stoppered to prevent diffusion of carbon dioxide into the mixture.
2. Obtain 100 mL of isopropyl alcohol to be used for quenching.
3. Rinse the burette with 2–3 mL of 0.1$N$ NaOH, and then fill it with the base *using a funnel.* Avoid bubbles in the burette.

4. Keep 1–2 mL of phenolphthalein in a test tube to be used as the indicator in titrating.

5. Add 5.0 mL of *t*-butyl chloride to the solvent mixture (from step 1) and note the time closest to the minute after a three-second vigorous mixing. This is the reaction mixture from which aliquots will be taken and titrated.

6. After 1 minute, pipette a 10-mL sample of the reaction mixture into an Erlenmeyer flask, and quench the reaction by adding 10 mL of isopropyl alcohol and some crushed ice. Add 2 drops of phenolphthalein and titrate with the standardized sodium hydroxide solution until a pale pink color persists for 10 seconds. Try to titrate as rapidly as possible.

7. Repeat step 6 at regular intervals of 2, 4, 6, 8, 10, 15, 20, and 30 minutes, and record the volumes of sodium hydroxide solution consumed.

8. Warm the last 10-mL sample reaction mixture in a 50°C water bath for about 5 minutes. Cool in ice *(do not add any isopropyl alcohol)* and titrate as before.

9. Drain the remaining NaOH solution, fill the burette with water (while still clamped!) and drain again.

10. Complete the Data and Calculations section (p. 159) and plot your data on the graph paper provided (p. 161).

**Questions**

1. What would be the effect on the titration values if the temperature were *higher* than room temperature?

2. How is the progress of the reaction followed in this experiment?

3. Using the first-order kinetics equation, prove that the time required for 99.9% of the reaction to take place is ten times that required for one-half of the reaction.

4. Can *n*-butyl chloride be substituted for *t*-butyl chloride in this experiment? Explain.

# DATA SHEET: SOLVOLYSIS OF *t*-BUTYL CHLORIDE: A KINETIC STUDY
## PAGE 1

Tabulate your results as follows:

[A] = NaOH volume at any time

$[A_o]$ = NaOH volume for infinity reading (from step 8) = _____ mL

$[A_t]$ = concentration at time $t$ = $[A_o]$ − [A]

| Time (min) | [A] (mL) | $[A_t]$ (mL) | $[A_o]/[A_t]$ | $\ln ([A_o]/[A_t])$ |
|---|---|---|---|---|
| 1 | | | | |
| 2 | | | | |
| 4 | | | | |
| 6 | | | | |
| 8 | | | | |
| 10 | | | | |
| 15 | | | | |
| 20 | | | | |
| 30 | | | | |

Plot $\ln ([A_o]/[A_t])$ vs. time on the graph paper. The slope of the straight line is equal to the rate constant ($k$). Calculate the half-life of the reaction by substituting $k$ in the equation

$$t_{1/2} = 0.693/k$$

Name: _____

## DATA SHEET: SOLVOLYSIS OF *t*-BUTYL CHLORIDE: A KINETIC STUDY
## PAGE 2

$\ln([A_0]/[A_t])$

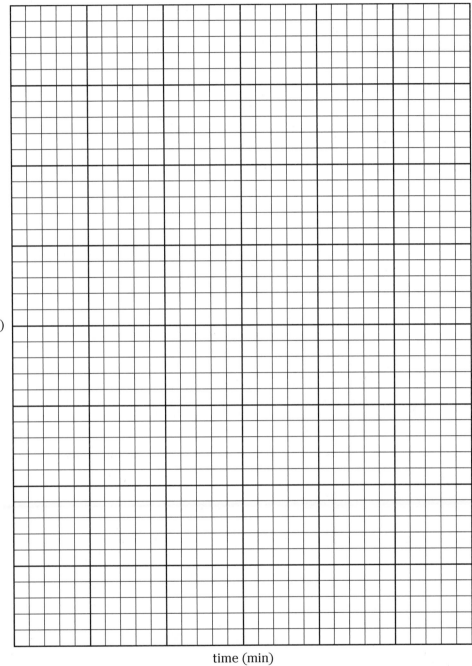

time (min)

## 13.3   STRUCTURAL, SOLVENT, AND TEMPERATURE EFFECTS ON NUCLEOPHILIC SUBSTITUTION REACTIONS

Nucleophilic substitution reactions constitute a very important class of organic processes. In a very general sense, these reactions involve:

- a **nucleophile:** an electron-rich species that forms a new bond to an electron-deficient carbon atom in a molecule;
- an **electrophile:** an electron-poor species that is attacked by the nucleophile and expels the leaving group;
- **leaving group;** the ion or molecule that is displaced from the electron-deficient carbon of the electrophile by the nucleophile.

There are two limiting mechanisms for nucleophilic substitution reactions; $S_N1$ (first-order), in which the leaving group departs *prior to* nucleophilic attack (often by the solvent; hence, this process is sometimes called solvolysis); and $S_N2$ (second order), where approach by the nucleophile occurs *at the same time as* the leaving group is departing. These reactions are highly dependent on the precise reaction conditions, as discussed in great detail in your textbook.

The purpose of this experiment is to observe *qualitatively* how factors such as electrophile structure, leaving group, solvent, and temperature influence the rate of nucleophilic substitution. The products will not be isolated, but instead observations will be made and conclusions will be drawn that will help the understanding of the mechanisms of these important reactions.

**Suggested References**   Carey, F. A., *Organic Chemistry.* 3rd ed., 309-42. New York: McGraw-Hill, 1996.

Fessenden, R. J., and J. S. Fessenden, *Organic Chemistry.* 5th ed., 185–228. Pacific Grove, CA: Brooks/Cole, 1994.

Loudon, G. M., *Organic Chemistry.* 3rd ed., 394–427. Redwood City, CA: Benjamin/Cummings, 1995.

McMurry, J., *Organic Chemistry.* 4th ed., 370-98. Pacific Grove, CA: Brooks/Cole, 1996.

Moore, J. A., and D. L. Dalrymple, *Experimental Methods in Organic Chemistry.* 2nd ed., 139–44. Philadelphia: W. B. Saunders, 1976.

Morrison, R. T., and R. N. Boyd, *Organic Chemistry.* 6th ed., 165–212, 249–71. Englewood Cliffs, NJ: Prentice-Hall, 1992.

Solomons, T. W. G. *Organic Chemistry.* 6th ed., 224–82. New York: John Wiley, 1996.

Vollhardt, K. P. C., and N. E. Schore, *Organic Chemistry.* 2nd ed., 178–230. New York: W. H. Freeman, 1994.

Wade, L. G., Jr., *Organic Chemistry.* 3rd ed., 240–86. Englewood Cliffs, NJ: Prentice-Hall, 1995.

 **Safety Tips**

✔   Alkyl halides are flammable and toxic. Avoid open flames, inhalation, or skin contact.

✔   Silver nitrate solutions will stain the skin. In case of contact, wash immediately with lots of water.

✔   Acetone and ethanol are highly flammable; avoid open flames or other sources of ignition.

## Part I. Structural Effects on S$_N$1 and the S$_N$2 Reactivity

In this part, the relative reactivities of a series of alkyl halides will be compared under two different sets of conditions: one favoring the S$_N$1 mechanism and another favoring the S$_N$2 mechanism. First, the S$_N$2 displacement of chloride or bromide by iodide in acetone will be studied:

$$R\text{—}X + NaI \longrightarrow R\text{—}I + NaX \text{ (s)}$$

Iodide is a good nucleophile, while acetone is an effective solvent for S$_N$2 displacements, particularly since the side-product (sodium chloride or sodium bromide) is insoluble in acetone. Thus, the course of the reaction may be monitored visibly by noting the appearance of turbidity, or a white, crystalline precipitate. This synthetically useful process is known as the *Finkelstein reaction.*

The same series of alkyl halides will then be treated with a solution of silver nitrate in aqueous ethanol, to observe the S$_N$1 reaction. Under these conditions, the unimolecular dissociation of the halide to form a carbocation is favored. This leads to the formation of the insoluble silver halide, while the carbocation reacts with the nucleophilic protic solvent (ethanol or water):

$$R\text{—}X \longrightarrow R^+ + X^- \xrightarrow[\text{(R}' = \text{H, CH}_2\text{CH}_3)]{Ag^+,\ R'OH} AgX \text{ (s)} + ROR' + H^+$$

---

**Experimental**

(May be performed by students working in pairs.)

1. Take ten clean, *dry* test tubes, and label them from 1 to 10. In tubes 1 to 5, place 0.2 mL (about 8 drops) of the following alkyl halides:
   (1) 1-chlorobutane
   (2) 1-bromobutane
   (3) 2-bromobutane
   (4) *t*-butyl chloride (i.e., 2-chloro-2-methylpropane)
   (5) crotyl chloride (i.e., (E)-1-chloro-2-butene).
2. Place the same amounts of the same alkyl halides in tubes 6 to 10 (e.g., 1-chlorobutane in tube 6, 1-bromobutane in tube 7, and so on). *Keep the tubes stoppered with corks before and after adding reagents.*
3. In a small conical flask or beaker, place 10 mL of 15% NaI in acetone. In a *separate* small conical flask or beaker, place 10 mL of 1% AgNO$_3$ in 90% aqueous ethanol. Make sure to label the containers.
4. Measure 2.0 mL of the 15% NaI solution with a 10 mL graduated cylinder. Add, quickly and in one portion, the NaI to test tube 1, cork immediately, swirl vigorously for 5 seconds to ensure complete mixing, and record the time elapsed before observing turbidity or precipitation.
5. Repeat step 4 for test tubes 2 through 5. Wash and *dry* the graduated cylinder thoroughly after you finish with test tube 5.
6. Repeat step 4 for test tubes 6 through 10, using the 1% AgNO$_3$ solution instead of the 15% NaI solution. If no change is evident after about 20 minutes, record this observation in the table and discard the contents of the tubes in the designated waste container.
7. If any of the tubes in the NaI series are still clear after about 20 minutes, loosen the corks slightly and place the tubes in a beaker containing warm (50°C) water; record your observations, noting the time that any changes may occur. Again, if no change is evident after about

20 minutes, record this observation in the table and discard the contents of the tubes in the waste container.

8. In your lab report, be sure to discuss how the structures of the alkyl halides used in this part influence their relative reactivities toward both $S_N1$ and $S_N2$ substitution.

## Part II. Solvent Effects on $S_N1$ Reactivity

The rate of a nucleophilic substitution reaction is dependent not only on the structure of the electrophile, but also on the reaction conditions. In particular, the solvent plays a major role in both $S_N1$ and $S_N2$ reactions. In this part, the effect of the solvent on the rate of $S_N1$ solvolysis of *tert*-butyl chloride will be studied. The most effective method for comparing relative rates of solvolysis relies on the fact that a strong acid is produced in this reaction—in this case, HCl:

$$t\text{-Bu-Cl} + \text{ROH} \longrightarrow t\text{-Bu-OR} + \text{HCl}$$

As the *t*-butyl chloride is consumed with time, an equivalent amount of hydrochloric acid is produced (as described in the equation above). This acid will be neutralized by the initial addition of a constant amount of sodium hydroxide solution that includes some phenolphthalein to serve as an indicator. The complete neutralization of the base will be reached once the phenolphthalein pink color disappears. Since in all cases a *constant* amount of base is added, the time of neutralization (i.e., the time needed to change the solution color from pink to colorless) will vary with each set of conditions.

The rate of formation of HCl by $S_N1$ solvolysis of an alkyl chloride is equal to the rate of disappearance of the alkyl chloride and is proportional only to the concentration of the alkyl chloride. This is an example of first-order kinetics:

$$d[\text{HCl}]/dt = -d[\text{RCl}]/dt = k[\text{RCl}]$$

Using the standard analysis (review your general chemistry notes on kinetics or the introduction of Experiment 13.2), one can determine the first-order rate constant by measuring the half-life of the reaction $t_{1/2}$ (i.e., the time required for half of the alkyl chloride to solvolyze):

$$k = \frac{0.693}{t_{1/2}}$$

Thus, if 0.5 equivalents of the hydroxide anion were present at the start of the reaction, $t_{1/2}$ would be the time at which the solution would become acidic. In this experiment, the amounts of alkyl halide and hydroxide will not be measured accurately, since only a relative comparison of the rates will be made. However, one could easily calculate the actual rate constant this way.

Appropriate solvent systems for this study are mixtures of water with methanol, ethanol, or acetone. The volume percent compositions indicated on the Data Sheet, Part II, will give conveniently measurable reaction rates (except for 50% MeOH and 70% acetone, which will not, so DO NOT use these mixtures!).

**Experimental**  (May be performed by students working in pairs.)

1. Prepare an insulated hot water bath (e.g., a mini-Dewar bath or a styrofoam cup) at about 30°C.
2. Prepare a series of test tubes with suggested relative components as described in the Data Sheet, Part II. Prepare 2.0 mL of the appropriate solvent mixture in each tube, using a graduated pipette to measure the volumes. To each test tube, add *exactly* 3 drops of 0.5 M NaOH solution containing phenolphthalein indicator. Cork the tubes and place them in the bath prepared in step 1, along with a thermometer.
3. After a few (1 to 2) minutes in the bath to allow temperature equilibration, add 3 drops of *t*-butyl chloride to each test tube. Note the time of addition, shake or swirl the test tubes, stopper them to prevent evaporation of the volatile (BP 48–51°C) chloride, and replace them in the bath. Add a few mL of hot water, if necessary, to maintain the temperature at 30±1°C.
4. Record the time required for the pink (basic) color to disappear in each solvent mixture. Tabulate your results in the Data Sheet, Part II.
5. Empty your test tubes into the designated waste container, and continue with Part III.

## Part III. Temperature Effects on $S_N1$ Reactivity

The rate of any chemical reaction increases with increasing temperature, since at higher temperatures the reacting molecules have greater kinetic energy. This phenomenon is quantitatively described by the Arrhenius equation:

$$k = Ae^{-(E_a/RT)}$$

where

   $E_a$ = activation energy (in cal)
   $T$ = temperature (K), held constant
   $A$ = a term related to the probability of the specific reaction
   $R$ = the gas constant, 1.99 cal/mol.K

The activation energy $E_a$ may be determined by measuring the rate constant at two or more temperatures. The logarithmic form of the Arrhenius equation shows that a plot of log $k$ versus $1/T$ should give a straight line, with a slope of $-E_a/2.303R$:

$$\log k = \log A - (E_a/2.303R)(1/T)$$

The effect of temperature on the rate of solvolysis of *t*-butyl chloride will be studied using the solvent systems and procedure from Part II, but at different temperatures. To obtain conveniently measurable rates, the solvent system from Part II that gives an end point after about 10 minutes should be used. Although no numerical values for the rate constants will be obtained, the neutralization times measured are *inversely proportional* to the rate constants. Thus, a plot of log t values versus $1/T$ (in K) should give a straight line, with a slope of $E_a/2.3R$, from which a value of $E_a$ can be obtained. As a rule of thumb, the rate constant approximately doubles with every temperature increase of 10°C.

**Experimental**

(May be performed by students working in pairs)

*Note:* In this part of the experiment, duplicate samples should be run and average times should be measured.

1. Select the solvent system that gives a neutralization time (at 30°C) closest to 10 minutes from the results in the Data Sheet, Part II.
2. Transfer the experimental results obtained using this particular solvent system to the Data Sheet, Part III, T = 303 K, Trial 1.
3. Repeat the experiment with the same particular solvent system, by preparing one 2-mL mixture of this solvent composition and adding 3 drops of NaOH solution. Add 3 drops of *t*-butyl choride, mix well, and measure the time required for the color to disappear, as in Part 2. The time should be close to the one obtained in Part II and should be placed in the Data Sheet, Part III, 303 K, Trial 2. Average the values of Trials 1 and 2.
4. Adjust the temperature of the water bath to 20±1°C using small pieces of ice.
5. Prepare two 2-mL tubes of the same solvent composition and add three drops of the NaOH indicator solution as in Part II.
6. Insert the test tubes from step 5 into the 20°C water bath and allow about 3 minutes for temperature equilibration.
7. Note the time, add three drops of *t*-butyl chloride to each tube, mix well, and measure the time required for the color to disappear. Transfer the values to the Data Sheet, Part III, 293 K, Trials 1 and 2, and average them.
8. Repeat the procedure with two other 2-mL samples after adjusting the bath temperature to 40±1°C. (You can obtain another styrofoam cup and start the 40°C run while you are waiting for the 20°C run to finish, as long as you watch all of the tubes carefully and note the time at which the color disappears.) Transfer the values to the Data Sheet, Part III, 313 K, Trials 1 and 2, and average them.
9. Using the average times required for the samples, at 20°, 30°, and 40°C, plot log t versus 1/T (in degrees Kelvin) and, from the slope, calculate the activation energy for the solvolysis reaction in that solvent system.

**Questions**

1. Write balanced equations for the expected $S_N2$ displacement reactions and $S_N1$ hydrolysis reactions for each of the alkyl halides employed in Part I.
2. Why should all of the test tubes be dry before any experiment?
3. In Part III, why should all test tubes be corked as soon as mixing of *t*-butyl chloride and NaOH takes place?
4. If you used 30% NaI in acetone in Part I, would you expect the $S_N2$ reaction of NaI and 1-bromobutane to proceed faster or slower? Explain.
5. In Part II, would the color change occur sooner, later, or at the same time if we used:
   a. twice as much *t*-butyl chloride (i.e., 6 drops)
   b. twice as much NaOH-phenolphthalein solution
   Explain.
6. Sketch a reaction coordinate diagram for the hydrolysis of *t*-butyl chloride, labeling the maxima and minima with appropriate structures, and indicating the activation energy measured in Part III.
7. Why is the hydrolysis of *t*-butyl chloride, rather than the reaction of *t*-butyl alcohol with HCl, used in this experiment?

# DATA SHEET: STRUCTURAL, SOLVENT, AND TEMPERATURE EFFECTS ON NUCLEOPHILIC SUBSTITUTION REACTIONS

## Part 1: Structural Effects on $S_N1$ and $S_N2$ Reactivity

| Test Tube # | Compound | Time Elapsed for Turbidity to Occur (minutes) |
|---|---|---|
| 1 | 1-chlorobutane | _____ |
| 2 | 1-bromobutane | _____ |
| 3 | 2-bromobutane | _____ |
| 4 | *t*-butyl chloride | _____ |
| 5 | crotyl chloride | _____ |

**Conclusion:** The descending order of reactivity of the above alkyl halides toward 15% NaI in acetone follows the sequence:

_____ > _____ > _____ > _____ > _____
(reactivity decrease ⟶ )

| Test Tube # | Compound | Time Elapsed for Precipitation to Occur (minutes) |
|---|---|---|
| 1 | 1-chlorobutane | _____ |
| 2 | 1-bromobutane | _____ |
| 3 | 2-bromobutane | _____ |
| 4 | *t*-butyl chloride | _____ |
| 5 | crotyl chloride | _____ |

**Conclusion:** The descending order of reactivity of the above alkyl halides toward 1% AgNO$_3$ (aqueous) follows the sequence:

_____ > _____ > _____ > _____ > _____
(reactivity decrease ⟶ )

## Part 2: Solvent Effects on $S_N1$ Reactivity

T = _____ °C = _____ K

| Solvent Mixture | Relative Ratio | Volumes Required for 2 mL Mixture | | Neutralization Time (minutes)* |
|---|---|---|---|---|
| | | Solvent | H$_2$O | |
| Methanol/H$_2$O | 55 : 45 | 1.1 mL | 0.9 mL | _____ |
| | 60 : 40 | 1.2 mL | 0.8 mL | _____ |
| | 65 : 35 | 1.3 mL | 0.7 mL | _____ |
| | 70 : 30 | 1.4 mL | 0.6 mL | _____ |
| Ethanol/H$_2$O | 50 : 50 | 1.0 mL | 1.0 mL | _____ |
| | 55 : 45 | 1.1 mL | 0.9 mL | _____ |
| | 60 : 40 | 1.2 mL | 0.8 mL | _____ |
| | 65 : 35 | 1.3 mL | 0.7 mL | _____ |
| | 70 : 30 | 1.4 mL | 0.6 mL | _____ |
| Acetone/H$_2$O | 50 : 50 | 1.0 mL | 1.0 mL | _____ |
| | 55 : 45 | 1.1 mL | 0.9 mL | _____ |
| | 60 : 40 | 1.2 mL | 0.8 mL | _____ |
| | 65 : 35 | 1.3 mL | 0.7 mL | _____ |

* Time required for NaOH to be consumed (i.e., pink color to disappear).

**Conclusion:** Based on the above data, the reactivity increases as the polarity of the medium (polarity of solvent OR water concentration) increases / decreases (circle one).

## Part 3: Temperature Effects on S$_N$1 Reactivity

Solvent mixture chosen from Part 2: _____ /H$_2$O, _____ : _____

| Relative Ratio of Solvent Mixture | Volumes Required for a 2 mL Mixture | | Temperature (K) | 1/T (K$^{-1}$) | Neutralization Time, *t* (min) | | | Log *t* |
|---|---|---|---|---|---|---|---|---|
| | Solvent | H$_2$O | | | Trial 1 | Trial 2 | Average | |
| _____ | _____mL | _____mL | 293 K | _____ | _____ | _____ | _____ | _____ |
| _____ | _____mL | _____mL | 303 K * | _____ | _____ | _____ | _____ | _____ |
| _____ | _____mL | _____mL | 313 K | _____ | _____ | _____ | _____ | _____ |

*Data on this row should be the same as the corresponding data in Part 2.

# chapter 14

# Alcohols and Ethers

## 14.1 THE GRIGNARD REACTION: SYNTHESIS OF 2-METHYL-2-HEXANOL

Organometallic compounds are characterized by the carbon-metal bond and are useful intermediates in organic synthesis. Such compounds include organolithium ($RLi$), organocadmium ($R_2Cd$), and the Grignard reagent ($RMgX$). The latter was discovered by **Victor Grignard,** the 1912 Chemistry Nobel Prize co-winner. The Grignard reagent is prepared from an alkyl (or aryl) halide and magnesium in ethereal solvents (e.g., diethyl ether, tetrahydrofuran).

$$R\text{—}X + Mg \xrightarrow{\text{ether}} RMgX$$

where R is an alkyl or aryl group and X is Cl, Br, or I.

The Grignard reagent is seldom isolated from solution because it is highly reactive toward any electrophile. Other reactants are normally added to the ether solution of the Grignard reagent in what is frequently termed a "one-pot synthesis."

The Grignard reagent, like many other organometallic reagents, is characterized by the considerable polarity of the carbon-metal bond. In alkyl (or aryl) halides, carbon is the more electropositive atom of the C—X bond (where X = halogen). When this carbon is bonded to magnesium, it becomes the more electronegative atom. This partial negative charge on the carbon of the Grignard reagent enables it to react;

i. With water or alcohol to give an alkane.

$$RMgX + H_2O \longrightarrow R\text{—}H + MgX(OH)$$

$$RMgX + R'O\text{—}H \longrightarrow R\text{—}H + MgX(OR')$$

ii. With a reactive alkyl halide to form a larger hydrocarbon.

$$RMgX + CH_2\text{=}CHCH_2\text{—}Cl \longrightarrow CH_2\text{=}CHCH_2\text{—}R + MgXCl$$

iii. With the polar carbonyl group to form a variety of products after acid/water treatment.

$$RMgX + \underset{\substack{\| \\ O}}{R'CH} \xrightarrow[\text{H}^+/\text{H}_2\text{O}]{\text{then}} \underset{\substack{| \\ OH}}{\overset{\substack{R \\ |}}{R'CH}} + MgX(OH)$$

aldehyde          2° alcohol

$$RMgX + \underset{\substack{\| \\ O}}{R' - C - R''} \xrightarrow[\text{H}^+/\text{H}_2\text{O}]{\text{then}} \underset{\substack{| \\ OH}}{\overset{\substack{R \\ |}}{R' - C - R''}} + MgX(OH)$$

ketone                3° alcohol

$$RMgX + O{=}C{=}O \xrightarrow[\text{H}^+/\text{H}_2\text{O}]{\text{then}} \underset{\substack{\| \\ O}}{R - C - OH} + MgX(OH)$$

carboxylic acid

$$RMgX + \underset{\substack{\| \\ O}}{R' - C - OR''} \longrightarrow \left[ \underset{\substack{\| \\ O}}{R' - C - R} \right] \xrightarrow[\text{2. H}^+/\text{H}_2\text{O}]{\text{1. RMgX}} \underset{\substack{| \\ OH}}{R'CR_2} + MgX(OH)$$

ester                    ketone                3° alcohol

iv. With the strained epoxide ring to give a 1° alcohol.

$$RMgX + \underset{O}{\triangle} \xrightarrow[\text{H}^+]{\text{then}} R - CH_2CH_2OH$$

These reactions are carried out under strictly anhydrous conditions to avoid reactions which lead to the formation of alkanes, as in reaction (i). In addition, the glassware used in the experiment should not be cleaned with industrial acetone to prevent side reactions which lead to the formation of alcohols (reaction iii) and alkanes (reaction i, due to alcohol contamination in industrial acetone). The magnesium metal should be finely divided in the form of filings to ensure a greater surface area. Extreme caution must be taken to avoid flames in the laboratory because ether is exceptionally flammable.

In this experiment, 1-chlorobutane will be reacted with magnesium in anhydrous ether to form the Grignard reagent, which will then be reacted with acetone to form the eventual product, 2-methyl-2-hexanol. The use of 1-bromobutane often gives faster initiation times.

$$CH_3CH_2CH_2CH_2Cl + Mg \xrightarrow{\text{ether}} CH_3CH_2CH_2CH_2MgCl$$

1-chlorobutane                    Grignard reagent          $(CH_3)_2C{=}O$

$$\underset{\substack{| \\ OH}}{CH_3CH_2CH_2CH_2 - C(CH_3)_2} \xleftarrow{\text{H}^+} \underset{\substack{| \\ O^-MgCl^+}}{CH_3CH_2CH_2CH_2 - C(CH_3)_2}$$

2-methyl-2-hexanol

In the last step, the alkoxide salt is hydrolyzed in a weakly acidic medium to avoid any unwanted side reactions of the product, a 3° alcohol (e.g., an elimination reaction to form an alkene). An ammonium chloride/ hydrochloric acid buffer is cautiously added until the pH of the mixture is clearly acidic and all the alkaline suspension is dissolved.

**Suggested References**     Carey, F. A., *Organic Chemistry.* 3rd ed., 569–75. New York: McGraw-Hill, 1996.

"Ether." In *McGraw-Hill Encyclopedia of Science and Technology,* vol. 6, 494–96. New York: McGraw-Hill, 1992.

"Grignard Reaction." In *McGraw-Hill Encyclopedia of Science and Technology,* vol. 8, 229. New York: McGraw-Hill, 1987.

"Grignard Reagents." In *Reagents for Organic Synthesis,* edited by L. F. Fieser and M. Fieser, vol. 1, 415–24. New York: John Wiley, 1967.

Loudon, G. Marc, *Organic Chemistry.* 3rd ed., 367–371. Redwood City, CA: Benjamin/ Cummings, 1995.

March, J., *Advanced Organic Chemistry.* 4th ed., 920. New York: John Wiley, 1992.

McMurry, J., *Organic Chemistry.* 4th ed., 357–59, 644–47. Pacific Grove, CA: Brooks/Cole, 1996.

Morrison, R. T., and R. N. Boyd, *Organic Chemistry.* 6th ed., 675, 685–94. Englewood Cliffs, NJ: Prentice-Hall, 1992.

Patnaik, P., *A Comprehensive Guide to the Hazardous Properties of Chemical Substances.* 290, 497–8. New York: Van Nostrand Reinhold, 1992.

Solomons, T. W. G., *Organic Chemistry.* 6th ed., 484–91. New York: John Wiley, 1996.

Vollhardt, K. P. C., and N. E. Schore, *Organic Chemistry.* 2nd ed., 256–76. New York: W. H. Freeman, 1994.

Wade, L. G., *Organic Chemistry.* 3rd ed., 406–15, 597, 835-36. Englewood Cliffs, NJ: Prentice-Hall, 1995.

Walborsky, H. M., "Mechanism of Grignard Reagent Formation. The Surface Nature of the Reaction." *Acc. Chem. Res.* 23 (1990): 286–93.

 **SAFETY TIPS**

✔  1-Chlorobutane, and especially ether, are flammable liquids and therefore all flames must be extinguished.

✔  Moreover, ether forms peroxides, which are extremely sensitive to shock, sparks, heat, and friction. Since the peroxides have a higher boiling point than ether, their residue remains in the distilling flask and can overheat and explode. Distillation to dryness, therefore, must be avoided.

✔  Containers of diethyl ether should be dated when opened, and should be disposed of if opened and unused for more than a few months. Finely divided magnesium metal reacts with water, liberating hydrogen, which can ignite or explode.

**Experimental**

1. Place 2.4 g of magnesium turnings (or magnesium ribbon cut in small pieces) and 15 mL of anhydrous ether in a clamped 100-mL round-bottomed flask. (For best and quickest results, the ether should be dried overnight with small pieces of sodium metal, calcium hydride, or lithium aluminum hydride, and filtered just before use).

2. Attach a Claisen adaptor, a separatory funnel, and a condenser equipped with a drying tube onto the round-bottomed flask as shown in Figure 14.1.

3. Add to the separatory funnel 10.5 mL of 1-chlorobutane (density = 0.89 g/mL; 9.3 g) that has been dried overnight over anhydrous calcium

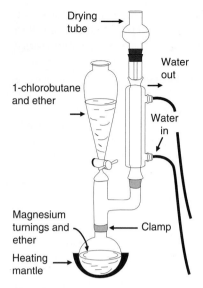

Drying tube

Water out

1-chlorobutane and ether

Water in

Magnesium turnings and ether

Clamp

Heating mantle

**Figure 14.1**   Set-up for the Grignard reaction

chloride and filtered just before use. Also add 30 mL of anhydrous ether. Swirl the funnel briefly to ensure proper mixing. Add in one portion approximately one-third of the volume to the round-bottomed flask, and swirl the flask for 2 minutes. The use of magnetic stirring, if available, is recommended.

4.  There is a very good chance that the reaction will not start immediately because of the magnesium oxide coating on the turnings. You may need to initiate the reaction as follows: To a test tube, add 15–20 magnesium turnings along with 2–3 mL of the 1-chlorobutane solution from the separatory funnel using a medicine dropper. Use a stirring rod to *crush* (not stir) the magnesium. Experience has shown that best results are obtained when a broken stirring rod is used, because the uneven surface is more efficient in removing the surface layer of magnesium oxide and exposing the pure magnesium metal. The solution should turn opalescent white, then gray (from clear) within a few minutes, and will finally bubble. Transfer the *entire contents* of the test tube (*especially* the crushed magnesium) to the 100-mL round-bottomed flask, and swirl the flask. Make sure that the joints are clean and that they fit perfectly.

5.  The reaction should start in the round-bottom flask after the initiation step, as indicated by the change of color of the solution and the reflux of the ether. Add the remaining contents of the separatory funnel dropwise, and replace the separatory funnel with a glass stopper.

6.  Once the reaction is going, the mixture will become progressively darker. When refluxing subsides, heat the mixture (using a heating mantle) for 10 minutes, and then allow the mixture to cool to room temperature.

7.  Immerse the round-bottom flask in an ice-water mixture for 5 minutes with occasional swirling. While the solution is cooling, transfer 7.6 mL (d = 0.79 g/mL, 6 g) of acetone in 15 mL of anhydrous ether to the dry (no need to wash) separatory funnel, and place the separatory funnel back on the Claisen adaptor.

8.  After the 5-minute cooling period, add the acetone/ether mixture dropwise from the separatory funnel while swirling or magnetically stirring the round-bottomed flask in the ice bath. *The reaction is highly exothermic, especially in the beginning, so special attention should be given to this step.* Let the solution cool to room temperature, stopper the round-bottomed flask with a *rubber (not a glass) stopper or cork,* and let stand until the next lab period.

9.  At the beginning of the next lab period, check to see if the ether has evaporated. If the mixture is dry, add 30 mL of ether. Pour the contents of the round-bottomed flask into a beaker containing a mixture of 25 g crushed ice and 20 mL of 20% ammonium chloride solution. Avoid pouring any unreacted magnesium, but make sure all other solids are included. Acidify with 20% hydrochloric acid until the solution is clear (pH = 4–5).

10.  Transfer the mixture to a separatory funnel, and separate the organic (top) layer from the water layer. Extract the organic layer once with 20 mL of 10% sodium carbonate (to remove any traces of hydrochloric acid), and once with 20 mL water (to remove any traces of carbonate).

11.  Transfer the organic layer to a 125-mL conical flask, add 1 g anhydrous magnesium sulfate, and let stand covered for about 5 minutes.

12.  Gravity-filter directly into a dry 100-mL round-bottomed flask, and perform a simple distillation. The ether (which should be collected in ice) distills at 34°–40°C. Transfer the distilled ether to a container labeled

"waste ether" to be properly disposed of by your instructor. Any traces of water that were not removed by the magnesium sulfate will distill off at about 100°C. They should also be included in the "waste ether" container.

13. Continue the distillation and collect the product (2-methyl-2-hexanol) in a preweighed vial (literature value 141°–142°C BP). Calculate the percent yield.

14. Run a gas chromatograph of your product and compare it with a gas chromatograph of a standard prepared by mixing equal quantities of ether, water, and 1-chlorobutane. Do any of the components of the mixture appear in the final product?

**Questions**

1. Explain the purpose of the following in the above experiment (include any pertinent equations):
   a. using anhydrous ether
   b. not rinsing glassware with acetone before the experiment
   c. using magnesium filings instead of "chunk" magnesium
   d. initiating the reaction in a test tube
   e. using an ammonium chloride/hydrochloric acid mixture to acidify the reaction mixture instead of using concentrated sulfuric acid.

2. Give the products of the reaction of methylmagnesium chloride with
   a. acetaldehyde
   b. acetone
   c. ethyl acetate (1 equivalent per Grignard)
   d. ethyl acetate (2 equivalents per Grignard)
   e. ethyl acetate (0.5 equivalents per Grignard)
   f. acetic acid
   g. benzyl chloride
   h. ethanol
   i. *n*-pentane
   j. 1-propyne
   k. propylene oxide.

3. Compare the IR spectra of 1-chlorobutane, acetone, and 2-methyl-2-hexanol. Identify the new peaks that appear in the product.

## 14.2  QUALITATIVE TESTS FOR ALCOHOLS

Alcohols are polar compounds that can be distinguished from all other oxygen-containing compounds by reactions characteristic of the hydroxyl group. In this experiment an unknown will be tested against known 1°, 2°, and 3° alcohols. The experimental results should give an idea of the general structure of the unknown. The reactions include **Jones' oxidation,** the **Lucas reaction,** and the **iodoform test.** A microboiling point determination will also be carried out.

### Jones' Oxidation

Primary and secondary alcohols can be oxidized with sodium dichromate in acid medium to aldehydes (and further to carboxylic acids) and ketones respectively.

$$R-CH_2OH \xrightarrow{Cr_2O_7^{-2}/H^+} R-CHO \xrightarrow{Cr_2O_7^{-2}/H^+} R-COOH$$

1° alcohol                    aldehyde              carboxylic acid

$$R_2CHOH \xrightarrow{Cr_2O_7^{-2}/H^+} R_2C=O \xrightarrow{Cr_2O_7^{-2}/H^+} \text{No reaction}$$

2° alcohol                    ketone

    This reaction gives positive results with 1° and 2° alcohols. The orange dichromate anion is reduced to the greenish-blue chromic ion. Tertiary alcohols do not react, and the color of the dichromate does not change. Prolonged heating of the test solution must be avoided, especially with tertiary alcohols, because this can lead to dehydration and the formation of alkenes that easily react with the dichromate anion.

 **SAFETY TIPS**

✔ Sodium dichromate is a toxic compound, and the use of plastic gloves is recommended.
✔ The Lucas Reagent ($ZnCl_2$ in concentrated hydrochloric acid) is extremely corrosive. It should be poured carefully with a funnel and with adequate ventilation.

**Experimental**

To each of four test tubes add 2 mL of a 1% sodium dichromate solution and 5 drops of concentrated sulfuric acid. Mix the contents thoroughly. Add 10 drops of the 1° alcohol to the first test tube, 10 drops of the 2° alcohol to the second test tube, 10 drops of the 3° alcohol to the third test tube, and, finally, 10 drops of your unknown to the fourth test tube. Cork the test tubes and warm gently in a 40°–50°C water bath for a minute; observe and record any color changes.

### Lucas Reaction

The Lucas reagent is a solution of zinc chloride in concentrated hydrochloric acid. The alcohols are differentiated by the rate of alkyl chloride formation. The lower alcohols (through $C_5H_{11}OH$) are soluble in the Lucas reagent.

$$R - OH + HCl \xrightarrow{ZnCl_2} R - Cl + H_2O$$

Tertiary alcohols react immediately to give the alkyl chloride, and a cloudy appearance develops. The alkyl chloride is generally insoluble in the reagent and soon forms a distinct upper layer. Secondary alcohols develop the cloudy appearance within 5 minutes of reaction. Saturated primary alcohols do not react at ordinary temperatures, and the solution remains clear after a long period of time. This trend is in accord with the expected stability of the intermediate carbocation ($3° > 2° > 1°$) formed during the reaction; it also makes it possible to distinguish between the three types of alcohols.

**Experimental**  In each of four different dry test tubes, place 5 mL of the Lucas reagent. Add 1 mL of the $1°$ alcohol to the first one, 1 mL of the $2°$ alcohol to the second one, 1 mL of the $3°$ alcohol to the third one, and, finally, 1 mL of the unknown alcohol to the fourth test tube. Cork each of the test tubes with a clean cork and observe the results. The $2°$ alcohol might not visibly react. In this case, warm the test tube for 30 seconds in a prewarmed $50°C$ water bath.

### The Iodoform Test

When an alcohol of the form

$$CH_3 - \overset{\overset{\displaystyle H}{|}}{\underset{\underset{\displaystyle R}{|}}{C}} - OH$$

is treated with a halogen (e.g., iodine) in basic medium, it is oxidized to the corresponding methyl ketone (or acetaldehyde if the alcohol is ethanol). Further reaction provides the $\alpha,\alpha,\alpha$-trihalo ketone (or aldehyde), which on hydrolysis yields haloform and the salt of a carboxylic acid.

$$CH_3 - \overset{\overset{\displaystyle H}{|}}{\underset{\underset{\displaystyle R}{|}}{C}} - OH \xrightarrow[\text{NaOH}]{KI/I_2} CH_3 - \underset{\underset{\displaystyle O}{\|}}{C} - R \longrightarrow CI_3 - \underset{\underset{\displaystyle O}{\|}}{C} - R \xrightarrow[\Delta]{NaOH/H_2O} CHI_3 + R - \underset{\underset{\displaystyle O}{\|}}{C} - O^-$$

methyl ketone                                           iodoform   carboxylate anion

Although all three halogens (Cl, Br, I) react in this manner, iodine is recommended for laboratory use because of safety, since both bromine and chlorine vapors are toxic and because of the ease of iodoform ($CHI_3$) isolation and identification. Chloroform and bromoform, although easily identifiable, are colorless liquids; iodoform is a yellow insoluble solid.

A positive iodoform test should involve *both* of the following experimental observations:

a. The brown color of iodine should be discharged as the reagent is added to the solution of the test compound.
b. The yellow precipitate of iodoform should form when the solution is heated briefly in a warm water bath while iodine solution is still in excess. In basic solution, all aldehydes and ketones with an $\alpha$

hydrogen can be substituted at the α position. Only those that can be trisubstituted will form iodoform. Thus, the loss of the brown iodine color does not *in itself* indicate a positive iodoform test. Both of these conditions (i.e., the loss of the brown color and the formation of the yellow precipitate) must be fulfilled for a positive test.

The addition of potassium iodide facilitates iodine's dissolution in water because iodine itself is sparingly soluble.

---

**Experimental**

*Perform the following test on a compound that is known to give a positive iodoform test and on the unknown.*

In a large test tube, place 2 mL of a 3*M* sodium hydroxide solution, 1 mL of the compound to be tested, and enough iodine solution ($I_2$/KI) to get a *persistent* brown color. Note that the color must be *brown* (not yellowish-brown). Warm the solution in a 60°C water bath for a few minutes. If the color discharges, add more $KI/I_2$ until the solution is persistently brown. A positive test for an alcohol of the type shown on p. 181 (or a methyl ketone or acetaladehyde) is indicated by a canary-yellow precipitate of iodoform.

## Microboiling Point Determination

Perform a microboiling point determination on the unknown alcohol. Use the setup described in chapter 3 (p. 25), and an oil bath instead of a water bath.

**Suggested References**

Carey, F. A., *Organic Chemistry.* 3rd ed., 737–40. New York: McGraw-Hill, 1996.
Loudon, G. Marc, *Organic Chemistry.* 3rd ed., 447–50, 467–69, 1051. Redwood City, CA: Benjamin/Cummings, 1995.
McMurry, J., *Organic Chemistry.* 4th ed., 874–76. Pacific Grove, CA: Brooks/Cole, 1996.
Morrison, R. T., and R. N. Boyd, *Organic Chemistry.* 6th ed., 697–98. Englewood Cliffs, NJ: Prentice-Hall, 1992.
Solomons, T. W. G., *Organic Chemistry.* 6th ed., 438–41, 473–77, 761. New York: John Wiley, 1996.
Vollhardt, K. P. C., and N. E. Schore, *Organic Chemistry.* 2nd ed., 253–54, 282–86, 681–82. New York: W. H. Freeman, 1994.
Wade, L. G., *Organic Chemistry.* 3rd ed., 364–415. Englewood Cliffs, NJ: Prentice-Hall, 1995.

**Questions**

1. Balance the following equations:
   a. $(CH_3)_2CHOH + Cr_2O_7^{-2}/H^+ \longrightarrow (CH_3)_2C=O + Cr^{+3}$
   b. $CH_3CH_2CH_2CH_2OH + MnO_4^-/OH^- \longrightarrow CH_3CH_2CH_2CO_2^- + MnO_2$.
2. How would you distinguish between the following compounds? State what you would do and what you would see.
   a. isoamyl alcohol and *sec*-amyl alcohol.
   b. isobutyl alcohol and diethyl ether.
   c. *n*-amyl alcohol and valeric acid.
   d. 2-pentanol and 3-pentanol.
3. Identify the following compounds:
   A  ($C_4H_{10}O$) positive iodoform
   B  ($C_9H_{12}O$) positive Lucas (immediate reaction); reaction with hot $KMnO_4$ gives benzoic acid; positive Jones
   C  ($C_9H_{12}O$) positive Jones; negative Lucas
   D  the only 1° alcohol to give a positive iodoform test
   E  ($C_9H_{12}O$) positive iodoform; reaction with hot $KMnO_4$ gives 1,4-benzenedicarboxylic acid
   F  ($C_4H_{10}O$) negative Jones; negative iodoform
   G  ($C_7H_8O$) negative Jones; negative Lucas; negative iodoform; no reaction with hot $KMnO_4$.

## DATA SHEET: QUALITATIVE TESTS FOR ALCOHOLS: UNKNOWN RESULTS

| Test | Observation | Conclusion |
|------|-------------|------------|
| 1. Jones | | |
| 2. Lucas | | |
| 3. Iodoform | | |

4. Microboiling point: _____ °C (if using a water bath and the BP is above 100°C, write " > 100°C").

| Alcohol | Formula | BP (°C) |
|---------|---------|---------|
| Methanol | $CH_3OH$ | 64 |
| Ethanol | | 78 |
| 1-Propanol | | 97 |
| 2-Propanol | | 82 |
| 1-Butanol | | 118 |
| 2-Butanol | | 100 |
| 2-Methyl-1-propanol | | 108 |
| 2-Methyl-2-propanol | | 83 |
| 1-Pentanol | | 138 |
| 2-Pentanol | | 120 |
| 3-Pentanol | | 116 |
| 3-Methyl-1-butanol | | 132 |
| 2-Methyl-2-butanol | | 102 |
| 2-Methyl-2-hexanol | | 142 |
| Cyclopentanol | | 140 |
| Cyclohexanol | | 162 |

*Conclusion:* Based on the above data, unknown # _____ is _____

## 14.3   THE WILLIAMSON ETHER SYNTHESIS: PREPARATION OF PHENACETIN FROM ACETAMINOPHEN

One of the most common methods used to prepare ethers is the **Williamson ether synthesis.** In this reaction, an alkoxide anion ($RO^-$) acts as a nucleophile in an $S_N2$ process, displacing a leaving group ($X^-$) from an electrophile ($R'—X$).

| an alkoxide anion | an alkyl halide | | an ether | a leaving group |
|---|---|---|---|---|
| (a nucleophile) | | | | |

The usual limitations on the $S_N2$ reaction apply: the alkoxide must not be too bulky; the electrophile is usually primary, secondary, or resonance-stabilized (e.g., allylic or benzylic), but never tertiary; X must be a good leaving group; and polar, aprotic solvents are preferred.

In this experiment, the Williamson ether synthesis will be demonstrated by converting one commonly used pharmaceutical product (acetaminophen) into another (phenacetin). Both of these compounds are found in over-the-counter pain relievers—acetaminophen is the active ingredient in Tylenol® and many other non-asprin analgesics. Phenacetin was a very popular analgesic until recently.

acetaminophen
(4-acetamidophenol)                 potassium carbonate        phenacetin
(4-acetamidoethoxybenzene)

The overall reaction is shown in this diagram. Acetaminophen (formally known as 4-acetamidophenol) is a **phenol,** and thus is sufficiently acidic ($pK_a$ around 10) to be deprotonated by potassium carbonate. The resulting phenoxide anion serves as a nucleophile in the $S_N2$ displacement of iodide (a good leaving group) from iodoethane (a primary alkyl halide). The solvent, 2-butanone, is polar and aprotic like acetone but has a higher boiling point. The side-products will be removed with an aqueous wash, and the organic product, an ethyl ether, will be purified by recrystallization.

**Suggested References**   Carey, F. A. *Organic Chemistry*. 3rd ed., 652–54. New York: McGraw Hill, 1996.

Fessenden, R. J., and J. S. Fessenden, *Organic Chemistry*. 5th ed., 316–17. Pacific Grove, CA: Brooks/Cole, 1994.

Loudon, G. M., *Organic Chemistry*. 3rd ed., 497–99, 840–41. Redwood City, CA: Benjamin/ Cummings, 1995.

McMurry, J., *Organic Chemistry*. 4th ed., 679–80. Pacific Grove, CA: Brooks/Cole, 1996.

Patai, S., ed., *The Chemistry of the Ether Linkage*. 446–67. London: Interscience, 1967.

Solomons, T. W. G. *Organic Chemistry*. 6th ed., 444–45. New York: John Wiley, 1996.

Volker, E. J., E. Pride, and C. Hough, "Drugs in the Chemistry Laboratory." *J. Chem. Educ.* 56. (1979):831.

Vollhardt, K. P. C., and N. E. Schore, *Organic Chemistry.* 2nd ed., 292–95, 879–80. New York: W. H. Freeman, 1994.

Wade, L. G., Jr., *Organic Chemistry.* 3rd ed., 432, 569–70. Englewood Cliffs, NJ: Prentice-Hall, 1995.

 **Safety Tips**

✔ 2-Butanone and ether are highly flammable—avoid open flames.
✔ Iodoethane is toxic, corrosive, and an irritant, and should be dispensed in the hood; avoid skin contact or inhalation.

## Experimental

1. Place acetaminophen (1.30 g) and powdered potassium carbonate (2.50 g) in a 50 mL round-bottom flask.
2. Suspend the solids in 2-butanone (15 mL), and add 1.0 mL iodoethane (d = 1.95 g/mL, 0.50 g) and a few boiling chips.
3. Attach a condenser and reflux the reaction mixture for 1 hour.
4. At the end of the reflux period, remove the heating mantle, turn off the cooling water, and allow the flask to cool slightly.
5. Gravity-filter the contents of the flask into a separatory funnel.
6. Rinse the solids that remain in the filter paper with two 5 mL portions of ether to carry any remnants of the product into the flask.
7. Extract the solution first with 20 mL of 5% aqueous NaOH (to remove any unreacted acetaminophen) and then with 20 mL of water.
8. Transfer the organic layer into a dry Erlenmeyer (conical) flask, and dry the solution with about 1 g of anhydrous $MgSO_4$.
9. Gravity-filter the dried solution into a small conical flask, add one boiling chip, and evaporate the solvents on a hot-water bath in the hood.
10. Recrystallize the crude product from a minimum amount of boiling water (10–15 mL).
11. Allow the solution to cool for a few minutes, then cool in an ice-water bath.
12. The crystallization process may be slow but can be initiated by scratching the sides of the flask with a stirring rod.
13. Collect the crystals by suction filtration, air dry them, and calculate the percent yield.
14. Determine the melting point (MP 134–136°C).
15. Determine the purity of your product by thin-layer chromatography (using silica gel plates and 4:1 ethyl acetate-dichloromethane); measure and record the $R_f$ values of acetaminophen and phenacetin.
16. If time permits, measure and compare the infrared spectra of acetaminophen and phenacetin.

## Questions

1. In the previous experiment, what is the purpose of:
   a. potassium carbonate?
   b. 2-butanone?
   c. ether (step 6)?
   d. extraction with 5% aqueous NaOH (step 7)?
   e. magnesium sulfate (step 8)?

2. Assign all of the important absorptions in the infrared spectra of acetaminophen and phenacetin. Which particular IR band(s) tell(s) you that the reaction has occurred?
3. Why is acetaminophen soluble in 5% NaOH, but phenacetin is not? Write an equation to illustrate your answer.
4. Suggest syntheses of the following ethers:
    a. phenyl *n*-propyl ether
    b. 2-methoxyheptane
    c. ethoxycyclohexane
    d. *t*-butyl methyl ether.
5. Identify the products of the reaction between:
    a. *t*-butyl chloride and sodium ethoxide
    b. sodium *t*-butoxide and ethyl chloride.
6. Acetaminophen was reacted with $K_2CO_3$ and iodoethane to yield phenacetin. Would a similar reaction take place if cyclohexanol, $K_2CO_3$, and iodoethane were mixed together? Explain.

# chapter 15

# Aromatic Reactions

## 15.1 NITRATION OF BROMOBENZENE

Nitration is a typical **electrophilic aromatic substitution** reaction in which the electrophile is the **nitronium ion,** $NO_2^+$, produced by the protonation and subsequent dehydration of nitric acid, $HNO_3$. Usually the proton is provided by sulfuric acid.

$$HNO_3 + H^+ \rightleftharpoons [H_2NO_3]^+ \longrightarrow H_2O + NO_2^+$$

nitric acid                    nitronium ion

With highly reactive aromatics such as phenol, $C_6H_5OH$, the formation of the nitronium ion is initiated by the autoprotonation of nitric acid, and sulfuric acid is not required.

$$HNO_3 + HNO_3 \rightleftharpoons NO_3^- + [H_2NO_3]^+ \longrightarrow NO_2^+ + H_2O$$

In this experiment, bromobenzene will be nitrated to give 2,4-dinitrobromobenzene (major product) and a mixture of *p*- and *o*-bromonitrobenzene (minor products). The mechanism involves the aromatic ring's attack on the strongly electrophilic nitronium ion to form a cyclohexadienyl carbocation. Eventually an aromatic system is regenerated by proton elimination.

2,4-dinitrobromobenzene
(major product)

*p*-bromonitrobenzene     *o*-bromonitrobenzene

(minor products)

Monosubstitution favors the *ortho* and *para* isomers over the *meta* because of the electron-releasing effect of bromine (via resonance), which stabilizes the intermediate cyclohexadienyl carbocation and the transition state leading to it.

Electron-withdrawing substituents, such as the nitro ($-NO_2$) and the carboxyl ($-COOH$) groups, favor *meta* substitution.

The second nitro group also substitutes at positions *o-* or *p-* to the bromine for the same reason.

The crystals should be thoroughly washed with water before the recrystallization step to remove any traces of acid.

Carey, F. A., *Organic Chemistry*. 3rd ed., 470–71. New York: McGraw-Hill, 1996.

Loudon, G. Marc, *Organic Chemistry*. 3rd ed., 752–53. Redwood City, CA: Benjamin/Cummings, 1995.

McMurry, J., *Organic Chemistry*. 4th ed., 576–77. Pacific Grove, CA: Brooks/Cole, 1996.

Morrison, R. T., and R. N. Boyd, *Organic Chemistry*. 6th ed., 525–27, 530–35. Englewood Cliffs, NJ: Prentice-Hall, 1992.

Solomons, T. W. G., *Organic Chemistry*. 6th ed., 660–61. New York: John Wiley, 1996.

Vollhardt, K. P. C., and N. E. Schore, *Organic Chemistry*. 2nd ed., 576–77. New York: W. H. Freeman, 1994.

Wade, L.G., *Organic Chemistry*. 2nd ed., 719–35. Englewood Cliffs, NJ: Prentice-Hall, 1991.

### ⚠ SAFETY TIPS

✔ Both concentrated sulfuric and nitric acids are corrosive and should be handled with care.

✔ All nitrated bromobenzene isomers are skin irritants and should also be handled with care. The use of plastic gloves is recommended.

✔ Ethanol is flammable, and all laboratory flames should be extinguished before the recrystallization step.

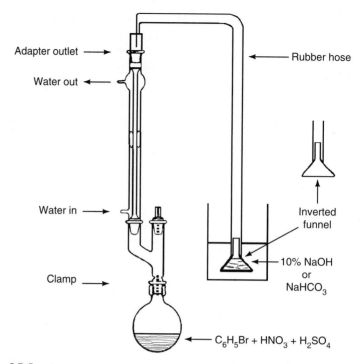

**Figure 15.1** Set-up for nitration of bromobenzene

**Experimental**

1. In a 100-mL round-bottomed flask, place 6 mL of concentrated nitric acid and 15 mL of concentrated sulfuric acid. Cool to room temperature by means of an ice bath.

2. Add dropwise 2.0 mL (d = 1.5 g/mL; 3.0 g) of bromobenzene while swirling the flask. Place an adaptor outlet containing a small piece of glass tubing on top of the condenser, and attach one end of a rubber hose to the glass tubing. Attach a glass funnel to the other end of the hose, and place the funnel in a beaker containing a 10% NaOH or NaHCO₃ solution. This will serve as a trap to neutralize nitrogen oxide fumes. Attach a condenser and stopper as seen in Figure 15.1, and reflux while swirling the flask at intervals.

3. After 30 minutes, the formation of the brown fumes of nitrogen oxides will cease. Pour the contents of the flask into a beaker containing 50 g of ice. Stir the mixture thoroughly with a stirring rod to promote crystallization of the yellow-white crystals of the product.

4. Vacuum-filter the crystals, and wash with about 50 mL of water.

5. Recrystallize the 2,4-dinitrobromobenzene from about 25–30 mL of hot ethanol. The crystals of the product will precipitate as the solution cools, while the monosubstituted isomers will remain in solution. If the 2,4-dinitrobromobenzene does not crystallize after the hot-filtration step, add about 5 or 10 mL of ice-cold water, and swirl the flask in an ice bath.

6. Determine the MP (lit. 75°C) and percent yield.

**Questions**

1. Explain the following:

   a. The nitration of nitrobenzene is much slower than the nitration of toluene.

   b. Direct nitration of aniline yields small amounts of *m*-nitroaniline, but no *o*- or *p*-nitroaniline.

2. Identify the major product in the mononitration of the following:
   a. *p*-nitrotoluene
   b. *o*-nitroanisole
   c. phenyl benzoate
   d. *p*-nitrobenzophenone
   e. *p-t*-butyltoluene
   f. phthalic anhydride.
3. Explain the purpose of the following in this experiment:
   a. sulfuric acid
   b. water washings.

## 15.2 FRIEDEL-CRAFTS ALKYLATION: SYNTHESIS OF 2,5-DI-*t*-BUTYL-1,4-DIMETHOXYBENZENE— A MICROSCALE SYNTHESIS

Friedel-Crafts reactions are useful for attaching an alkyl or acyl group to an aromatic ring. The reaction involves the formation of a strongly electrophilic carbocation ($R^+$, Friedel-Crafts alkylation) or acylium ion ($RC\equiv O^+$, Friedel-Crafts acylation), which reacts with an aromatic ring via a typical electrophilic aromatic substitution mechanism.

In this experiment, a Friedel-Crafts alkylation will be carried out. The electrophile is the *t*-butyl carbocation (which does not rearrange) and the aromatic species is 1,4-dimethoxybenzene. Thus,

$$(CH_3)_3C - OH + H^+ \longrightarrow (CH_3)_3C - \underset{+}{O}H_2 \longrightarrow (CH_3)_3C^+ + H_2O$$

2, 5-di-*t*-butyl-1, 4-dimethoxybenzene

The dehydrating agent is concentrated $H_2SO_4$. 1,4-Dimethoxybenzene is a reactive aromatic compound that easily undergoes dialkylation at positions 2 and 5. No further reaction occurs because the two remaining positions are sterically hindered by the bulky *t*-butyl groups.

The reaction is performed at low temperatures to prevent possible unwanted side reactions between the strong acid and the activated aromatic ring, as well as elimination and subsequent polymerization of the *t*-butyl carbocation.

**Suggested References**    Carey, F. A., *Organic Chemistry*. 3rd ed., 474–77. New York: McGraw-Hill, 1996.

Blatchy, T. M., and N. H. Hartshorne, "Curling of Crystals of 2,5-Di-*t*-butyl-1,4-dimethoxybenzene During Growth." *Trans. Faraday Soc.* 62. (1966): 512.

Loudon, G. Marc, *Organic Chemistry*. 3rd ed., 757–59, 774, 75. Redwood City, CA: Benjamin/Cummings, 1995.

McMurry, J., *Organic Chemistry*. 4th ed., 579–83. Pacific Grove, CA: Brooks/Cole, 1996.

Morrison, R. T., and R. N. Boyd, *Organic Chemistry*. 6th ed., 556–61. Englewood Cliffs, NJ: Prentice-Hall, 1992.

Oesper, P. F., C. P. Smyth, and M. S. Kharasch, "The Reduction of Dipole Moment by Steric Hindrance in Di-*t*-butylhydroquinone and its Dimethyl Ether." *J. Am. Chem. Soc.* 64. (1942): 937.

Olah, G. A., *Friedel-Crafts Chemistry.* New York: John Wiley, 1973.

Olah, G. A., *Friedel-Crafts and Related Reactions.* New York: John Wiley, 1963.

Roberts, R. M., "Friedel-Crafts Chemistry." *Chem. Eng. News.* (Jan. 25, 1965): 96.

Solomons, T. W. G., *Organic Chemistry.* 6th ed., 663-69. New York: John Wiley, 1996.

Stolow, R. D., and J. W. Larsen, "The Transient Cylindrical 'Conformation' of Growing Crystals of 2,5-Di-*t*-butyl-1,4-dimethoxybenzene." *Chemistry and Industry.* 449. (1963).

Vollhardt, K. P. C., and N. E. Schore, *Organic Chemistry.* 2nd ed., 579-89. New York: W. H. Freeman, 1994.

Wade, L. G., *Organic Chemistry.* 3rd ed., 757-74. Englewood Cliffs, NJ: Prentice-Hall, 1995.

 **SAFETY TIPS**

✔ Concentrated sulfuric acid is extremely corrosive and should be handled with care.
✔ Methanol is a flammable liquid and is hightly toxic if ingested.

**Experimental**

1. In an 18 × 150 mm test tube, place 0.30 g of 1,4-dimethoxybenzene, 0.5 mL (d = 0.79 g/mL; 0.40 g) of *t*-butyl alcohol, and 1.0 mL of glacial acetic acid. After swirling the mixture to dissolve the arene, immerse the test tube in an ice-water bath.
2. While this solution is cooling, add 2.0 mL of concentrated sulfuric acid to another test tube, and place the tube in an ice-water bath.
3. When the temperatures of both solutions are between 0° and 5°C, use a disposable pipette to start adding the sulfuric acid to the mixture from step 1 while swirling the test tube and occasionally stirring with a glass stirring rod. *Watch for any spilled acid. If any drops fall on your fingers, wash them with lots of cold water.*
4. After the addition is complete, remove the mixture from the ice bath. Let it warm up to room temperature while you continue to stir with a stirring rod. *(Do not use the thermometer since you might break it as you stir!)* Let the temperature of the mixture rise to 25°C and continue stirring for an additional 5 minutes.
5. Add a few chips of ice to the mixture, stir, add water up to about 1 inch from the top of the test tube, and stir again. Vacuum filter the mixture using a Hirsch funnel, wash the white solid with 10 mL of water, and let dry while still on suction for (at least) 5 minutes.
6. Transfer the white solid to a 25 mL conical flask, and recrystallize from a minimum amount (about 3–5 mL) of methanol. After drying, record the MP (lit. 104–105°C) and the percent yield.

**Questions**

1. What is the purpose of the following in this experiment?
   a. concentrated sulfuric acid
   b. acetic acid
2. Give the major products in the reactions of the following:
   a. anisole with isobutyl alcohol (in acid medium)
   b. anisole with 2,3-dimethyl-2-butene (in acid medium)

c. *m*-dimethoxybenzene with propylene oxide (in acid medium)
d. *m*-dimethoxybenzene with 1-chloro-3,3-dimethylbutane (in the presence of AlCl$_3$).

3. Describe the $^1$H NMR spectrum of this experiment's product.
4. What are the main differences between the Friedel-Crafts alkylation and Friedel-Crafts acylation reactions?
5. After reading about the Friedel-Crafts acylation reaction, predict the products of the Lewis acid-catalyzed reaction of acetyl chloride with the following:
   a. anisole
   b. *p*-dichlorobenzene
   c. *p*-methylacetophenone.

## 15.3  SYNTHESIS OF 2,4-DINITROPHENYLANILINE: A NUCLEOPHILIC AROMATIC SUBSTITUTION

Aromatic compounds are normally susceptible to electrophiles, and they undergo electrophilic aromatic substitution reactions. When the benzene ring is properly substituted with electron-withdrawing groups (such as nitro, $-NO_2$, carboxyl, $-COOH$, etc.), a **nucleophilic aromatic substitution** can take place. In these cases, the nucleophile will displace a good leaving group (such as a halide or a methoxy group, $-OCH_3$). The position of the electron-withdrawing group relative to the leaving group is extremely important because the intermediate is stabilized by resonance. In all nucleophilic aromatic substitutions, the electron-withdrawing group (or better, groups) are located *ortho* and/or *para* to the leaving group. They activate the aromatic ring toward nucleophilic attack in two ways: (1) by lowering the electron density of the ring, and (2) by stabilizing the intermediate cyclohexadienyl anion by resonance.

In this experiment, 2,4-dinitrochlorobenzene will be reacted with a weak nucleophile (aniline) to produce N-(2,4-dinitrophenyl) aniline. The reaction is uncatalyzed and proceeds smoothly without side products.

2, 4-dinitrochlorobenzene          aniline          N-(2, 4-dinitrophenyl) aniline

The use of a strong base is generally avoided since it promotes the formation of a benzyne and leads to more complicated reactions.

benzyne

**Suggested References**   Carey, F. A., *Organic Chemistry.* 3rd ed., 980–85. New York: McGraw-Hill, 1996.

Loudon, G. Marc, *Organic Chemistry.* 3rd ed., 829–32. Redwood City, CA: Benjamin/ Cummings, 1995.

McMurry, J., *Organic Chemistry.* 4th ed., 579–99. Pacific Grove, CA: Brooks/Cole, 1996.

Miller, J., *Aromatic Nucleophilic Substitution.* New York: Elsevier, 1968.

Morrison, R. T., and R. N. Boyd, *Organic Chemistry.* 6th ed., 952–61. Englewood Cliffs, NJ: Prentice-Hall, 1992.

Solomons, T. W. G., *Organic Chemistry.* 6th ed., 980–85. New York: John Wiley, 1996.

Wade, L. G., *Organic Chemistry.* 3rd ed., 783–84. Englewood Cliffs, NJ: Prentice-Hall, 1995.

**SAFETY TIPS**

✔ 2,4-Dinitrochlorobenzene, aniline and N-(2,4-dinitrophenyl) aniline are all skin irritants and the use of gloves is recommended.

✔ Ethanol is a highly flammable liquid. No open flames should be used.

**Experimental**

1. Place 15 mL of ethanol, 0.8 g of 2,4-dinitrochlorobenzene, and 1.0 mL (d = 1.02 g/mL, 1.0 g) of aniline in a 50-mL round-bottomed flask, and reflux for 30 minutes.
2. Cool the solution in ice, and vacuum-filter the red crystals.
3. Recrystallize the solid from approximately 60–70 mL of ethanol. Let the solution cool slowly.
4. Isolate the red crystals by vacuum filtration, determine the MP (lit. 156°C), and calculate the percent yield.

**Questions**

1. What would be the effect if hydrochloric acid was added to the reaction mixture accidentally?
2. Give the major product in the reactions of the following:
   a. 2,4-dinitrochlorobenzene and hydrazine
   b. 3,5-dinitrochlorobenzene and aniline
   c. 2,4,6-trinitrochlorobenzene and aniline.
3. Why isn't benzyne formed in this reaction?
4. Write a mechanism for the formation of N-(2,4-dinitrophenyl) aniline via the above procedure. Show all resonance forms that stabilize the intermediate in this reaction.

## 15.4 OXIDATION OF THE SIDE CHAIN OF AN ARENE: SYNTHESIS OF 2-CHLOROBENZOIC ACID FROM 2-CHLOROTOLUENE

The side chain of an arene can be oxidized to a carboxylate group, forming the corresponding benzoic acid, by means of a strong oxidizing agent such as basic permanganate solution or acidic dichromate solution. For this method to succeed, the substituent must contain at least one benzylic hydrogen atom.

R = alkyl chain with benzylic hydrogen
[O] = $MnO_4^-/OH^-$ followed by $H^+$, or $Cr_2O_7^{2-}/H^+$

The mechanism, which is not well understood, is believed to involve the removal of a benzylic hydrogen in a free-radical process, which leads first to the formation of the alcohol, then to the ketone (or alkene), and finally to the carboxylic acid. Consistent with this proposal is the observation that benzylic alcohols, aryl ketones, and vinyl arenes, which are possible intermediates in this mechanism, are also efficiently oxidized to benzoic acids. Note that all alkyl-substituted arenes always yield benzoic acid derivatives, as long as they carry a benzylic hydrogen. The rest of the side-chain is converted into carbon dioxide and water.

In this experiment, 2-chlorotoluene will be converted to 2-chlorobenzoic acid via hot alkaline $KMnO_4$.

2-chlorotoluene   potassium permanganate   2-chlorobenzoic acid   mangane dioxide

The excess permanganate will be destroyed with sodium thiosulfate prior to filtration through Celite (Experiment 10.2) to remove the gelatinous $MnO_2$.

$$MnO_4^- + S_2O_3^{2-} \longrightarrow MnO_2 + SO_4^{2-}$$

Finally, the clear solution will be acidified to precipitate out the 2-chlorobenzoic acid.

**Suggested References**    Carey, F. A., *Organic Chemistry.* 3rd ed., 438–40. New York: McGraw-Hill, 1996.

Clarke, H. T., and Taylor, E. R., "2-Chlorobenzoic Acid." In *Organic Syntheses,* vol. 2, 135–36. 1943.

McMurry, J., *Organic Chemistry.* 4th ed., 602–03. Pacific Grove, CA: Brooks/Cole, 1996.

Morrison, R. T., and R. N. Boyd, *Organic Chemistry.* 6th ed., 563–64, 720–50. Englewood Cliffs, NJ: Prentice-Hall, 1992.

"Potassium Permanganate." In *Reagents for Organic Synthesis.* ed. L. F. Fieser and M. Fieser, vol. 1, 942–52. New York: John Wiley, 1967.

Solomons, T. W. G., *Organic Chemistry.* 6th ed., 690–91. New York: John Wiley, 1996.

Vollhardt, K. P. C., and N. E. Schore, *Organic Chemistry.* 2nd ed., 866–67. New York: W. H. Freeman, 1994.

Wade, L. G., *Organic Chemistry.* 3rd ed., 790–91. Englewood Cliffs, NJ: Prentice-Hall, 1995.

 **SAFETY TIPS**

✔ 2-Chlorotoluene is flammable and an irritant; avoid open flames, skin contact, or inhalation.

✔ Sodium hydroxide pellets are extremely corrosive; in case of skin contact, wash with dilute acetic acid and water.

✔ Potassium permanganate is a strong oxidant and skin stainer; in case of skin contact, wash with dilute sodium thiosulfate and water.

✔ Concentrated hydrochloric acid is extremely corrosive; it should be poured carefully via funnel and with adequate ventilation.

**Experimental**

1. To a 250 mL round-bottomed flask, add 6.0 g of potassium permanganate, 0.46 g of sodium hydroxide (about 5 pellets), several boiling chips (or a magnetic stir bar, if magnetic stirrers are available), and 70 mL of water; stir to dissolve.

2. Add 2.0 mL (d = 1.08 g/mL; 21.2 g) of 2-chlorotoluene, attach a reflux condenser, and heat the reaction mixture to boiling. Reflux for 1.5 to 2 hours, depending on the time available.

3. After the indicated reflux period, remove the heating mantle and allow the mixture to cool briefly.

4. Remove the reflux condenser, and carefully add saturated aqueous sodium thiosulfate solution ($Na_2S_2O_3$) until the purple color due to excess permanganate has disappeared (this should only require a small amount of thiosulfate solution).

5. Remove the precipitated manganese dioxide by suction filtration through a pad of Celite, using a Büchner funnel; the filtrate should be clear and colorless.

6. Wash the filter cake with 10 mL of hot water.

7. Transfer the filtrate to a 250 mL beaker, and boil away approximately half of the water, using a hotplate or a Bunsen burner in a well-ventilated fume hood.

8. Cool the concentrated solution in an ice-water bath, and (in a fume hood) add concentrated hydrochloric acid until the pH is below 4.

9. Continue cooling the resulting suspension with stirring in the ice-water bath to complete the precipitation, and isolate the product by suction filtration. Wash with 5 mL of ice-cold water, and allow it to air-dry with suction for a few minutes.

10. Dry the product in an oven for an hour, if time permits, or at room temperature overnight.

11. Determine the melting point (lit. 138–140°C), weigh the product, and calculate the percent yield.

12. If desired, the product may be further purified by recrystallization from toluene.

13. The brown manganese dioxide stains can be removed from glassware by brief treatment with concentrated hydrochloric acid (in the hood) and use of a test tube brush.

**Questions**

1. Give balanced equations for the oxidation of toluene by the following:
   a. $MnO_4^-$ in base, to yield $MnO_2$ and benzoate anion
   b. $Cr_2O_7^{2-}$ in acid, to yield $Cr^{3+}$ and benzoic acid
   c. $MnO_4^-$ in acid, to yield $Mn^{2+}$ and benzoic acid.

2. Write a balanced equation for the reaction of permanganate and thiosulfate in basic medium. Assume that permanganate is reduced to manganese dioxide, and thiosulfate is oxidized to sulfate.

3. In the experiment you just performed, which reagent, permanganate or 2-chlorotoluene, was the limiting reagent? By what mole percentage was the other reagent in excess?

4. Would it be feasible to attempt the synthesis of benzaldehyde from toluene via the above procedure? Explain.

5. What are the products of the hot permanganate oxidation of the following:
   a. *p*-methoxytoluene
   b. 1-chloro-3-ethylbenzene
   c. benzyl alcohol
   d. *p*-xylene
   e. *p-t*-butyltoluene
   f. cyclohexylbenzene

## 15.5  BENZYNE: SYNTHESIS OF TRIPTYCENE

Arynes, or dehydrobenzenes, are very unstable because of the poor over-lap between the orbitals that make up the $C \equiv C$ formal bond, and are therefore, very reactive species. The simplest member of this family of intermediates is **benzyne,** which can be formed by the loss of the substituents A and B:

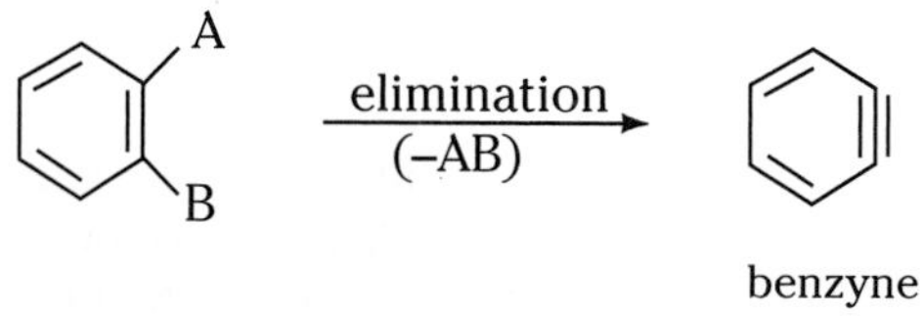

For example, chlorobenzene (Cl⁻ being a good leaving group) can form benzyne when treated with a strong base, such as $NaNH_2$.

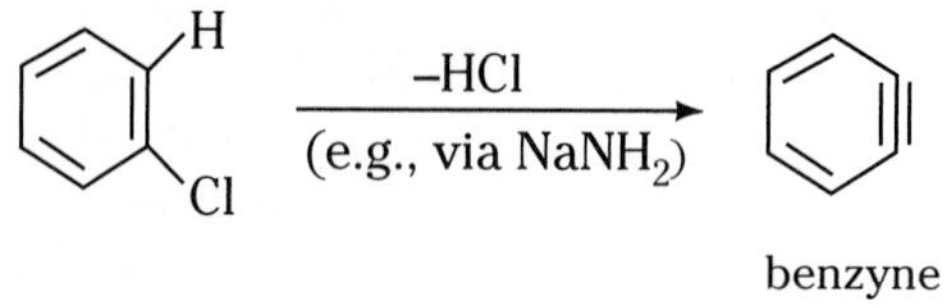

In this experiment, benzyne will be formed via diazotization of anthranilic acid (*o*-aminobenzoic acid) using isoamyl nitrite, and by sub-sequent elimination of two gases (nitrogen and carbon dioxide). The for-mation of these gases causes a favorable shift of the equilibrium to the right-hand side.

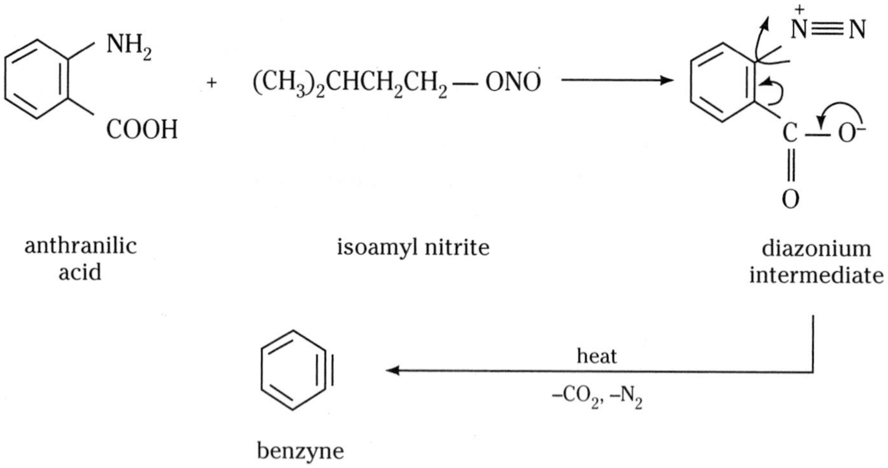

| anthranilic acid | isoamyl nitrite | diazonium intermediate |

benzyne

The choice of solvents is very crucial. The solvent cannot be protic, because benzyne will quickly react via addition.

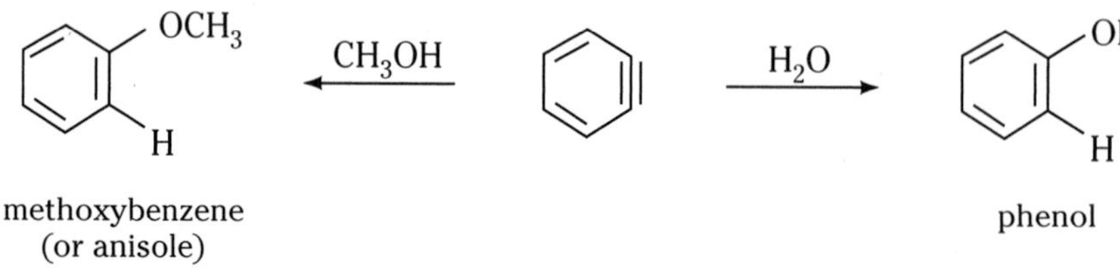

methoxybenzene          phenol
(or anisole)

Moreover, the solvent should be polar so that reactants and interme-diates are dissolved. If the diazonium intermediate $[o - C_6H_4(N_2^+)COO^-]$ does not dissolve, the possibility of an explosion is *very high* because more and more of the explosive solid will accumulate.

Finally, the solvent should have a high boiling point so its temperature can be high enough to decompose the diazonium intermediate as soon as it forms, thereby avoiding its accumulation. Diethylene glycol diethyl ether is a solvent that fulfills all these requirements.

Once the benzyne is formed, it will be trapped by anthracene in a Diels-Alder reaction to form triptycene, the desired product.

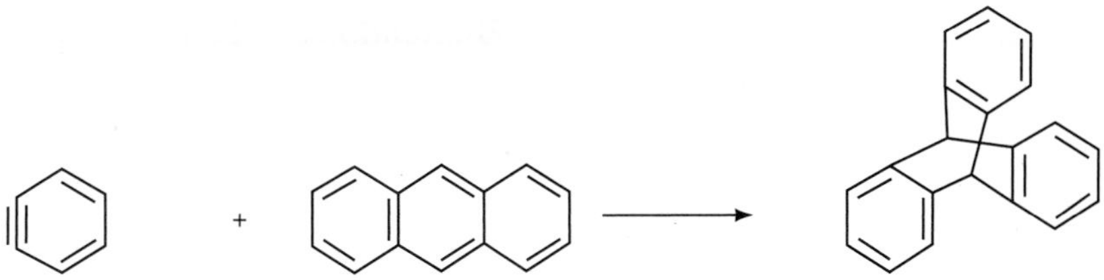

benzyne                    anthracene                              triptycene

The Diels-Alder reaction is energenetically favorable, and leads to the formation of a more stable product (see Chapter 12, pp. 147–50). Anthracene has a resonance energy *per ring* lower than that of benzene, and in this sense, it is "less aromatic." The product, triptycene, has three "unfused" benzene rings and is actually more stable (i.e., more aromatic) than the polynuclear aromatic anthracene.

Problems arise with the purification of triptycene. Anthracene is added in excess to prevent dimerization of benzyne. As a result anthracene will constitute an impurity in the isolated triptycene.

dimerization product

The Diels-Alder reaction comes to the rescue again! The excess anthracene is removed by reacting it with maleic anhydride, which results in a product that can be hydrolyzed by a strong base to a water-soluble diacid.

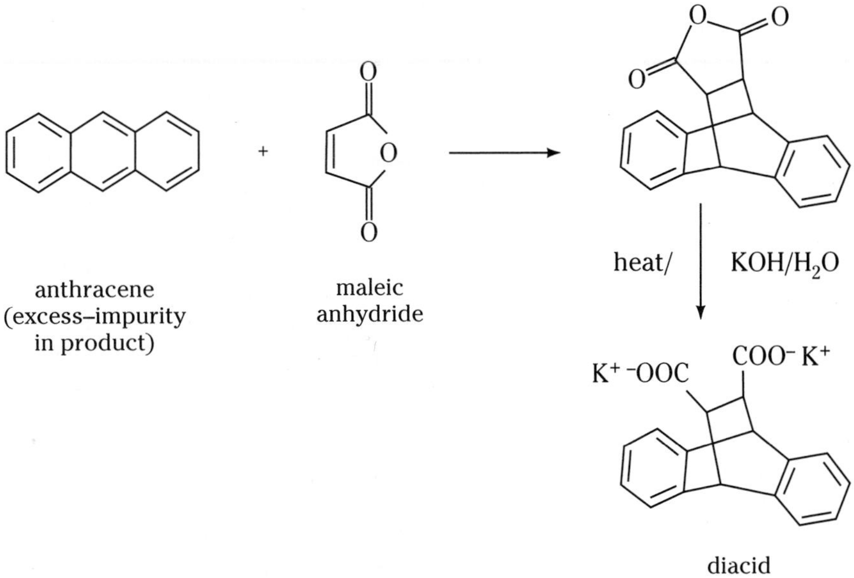

anthracene              maleic
(excess–impurity        anhydride
in product)

heat/ | KOH/H$_2$O

$K^+$ $^-$OOC   COO$^-$ $K^+$

diacid
(water soluble)

In summary, the overall procedure features the following reactions and can be considered as a good all-around experiment:

a. **Diazotization**

$$R\!-\!NH_2 + R'\!-\!ONO \longrightarrow [R\!-\!N\!\equiv\!N]^+$$

b. **Elimination** and benzyne formation

c. **Diels-Alder reaction** of a benzyne with a "less" aromatic compound to form a "more" aromatic compound
d. **Hydrolysis** of an anhydride to a carboxylic acid using a base

**Suggested References**     Carey, F. A., *Organic Chemistry*. 3rd ed., 965–70. New York: McGraw-Hill, 1996.

Friedman, L., and F. M. Logullo, "Arynes via Aprotic Diazotization of Anthranilic Acids." *J. Org. Chem.* 34 (1969): 3089.

Heaney, H., "The Benzyne and Related Intermediates." *Chem. Rev.* 62 (1962): 81.

Loudon, G. Marc, *Organic Chemistry*. 3rd ed., 832–35. Redwood City, CA: Benjamin/Cummings, 1995.

McMurry, J., *Organic Chemistry*. 4th ed., 600–02. Pacific Grove, CA: Brooks/Cole, 1996.

Morrison, R. T., and R. N. Boyd, *Organic Chemistry*. 6th ed., 962–67. Englewood Cliffs, NJ: Prentice-Hall, 1992.

Patnaik, P., *A Comprehensive Guide to the Hazardous Properties of Chemical Substances*. 475. New York: Van Nostrand Reinhold (1992).

Solomons, T. W. G., Organic Chemistry. 6th ed., 872–73, 983–85. New York: John Wiley, 1996.

Vollhardt, K. P. C., and N. E. Schore, *Organic Chemistry*. 2nd ed., 875–76. New York: W. H. Freeman, 1994.

Wade, L. G., *Organic Chemistry*. 3rd ed., 785–86. Englewood Cliffs, NJ: Prentice-Hall, 1995.

Wittig, G., and K. Niethammer, "Dehydrobenzol und Acridin." *Chem. Ber.* 93 (1960): 944.

Wittig, G., H. Härle, E. Knaus, and K. Niethammer, "Synthese von Derivaten des Triptycens und Iso-aza-triptycens." *Chem. Ber.* 93 (1960): 9514.

---

⚠ **SAFETY TIPS**

✔ Isoamyl nitrite is a flammable and toxic liquid. 1,2-Dichloroethane is a suspected carcinogen and methyl ethyl ketone is flammable. Prolonged exposure to these liquids should be avoided and their transfer should be done in a well-ventilated fume hood.

✔ The dust of anthracene and anthranilic acid are irritants, while maleic anhydride is corrosive.

✔ All flames should be extinguished in the lab while the experiment is being done.

✔ Prolonged inhalation of methyl ethyl ketone can cause irritation of the eyes and nose.

✔ Wearing plastic gloves during the experiment is highly recommended.

**Experimental**

1. Place 5.4 g of anthracene, 3.1 mL (d = 0.87 g/mL; 2.7 g) isoamyl nitrite, and 40 mL of 1,2-dichloroethane in a 100-mL round-bottomed flask equipped with a condenser and separatory funnel, as seen in Figure 15.2.
2. Add a few boiling chips, and reflux using a heating mantle.
3. After the mixture has started to reflux, lower the heating mantle and add a solution of 3.0 g of anthranilic acid in 15 mL of diethylene glycol diethyl ether dropwise via the separatory funnel. The addition should take about 5 minutes, while the flask is frequently agitated.
4. After the addition is complete, reflux for 15 minutes. Re-fit the flask for a simple distillation and distill until a head temperature of 150°C is reached.
5. While the mixture is cooling, transfer the distillate from step 4 (*use a funnel!*) to a designated waste bottle in the hood.
6. Add 3.0 g of maleic anhydride all at once, and reflux for about 5 minutes.
7. Cool in an ice-water bath for 5 minutes. Add a solution of 6 g potassium hydroxide pellets dissolved in 20 mL of methanol and 10 mL of water, and reflux for 5 more minutes.
8. Cool the solution again, filter via suction, and wash the crystals with a 4:1 methanol-water solution until the washings (*not the solution!*) are colorless. Check the filtrate by collecting a few drops periodically in a test tube that is inserted in the filter flask.
9. Purify the product as follows: Dissolve the collected solid in approximately 20 mL of butanone (methyl ethyl ketone) by heating the mixture in a hot-water bath. Add a pinch of decolorizing carbon, and vacuum filter the boiling solution. Evaporate the solvent to roughly two-thirds of its original volume, add about 10 mL of methanol, and cool in an ice-water bath. Collect the product by vacuum filtration, wash it with approximately 2–3 mL of ice-cold methanol, and determine the MP (literature value 224–25°C). Calculate the percent yield.

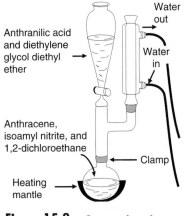

Anthranilic acid and diethylene glycol diethyl ether →

Water out

Water in

Anthracene, isoamyl nitrite, and 1,2-dichloroethane

Clamp

Heating mantle →

**Figure 15.2** Set-up for the synthesis of triptycene

**Questions**

1. What is the purpose of the following in this experiment:
   a. not using protic solvents (give a pertinent equation)
   b. using decolorizing carbon (step 9)
   c. refluxing the product with potassium hydroxide solution
   d. heating the reaction mixture *before* adding anthranilic acid.
2. Identify the compound that is
   a. distilled in step 4.
   b. left in the filtrate of step 8.
   c. formed if no anthracene is added.
3. Identify the compounds A through D below

$$\text{(structure: benzene ring with OCH}_3\text{, NH}_2\text{, COOH)} + \text{iso–C}_5\text{H}_{11}\text{—ONO} \longrightarrow \textbf{A} \ (\text{C}_7\text{H}_6\text{O}) \xrightarrow{\text{CH}_3\text{OH}} \textbf{B} + \textbf{C}$$

isomers of $\text{C}_8\text{H}_{10}\text{O}_2$

$$\text{(structure: benzene ring with two CH}_3\text{ and Cl)} + \text{NaNH}_2/\text{heat} \longrightarrow \textbf{D} \ (\text{C}_{16}\text{H}_{16})$$

(dimerization product)

4. Explain the following:
   a. When *m*-chlorotoluene is heated with KOH at 350°C, both *m*- and *p*-cresol are formed ($\text{C}_7\text{H}_8\text{O}$).
   b. When treated with $\text{NaNH}_2/\text{NH}_3$, both *o*-bromoanisole and *m*-bromoanisole yield *m*-methoxyaniline.

# chapter 16

# Carboxylic Acids and Derivatives

## 16.1 EQUIVALENT WEIGHT OF ORGANIC ACIDS—A MICRO OR MACROSCALE DETERMINATION

As a class, the carboxylic acids constitute one of the most acidic families in organic chemistry. They dissolve in dilute alkali (e.g., sodium hydroxide) as do phenols, but unlike most phenols and compounds of other organic families, they are also soluble in weakly basic sodium bicarbonate solutions. The products of these reactions are the carboxylate salt and water.

$$RCOOH \xrightarrow{\begin{array}{c}NaOH\end{array}} RCOO^-Na^+ + H_2O$$

carboxylic acid

sodium carboxylate

$$\xrightarrow{NaHCO_3} RCOO^-Na^+ + H_2O + CO_2$$

In this experiment, a known quantity of acid will be titrated against a standardized solution of sodium hydroxide. The equivalent weight (EW) of the acid will be calculated from the formula

$$EW_{acid} = \frac{(mass)_{acid}}{N_b \times V_b}$$

where $N_b$ is the normality of the base, and $V_b$ is the volume of the base.

The experimentally calculated equivalent weight will be compared to the equivalent weights of a series of organic acids listed in Table 16.1. Note that the usual relationship between an acid's molecular weight (MW) and equivalent weight (EW) is given as

$$EW = \frac{MW}{\text{number of titratable hydrogens}}$$

From this relationship, the equivalent weight of a monocarboxylic acid is equal to its molecular weight; while for a dicarboxylic acid, it is one-half its molecular weight; for a tricarboxylic acid, one-third its molecular weight, etc.

Suggested References

Carey, F. A., *Organic Chemistry.* 3rd ed., 770–73. New York: McGraw-Hill, 1996.
Loudon, G. Marc, *Organic Chemistry.* 3rd ed., 937–41. Redwood City, CA: Benjamin/Cummings, 1995.
McMurry, J., *Organic Chemistry.* 4th ed., 778–81. Pacific Grove, CA: Brooks/Cole, 1996.

Morrison, R. T., and R. N. Boyd, *Organic Chemistry.* 6th ed., 744–45. Englewood Cliffs, NJ: Prentice-Hall, 1992.

Solomons, T. W. G., *Organic Chemistry.* 6th ed., 795–96. New York: John Wiley, 1996.

Vollhardt, K. P. C., and N. E. Schore, *Organic Chemistry.* 2nd ed., 726–28. New York: W. H. Freeman, 1994.

Wade, L. G., *Organic Chemistry.* 3rd ed., 73, 942–43. Englewood Cliffs, NJ: Prentice-Hall, 1995.

## Experimental

*Note:* This experiment may also be performed under microscale conditions using the following:

- a microburette
- 2 mL methanol
- *one* drop of phenolphthalein delivered with a fine-tipped pipette
- a 0.5–1.0 *N* sodium hydroxide solution, and
- 50 mg samples of the unknown acid.

1. Accurately weigh 0.1 to 0.2 g of your unknown acid, and transfer it into a 125-mL Erlenmeyer flask. Add 10 mL of methanol and 2 drops of phenolphthalein indicator, and swirl the contents to ensure proper mixing.
2. Titrate against the standardized (0.10 to 0.15 *N*) sodium hydroxide solution, remembering to place a white paper under the flask so that the change of color (from colorless to pink) can be noted once the equivalence point has been reached.
3. Repeat the titration with a new weighed sample of the *same* carboxylic acid. The mass used for the second titration should be different from the mass used for the first determination. Calculate the equivalent weight. The two values should agree within ±2 amu. If they do not, perform a third titration. Remember to wash the flask thoroughly after each titration, first with tap water and then with distilled water.
4. Wash the burette after the last titration, *while still clamped.* Drain the burette, then slowly add about 10 mL of water and drain; then 10 mL of dilute hydrochloric acid and drain; finally 50 mL of distilled water and drain. *Do not place the burette under the faucet in order to wash it.*
5. Identify your unknown carboxylic acid by matching your result with the possibilities listed in Table 16.1. As is done for the first entry (formic acid), complete the table by writing in the formula and the equivalent weight (EW) for each of the acids listed.

## Questions

1. 0.11 g of a monocarboxylic acid is neutralized by 12.5 mL of a 0.1*N* sodium hydroxide solution.
   a. What is the molecular weight of the acid?
   b. Suggest a molecular formula for this acid.
   c. How many carboxylic acids of this molecular formula are possible?
2. What would be the effect on the equivalent weight calculations, if during the titration,
   a. some of the carboxylic acid adhered to the walls and was not dissolved in methanol?
   b. a defective burette was used and a higher value for the volume of sodium hydroxide was recorded?
   c. a 2 *N* sodium hydroxide solution was used instead of a 0.1 *N* sodium hydroxide solution?
3. How many grams of phthalic acid o—$C_6H_4(COOH)_2$ would be neutralized in 16.2 mL of a 0.100N NaOH solution?

**Table 16.1** Carboxylic acids

| Name | Solid/Liquid | Formula | Equivalent Weight (EW) |
|---|---|---|---|
| Formic | Liquid | $HCO_2H$ | 46 |
| Acetic | | | |
| Propionic | | | |
| Butyric | | | |
| Valeric | | | |
| Caproic | | | |
| Caprylic | | | |
| Capric | | | |
| *n*-Decanoic | | | |
| Benzoic | | | |
| *p*-Nitrobenzoic | | | |
| *o*-Chlorobenzoic | | | |
| *p*-Anisic | | | |
| Sorbic | | | |
| *p*-Toluic | | | |
| Phenylacetic | | | |
| Oxalic | Solid | $(COOH)_2$ | 45 |
| Malonic | | | |
| Maleic | | | |
| Phthalic | | | |
| Adipic | | | |
| Succinic | | | |

## DATA SHEET: EQUIVALENT WEIGHT OF ORGANIC ACIDS

|  | Attempt #1 | Attempt #2 | Attempt #3 |
|---|---|---|---|
| 1. Mass of carboxylic acid | _____g | _____g | _____g |
| 2. Initial burette reading | _____mL | _____mL | _____mL |
| 3. Final burette reading | _____mL | _____mL | _____mL |
| 4. Volume of NaOH needed for neutralization | _____mL | _____mL | _____mL |
| 5. Normality of NaOH (standardized) | _____N | _____N | _____N |
| 6. Equivalent weight of acid | _____ | _____ | _____ |

7. Equivalent weight of unknown # _____ is _____ amu (average)

*Conclusion:* My unknown carboxylic acid (# _____ ) is _____

## 16.2  HYDROLYSIS OF BENZONITRILE

A carboxylic acid can generally be prepared via the following methods:
Method A. *Oxidation of a primary alcohol or aldehyde.*

$$R - CH_2OH \atop (1° \text{ alcohol})$$
$$R - CHO \atop (\text{aldehyde})$$
$$\xrightarrow{[O]} R - COOH \quad \text{carboxylic acid}$$

where [O] is $Cr_2O_7^{-2}$, or $MnO_4^-/OH^-$ followed by $H^+$
Method B. *Carbonation of a Grignard reagent.*

$$RMgX + CO_2 \xrightarrow{\text{then } H^+} R - COOH$$

Method C. *Oxidation of a side chain of a aryl group.*

$$Ar - R + MnO_4^-/OH^- \xrightarrow{\text{then } H^+} Ar - COOH + MnO_2$$

(plus $CO_2$ and $H_2O$ if R carries more than one carbon)

Method D. *Hydrolysis of a nitrile.*

$$R - C \equiv N + OH^-/H_2O \longrightarrow R - COO^- + NH_3$$
$$\xrightarrow[H^+]{} R - COOH$$

Comparing the four methods, one concludes that

1. Methods A and D result in the formation of a carboxylic acid with the same number of carbons as the starting 1° alcohol, aldehyde, or nitrile.
2. Method C yields an acid with fewer carbon atoms than the starting arene (unless Ar—R is a toluene derivative).
3. Method B yields an acid with one carbon atom more than the starting material (usually an alkyl halide) contained.

Nitriles ($R - C \equiv N$) can be prepared from alkyl halides (R—X) by an $S_N2$ reaction. The method can generally be used only with 1° alkyl halides because 3° (and often 2°) alkyl halides undergo elimination (rather than substitution) when they react with the basic cyanide ion. Thus,

$$R - CH_2X \xrightarrow{^-C \equiv N} R - CH_2 - C \equiv N + X^-$$
$$R_3C - X \xrightarrow{^-C \equiv N} \text{alkene}$$

Alkaline hydrolysis of nitriles usually gives higher yields than acid hydrolysis because the reaction is driven to the right by the loss of a gaseous product, ammonia (Le Chatelier's principle). In acid medium, ammonia is converted into the ammonium ion (not a gaseous product).

$$R - C \equiv N + H_2O + OH^- \longrightarrow R - COO^- + NH_3 \quad \text{(alkaline medium)}$$
$$R - C \equiv N + 2H_2O + H^+ \longrightarrow R - COOH + NH_4^+ \quad \text{(acidic medium)}$$

There is also a small entropy effect that favors alkaline hydrolysis.

**Suggested References**

Carey, F. A., *Organic Chemistry.* 3rd ed., 783-84. New York: McGraw-Hill, 1996.

Loudon, G. Marc, *Organic Chemistry.* 3rd ed., 994-95. Redwood City, CA: Benjamin/Cummings, 1995.

McMurry, J., *Organic Chemistry.* 4th ed., 787, 825-26. Pacific Grove, CA: Brooks/Cole, 1996.

Morrison, R. T., and R. N. Boyd, *Organic Chemistry.* 6th ed., 678–79, 724, 870. Englewood Cliffs, NJ: Prentice-Hall, 1992.

Solomons, T. W. G., *Organic Chemistry.* 6th ed., 804, 826–27, 1150. New York: John Wiley, 1996.

Vollhardt, K. P. C., and N. E. Schore, *Organic Chemistry.* 2nd ed., 723–33. New York: W. H. Freeman, 1994.

Wade, L. G., *Organic Chemistry.* 3rd ed., 957, 1009–10. Englewood Cliffs, NJ: Prentice-Hall, 1995.

---

 **SAFETY TIPS**

✔  Benzonitrile is an irritant, and it should be transferred with a pipette in a well-ventilated area. It is also a combustible liquid (flash point 71°C) and all flames should be extinguished.

✔  Sodium hydroxide is caustic and should be treated with respect.

✔  Concentrated hydrochloric acid is extremely corrosive; it should be poured carefully and with adequate ventilation.

---

**Experimental**

1. In a 100-mL round-bottomed flask, place 2.0 mL (d = 1.01 g/mL; 2.0 g) of benzonitrile and 25 mL of a 10% sodium hydroxide solution.

2. Reflux the mixture for about 40 minutes. During this time, the ammonia vapors will escape from the top of the condenser. The presence of ammonia in these vapors can be tested for with a piece of moist red litmus paper held over the top of the condenser.

3. After 40 minutes, pour the contents of the flask into a conical flask, cool in an ice-water bath, and acidify with concentrated hydrochloric acid until precipitation of benzoic acid is complete (pH = 4).

4. Collect the fluffy benzoic acid via vacuum filtration, rinse with 5 mL of ice-cold water, and recrystallize from water (approximately 15 mL). Determine the percent yield and the MP (literature value 122°C).

5. Run IR spectra for both benzonitrile and benzoic acid, and compare them.

**Questions**

1. Infrared spectroscopy can be used to monitor the reaction: i.e., the formation of benzoic acid, and consumption of benzonitrile. Which major absorption peaks characteristic of the product and the starting material should change?

2. The procedure calls for 25 mL of a 10% (w/w) NaOH solution. Assuming that pure pellets of sodium hydroxide (MW 40) are used, and that the density of water is 1 g/mL, explain how to make this solution.

3. Litmus paper is used to determine the acidity of the medium (step 3) and the presence of ammonia (step 2). What color of litmus paper should be used in each case, and what color changes will be observed?

4. Identify the products of the following reactions:

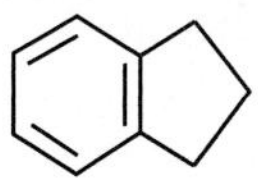

+ MnO$_4^-$/OH$^-$ ⟶ **A**

(CH$_3$)$_2$CHCH$_2$CH$_2$Cl + $^-$C≡N ⟶ **B** $\xrightarrow[\text{then H}^+]{\text{OH}^-/\text{H}_2\text{O}}$ **C**

CH$_3$CH$_2$CH$_2$Br + Mg/ether ⟶ **D** $\xrightarrow[\text{then H}^+]{\text{CO}_2}$ **E**

## 16.3   SYNTHESIS OF ASPIRIN

Aspirin is the most widely used drug in modern society. Its name is derived from its structure and chemical name, **acetylsalicylic acid**. In earlier times, salicylic acid, from which all the salicylates are derived, was known as **spiraeic acid** (isolated from the meadowsweet family), and aspirin was called **acetylspiraeic acid**. Aspirin is a powerful antipyretic (fever-reducing) agent, anti-inflammatory (swelling-reducing) agent, as well as an analgesic (pain killer).

Salicylic acid itself has analgesic properties, and in the past, it was used in the form of its sodium salt. Its unpleasant side effects, principally serious gastric disturbances, prompted a modification that involved acetylation of the phenolic group, which was found to be beneficial.

In this experiment, salicylic acid will be esterified at the phenolic group with acetic anhydride, a commonly used acylating agent. Phosphoric acid catalyzes the reaction by protonating the carbonyl group of the anhydride.

|           |          |            |         |        |
| salicylic | acetic   | phosphoric | aspirin | acetic |
| acid      | anhydride | acid      |         | acid   |

At the end of the reaction time, the excess acetic anhydride is destroyed with water. Ice is added to precipitate the aspirin, which is then recrystallized from water.

It should be pointed out that salicylic acid is difunctional—it can serve as either an acid or an alcohol in esterification reactions. For example, in the presence of methyl alcohol and an acid catalyst, salicylic acid is esterified at the carboxyl group and methyl salicylate (commonly known as oil of wintergreen) is produced.

|                |           |                   |
| salicylic acid | methanol  | methyl salicylate |

In the synthesis of aspirin as described here, salicylic acid reacts with the strong acylating agent acetic anhydride, and esterification occurs at the phenolic position.

**Suggested References**

Carey, F. A., *Organic Chemistry*. 3rd ed., 989–91. New York: McGraw-Hill, 1996.

Loudon, G. Marc, *Organic Chemistry*. 3rd ed., 1004–05. Redwood City, CA: Benjamin/Cummings, 1995.

McMurry, J., *Organic Chemistry*. 4th ed., 822–23, 1020. Pacific Grove, CA: Brooks/Cole, 1996.

Morrison, R. T., and R. N. Boyd, *Organic Chemistry*. 6th ed., 914. Englewood Cliffs, NJ: Prentice-Hall, 1992.

Solomons, T. W. G., *Organic Chemistry*. 6th ed., 814–15, 976. New York: John Wiley, 1996.

Vollhardt, K. P. C., and N. E. Schore, *Organic Chemistry.* 2nd ed., 752, 780. New York: W. H. Freeman, 1994.

Wade, L. G., *Organic Chemistry.* 3rd ed., 795. Englewood Cliffs, NJ: Prentice-Hall, 1995.

### ⚠ SAFETY TIPS

✔ Acetic anhydride is irritating to the eyes and nasal membranes. Because of its lachrymatory properties, it should be poured in the fume hood, and skin contact should be avoided.

✔ Concentrated phosphoric acid is extremely corrosive and should be handled with care.

✔ Ethanol is a highly flammable liquid; no open flames should be used.

**Experimental**

1. Place 2.0 g of salicylic acid in a 125-mL conical flask.
2. Add 4.0 mL (d = 1.08 g/mL; 4.3 g) of acetic anhydride and 6–8 drops of 85% phosphoric acid.
3. Swirl the flask to mix all reagents, and place it in a 70°C water bath. Heat for about 10 minutes.
4. Remove the flask from the water bath and add 2 mL of cold water dropwise, swirling the flask after each addition.
5. Add 10 g of ice, and cool in an ice water bath until precipitation is complete.
6. Isolate the crude product by vacuum filtration.
7. Recrystallize the product as follows: Place the crude aspirin in a 125-mL conical flask, and add 5–6 mL of ethanol. Heat the flask on a steam bath until all crystals dissolve, add 10 mL of water, and re-heat until the solution is almost boiling. Remove the flask from the steam bath, and let it cool to room temperature.
8. Cool in an ice-water bath until precipitation is complete and isolate the product by vacuum filtration. Determine the MP (literature value 138°C), and calculate the percent yield.

**Questions**

1. Give equations for the reaction of salicylic acid with the following:
   a. acetyl chloride/pyridine
   b. methanol/HCl/heat
   c. sodium bicarbonate, room temperature
   d. thionyl chloride
   e. 10% sodium hydroxide, room temperature
   f. excess sodium hydroxide, reflux
   g. ammonium hydroxide, room temperature.
2. Give equations for the reaction of aspirin with the reagents in question 1.
3. Give a step-by-step mechanism for the reaction of the following:
   a. salicylic acid with acetic anhydride/phosphoric acid
   b. salicylic acid with methanol/hydrochloric acid
   c. aspirin with 10% aqueous sodium hydroxide/heat.
4. What is the purpose of the following in the above experiment?
   a. phosphoric acid
   b. adding of water (step 4)
   c. ice (step 5)
   d. ethanol (step 7).

## 16.4   SYNTHESIS OF ISOAMYL ACETATE

Esters ($R-\overset{\overset{\displaystyle O}{\|}}{C}-OR'$; R=H, alkyl or aryl group) are derivatives of acids. They tend to have pleasant odors and tastes and they contribute to the characteristic fragrances and flavors of many flowers and fruits. Some examples are given in Table 16.2.

**Table 16.2**   Some common esters

| Ester | Formula | Fragrance |
|-------|---------|-----------|
| Isoamyl acetate | $(CH_3)_2CHCH_2CH_2O-\underset{\underset{\displaystyle O}{\|}}{C}-CH_3$ | Banana |
| Benzyl acetate | $\langle\!\!\langle\bigcirc\rangle\!\!\rangle - CH_2-O-\underset{\underset{\displaystyle O}{\|}}{C}-CH_3$ | Peach |
| Methyl butyrate | $CH_3O-\underset{\underset{\displaystyle O}{\|}}{C}-CH_2CH_2CH_3$ | Apple |
| Ethyl butyrate | $CH_3CH_2-O-\underset{\underset{\displaystyle O}{\|}}{C}-CH_2CH_2CH_3$ | Pineapple |
| Isobutyl propionate | $(CH_3)_2CHCH_2-O-\underset{\underset{\displaystyle O}{\|}}{C}-CH_2CH_3$ | Rum |
| Benzyl butyrate | $\bigcirc - CH_2O-\underset{\underset{\displaystyle O}{\|}}{C}CH_2CH_2CH_3$ | Cherry |
| Methyl anthranilate | $CH_3O-\underset{\underset{\displaystyle O}{\|}}{C}-\bigcirc\!\!-NH_2$ | Grape |
| Ethyl phenylacetate | $CH_3CH_2O-\underset{\underset{\displaystyle O}{\|}}{C}-CH_2-\bigcirc$ | Honey |
| Methyl salicylate | $\bigcirc\!\!\begin{smallmatrix}\overset{\displaystyle O}{\|}\\ -COCH_3\\ -OH\end{smallmatrix}$ | Wintergreen |

Esters can be prepared from the corresponding carboxylic acids by two popular methods:

Method A: $R-COOH + R'OH \underset{\longleftarrow}{\overset{H^+}{\rightleftharpoons}} R-COOR' + H_2O$

Method B:
$$R-\overset{\overset{\displaystyle O}{\|}}{C}-OH \xrightarrow{SOCl_2} SO_2 + HCl + R-\overset{\overset{\displaystyle O}{\|}}{C}-Cl$$

$$\downarrow R'OH$$

$$\longrightarrow RCOOR' + HCl$$

Often method A does not give high yields because the reaction is reversible. Although method B is a two-step method, it gives higher yields because gaseous products (HCl and $SO_2$) are formed and are removed from the system, thus driving the equilibrium to the right side. To increase the yield of ester via method A, a large excess of one of the reagents is used— preferably the more water-soluble reactant, which can then be removed from the product mixture during the water work-up process.

In this experiment, isoamyl acetate will be prepared via the acid-catalyzed esterification of acetic acid with isoamyl alcohol.

$$CH_3-\overset{\overset{\displaystyle O}{\|}}{C}-OH + (CH_3)_2CH(CH_2)_2OH \underset{\longleftarrow}{\overset{H_2SO_4}{\rightleftharpoons}} CH_3-\overset{\overset{\displaystyle O}{\|}}{C}-O(CH_2)_2CH(CH_3)_2 + H_2O$$

acetic acid      isoamyl alcohol          isoamyl acetate

This ester is an ingredient of banana oil, and it is also the alarm pheromone of the honeybee. **Pheromones** are communicatory chemicals secreted by organisms that cause a response in other organisms of the same type. After refluxing, the mixture is poured in water and then extracted with sodium bicarbonate to remove the unreacted acid. A second extraction with water is necessary to remove any traces of the basic bicarbonate which might reverse the reaction during the final distillation.

**Suggested References**    Carey, F. A., *Organic Chemistry.* 3rd ed., 619–20. New York: McGraw-Hill, 1996.

Loudon, G. Marc, *Organic Chemistry.* 3rd ed., 946–49. Redwood City, CA: Benjamin/Cummings, 1995.

McMurry, J., *Organic Chemistry.* 4th ed., 812–15. Pacific Grove, CA: Brooks/Cole, 1996.

Morrison, R. T., and R. N. Boyd, *Organic Chemistry.* 6th ed., 737–39, 761–62. Englewood Cliffs, NJ: Prentice-Hall, 1992.

Svoronos, P., E. Sarlo, and P. Kulas, "Equilibrium Constants of Esterification Reactions." *J. Chem. Education.* 67. (1990): 796.

Solomons, T. W. G., *Organic Chemistry.* 6th ed., 812–15. New York: John Wiley, 1996.

Vollhardt, K. P. C., and N. E. Schore, *Organic Chemistry.* 2nd ed., 740–43. New York: W. H. Freeman, 1994.

Wade, L. G., *Organic Chemistry.* 3rd ed., 962–64, 1003–05. Englewood Cliffs, NJ: Prentice-Hall, 1995.

 **SAFETY TIPS**

✔ Isoamyl alcohol is a flammable and irritating liquid. It should be dispensed with care in the fume hood, and all flames should be extinguished.

✔ Concentrated sulfuric acid is extremely corrosive and should be handled with care.

**Experimental**

1. Place 8.5 mL (d = 0.81 g/mL; 6.9 g) of isoamyl alcohol and 15.0 mL of acetic acid in a 100-mL round-bottomed flask.
2. Add dropwise 5 mL of concentrated sulfuric acid and a few boiling chips, and reflux for one hour.
3. Cool the round-bottomed flask in a cold-water bath, and then pour its contents into a beaker containing 40 mL of ice-cold water.
4. Transfer the mixture to a separatory funnel and mix thoroughly with occasional venting. If the two layers do not separate, add 2–3 g of solid sodium chloride, shake, and separate. The sodium chloride will reduce the water solubility of the ester (a process called **salting out**) and at the same time increase the density of the aqueous layer, thereby facilitating the separation of the two layers.
5. Extract the organic layer with 15 mL of 5% sodium bicarbonate (**Caution:** *Carbon dioxide gas is formed and must be vented frequently during the extraction*). Shake the funnel as before with proper ventilation of the evolving carbon dioxide gas.
6. Repeat the extraction with a second 15-mL portion of the bicarbonate solution. Check the aqueous layer to make sure it is neutral or slightly alkaline. If it is not, perform a third 15-mL bicarbonate solution extraction.
7. Finally, extract the organic layer with 15 mL of water, and then transfer the organic layer to a dry flask containing 1 g of anhydrous magnesium sulfate. Stopper the flask and occasionally stir.
8. Set up a simple distillation apparatus using a 100-mL round-bottomed flask. Gravity-filter the dried product into the distilling flask, and collect the portion that distills between 130° and 145°C in a preweighed vial or test tube.
9. Record the boiling point range, the mass, and the percent yield.
10. Run IR spectra for isoamyl alcohol, acetic acid and the product and compare.

**Questions**

1. Justify the purpose of the following in this experiment:
   a. using concentrated (and not dilute) sulfuric acid
   b. adding excess acetic acid
   c. extracting with sodium bicarbonate (step 5)
   d. extracting with water after extracting with sodium bicarbonate (step 7)
   e. using magnesium sulfate (step 7).
2. Would the reaction work if concentrated hydrochloric acid was used instead of sulfuric acid?
3. Why wouldn't concentrated sodium hydroxide (instead of sulfuric acid) give good results?
4. Give the structure of the product in the reaction below:

$$\text{(structure with } C\text{—}OH \text{ and } CH_2OH \text{ on benzene ring)} \xrightarrow{\text{SOCl}_2/\text{heat}} C_8H_6O_2$$

# 16.5   IMIDES: SYNTHESIS OF N-PHENYLPHTHALIMIDE

With the exception of acyl halides, anhydrides are by far the most reactive of the carboxylic acid derivatives. They are powerful acylating agents, and react with many different reagents to yield a variety of products.

$$\left( R - \underset{\underset{O}{\|}}{C} - \right)_2 O \; + \; H_2O \longrightarrow 2\, R - \underset{\underset{O}{\|}}{C} - OH$$

anhydride        water        carboxylic acid

$$\left( R - \underset{\underset{O}{\|}}{C} - \right)_2 O \; + \; R'OH \longrightarrow R - \underset{\underset{O}{\|}}{C} - OR' \; + \; R - \underset{\underset{O}{\|}}{C}OH$$

anhydride        alcohol        ester        carboxylic acid

$$\left( R - \underset{\underset{O}{\|}}{C} - \right)_2 O \; + \; 2R'NH_2 \longrightarrow R - \underset{\underset{O}{\|}}{C} - NHR' \; + \; R - \underset{\underset{O}{\|}}{C} - O^- \overset{+}{N}H_3R$$

anhydride        1° amine        amide        carboxylic acid salt

In this experiment, phthalic anhydride will be reacted with aniline, a weak nucleophile in acetic acid as the solvent, to give an N-phenylphthalimide.

phthalic anhydride   aniline      acetic      N–phenylphthalimide
                                  acid

Acetic acid not only acts as a solvent, but also promotes the nucleophilic attack by protonating the carbonyl group of the anhydride.

**Suggested References**   Carey, F. A., *Organic Chemistry.* 3rd ed., 836–37. New York: McGraw-Hill, 1996.

Loudon, G. Marc, *Organic Chemistry.* 3rd ed., 1005. Redwood City, CA: Benjamin/ Cummings, 1995.

McMurry, J., *Organic Chemistry.* 4th ed., 947–48. Pacific Grove, CA: Brooks/Cole, 1996.

Morrison, R. T., and R. N. Boyd, *Organic Chemistry.* 6th ed., 764–65. Englewood Cliffs, NJ: Prentice-Hall, 1992.

Solomons, T. W. G., *Organic Chemistry.* 6th ed., 821–22. New York: John Wiley, 1996.

Vollhardt, K. P. C., and N. E. Schore, *Organic Chemistry.* 2nd ed., 744. New York: W. H. Freeman, 1994.

**SAFETY TIPS**

✔ Aniline is a toxic chemical. Its vapors should not be inhaled, and skin contact should be avoided.

✔ Phthalic anhydride is a moisture-sensitive irritant, and caution should be used in handling it.

---

**Experimental**

1. Reflux 1.0 mL (d = 1.02 g/mL; 1.0 g) of aniline and 1.0 g phthalic anhydride in 15 mL of glacial acetic acid for 30 minutes.
2. Cool in ice and isolate the crystals via vacuum filtration.
3. Recrystallize from ethanol (approximately 50-60 mL) and determine the MP (literature value 210 °C), and the percent yield.
4. Record the IR spectra for both phthalic anhydride and N-phenylphthalimide and compare.

**Questions**

1. Complete the following reactions:

a.

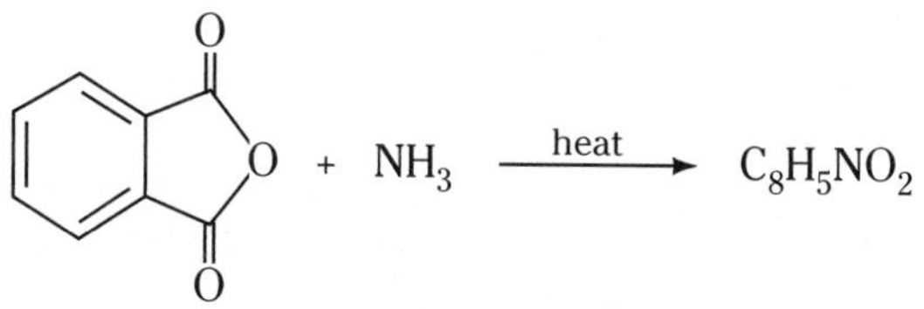

+ $NH_3$  $\xrightarrow{\text{heat}}$  $C_8H_5NO_2$

b.

+ $OH^-/H_2O$  $\xrightarrow[\text{then } H^+]{\text{heat}}$  $C_8H_6O_4$

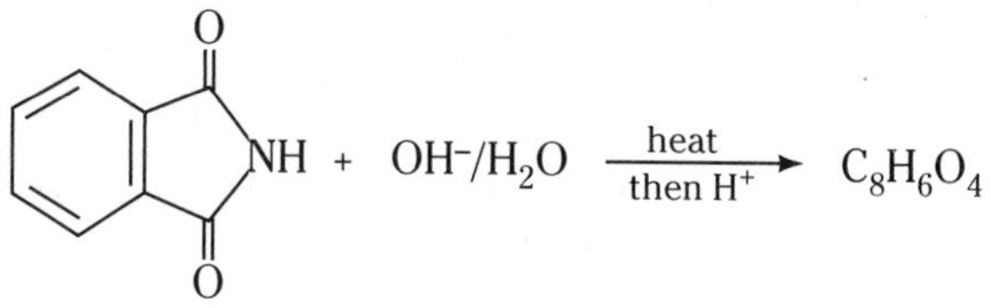

2. Suggest a mechanism for the synthesis of N-phenylphthalimide as performed in this experiment.

## 16.6 SAPONIFICATION OF AN ESTER

Esters can be converted to their constituent acids and alcohols by hydrolysis. **Saponification** is the hydrolysis of an ester in basic medium. It is an irreversible process, and for this reason, it is usually superior to acid-catalyzed hydrolysis which is reversible and gives smaller yields of product. In this experiment, methyl benzoate will be refluxed with a sodium hydroxide solution to yield the sodium salt of benzoic acid, and methanol. Because the solution is rather dense, boiling chips and careful heating are necessary to prevent severe bumping. Hydrochloric acid will be added slowly to neutralize the excess base as well as to precipitate the carboxylic acid.

**Suggested References**

Carey, F. A., *Organic Chemistry*. 3rd ed., 826-31. New York: McGraw-Hill, 1996.

Loudon, G. Marc, *Organic Chemistry*. 3rd ed., 989–90. Redwood City, CA: Benjamin/Cummings, 1995.

McMurry, J., *Organic Chemistry*. 4th ed., 825-28. Pacific Grove, CA: Brooks/Cole, 1996.

Morrison, R. T., and R. N. Boyd, *Organic Chemistry*. 6th ed., 772–75. Englewood Cliffs, NJ: Prentice-Hall, 1992.

Solomons, T. W. G., *Organic Chemistry*. 6th ed., 815–18. New York: John Wiley, 1996.

Vollhardt, K. P. C., and N. E. Schore, *Organic Chemistry*. 2nd ed., 781. New York: W. H. Freeman, 1994.

Wade, L. G., *Organic Chemistry*. 3rd ed., 1005–07. Englewood Cliffs, NJ: Prentice-Hall, 1995.

 **SAFETY TIPS**

✔ Methyl benzoate is an irritant and should be poured in a fume hood.
✔ Sodium hydroxide is caustic and should be handled with care.

**Experimental**

1. Place 2 g of sodium hydroxide pellets and 20 mL of water in a 100-mL round-bottomed flask, together with 0.9 mL (d = 1.09 g/mL; 1.0 g) of methyl benzoate and several (10 or 12) boiling chips.
2. Reflux for 45 minutes and let cool to room temperature.
3. Pour the contents into a 250-mL beaker, and acidify with concentrated hydrochloric acid until precipitation no longer occurs (pH = 4).
4. Cool in an ice-water bath for about 5 minutes.

5. Isolate the solid by vacuum filtration, and then recrystallize as follows: Transfer the solid to a 125-mL conical flask, add about 15–20 mL of water and 2 or 3 boiling chips. Heat to boiling and then gravity-filter the *hot* solution through a filter paper. Allow the filtrate to cool in an ice bath, and then collect the crystals by vacuum filtration.
6. Determine the MP (literature value 122°C) and the percent yield.
7. Run IR spectra for both methyl benzoate and benzoic acid and compare.

**Questions**

1. Suggest a mechanism for the basic hydrolysis of methyl benzoate.
2. What is the general formula for a soap? Write an equation for the reaction of ethyl stearate and sodium hydroxide. Explain briefly how a soap works.
3. Is sodium hydroxide a catalyst or a reactant? Explain.
4. Why is basic hydrolysis irreversible?
5. Provide the hydrolysis products in the reactions below:

a.

b.

c.

d.

# 17

# Aldehydes and Ketones

## 17.1 OXIDATION OF CYCLOHEXANOL TO CYCLOHEXANONE

Aldehydes and ketones can be prepared by oxidizing primary and secondary alcohols, respectively. Unless special precautions are taken, oxidation may proceed past the desired stage. Aldehydes must be removed from the reaction vessel as soon as they are formed because they can easily be oxidized to carboxylic acids. On the other hand, ketones are generally stable under the usual reaction conditions except when *enolization* is favored.

$$R - CH_2OH \xrightarrow{[O]} R - \overset{\overset{\displaystyle O}{\|}}{C} - H \xrightarrow{[O]} R - \overset{\overset{\displaystyle O}{\|}}{C} - OH$$

1° alcohol          aldehyde          carboxylic acid

$$\overset{R}{\underset{R'}{>}}CHOH \xrightarrow{[O]} \overset{R}{\underset{R'}{>}}C=O$$

2° alcohol          ketone

The most common oxidizing agent in the laboratory is acidified sodium dichromate, especially for synthesizing ketones from 2° alcohols. The solvent is usually water, but acetic acid or acetone can be used for water-insoluble alcohols.

In this experiment, cyclohexanone will be prepared by oxidizing cyclohexanol with sodium dichromate and sulfuric acid.

$$\text{cyclohexanol} \ -OH \ + \ Na_2Cr_2O_7 \xrightarrow{H^+} \ =O \ + \ Cr^{+3}$$

cyclohexanol   sodium dichromate          cyclohexanone

The reaction conditions call for a controlled temperature to prevent enolization and subsequent opening of the ring to give adipic acid.

$$=O \ \rightleftharpoons \ -OH \xrightarrow{[O]} \overset{COOH}{\underset{COOH}{}}$$

keto form          enol form          adipic acid

Oxalic acid is added at the end of the reaction to destroy any excess dichromate.

$$(COOH)_2 + Cr_2O_7^{-2}/H^+ \longrightarrow CO_2 + Cr^{+3}$$
oxalic acid

The mixture is then distilled, and the distillate (which consists of cyclohexanone and water) is treated with solid sodium chloride to saturate the aqueous layer and thus "salt out" the cyclohexanone. This "salting out" step is important for the recovery of the product, especially since the density of cyclohexanone (0.95 g/mL) is close to that of water.

**Suggested References**

Carey, F. A., *Organic Chemistry*. 3rd ed., 622–26. New York: McGraw-Hill, 1996.

Loudon, G. Marc, *Organic Chemistry*. 3rd ed., 467–69. Redwood City, CA: Benjamin/Cummings, 1995.

March, J., *Advanced Organic Chemistry*. 4th ed., 1167. New York: John Wiley, 1992.

McMurry, J., *Organic Chemistry*. 4th ed., 653–56. Pacific Grove, CA: Brooks/Cole, 1996.

Patnaik, P., *A Comprehensive Guide to the Hazardous Properties of Chemical Substances*. 80. New York: Van Nostrand Reinhold, 1992.

"Sodium Dichromate Dihydrate." In *Reagents for Organic Synthesis*, edited by L. F. Fieser and M. Fieser, 1059–61. New York: John Wiley, 1967.

Solomons, T. W. G., *Organic Chemistry*. 6th ed., 475. New York: John Wiley, 1996.

Vollhardt, K. P. C., and N. E. Schore, *Organic Chemistry*. 2nd ed., 253–54, 637–40, 731. New York: W. H. Freeman, 1994.

Wade, L. G., *Organic Chemistry*. 3rd ed., 433–37. Englewood Cliffs, NJ: Prentice-Hall, 1995.

---

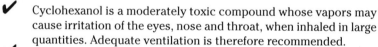 **SAFETY TIPS**

✔ Cyclohexanol is a moderately toxic compound whose vapors may cause irritation of the eyes, nose and throat, when inhaled in large quantities. Adequate ventilation is therefore recommended.

✔ Sodium dichromate is a toxic compound and suspected carcinogen. The use of gloves is highly recommended.

✔ All chromium-containing waste solutions should be placed in a container to be properly disposed by the instructor.

✔ Concentrated sulfuric acid is corrosive and should be handled with care.

✔ Oxalic acid is toxic when ingested.

---

**Experimental**

1. Prepare a mixture of sulfuric acid and ice by carefully adding 10 mL of concentrated sulfuric acid to 40 g of ice in a 125-mL Erlenmeyer flask.

2. Add 10.4 mL (d = 0.96 g/mL; 10.0 g) of cyclohexanol portionwise, keeping the temperature of the mixture below 25°C.

3. Dissolve 11.0 g of sodium dichromate in 15 mL of water in a second 125-mL conical flask.

4. Place the ice-acid-cyclohexanol flask (from step 2) in a beaker of ice and water, and *slowly* add the sodium dichromate solution 1 mL at a time (using a medicine dropper or disposable pipette) so that the temperature remains within the range of 30°–35°C. Note the change in color: The orange solution of dichromate turns into the green solution of chromic salts.

5. Add the last 2–3 mL of the dichromate solution in one portion, remove the flask from the ice-water bath and allow the temperature of the mixture to rise to 40°–50°C.
6. Let the mixture stand at room temperature until the temperature falls back to 30°–35°C, and then add 1 g of oxalic acid.
7. Transfer the mixture to a 250-mL round-bottomed flask, and set it up for a simple distillation. Rinse the Erlenmeyer flask with about 30 mL of water, and transfer the washings to the distilling flask using a funnel. Distill the mixture and collect the fraction that distills between 95° and 100°C (approximately 40 mL).
8. Transfer the distillate to a 125-mL Erlenmeyer flask, add 8 g of solid sodium chloride, and stir. Decant the water-cyclohexanone mixture into a separatory funnel without transferring any solid material along with the liquid. Separate and discard the bottom water layer.
9. Pour the upper cyclohexanone layer into a *dry* conical flask containing about 1 g of anhydrous magnesium sulfate. Cover the flask and let it stand for about 5 minutes.
10. Gravity-filter the mixture into a dry 100-mL round-bottomed flask (do not use a 50-mL or 25-mL round-bottomed flask, because a great degree of foaming will occur during the distillation). Perform a simple distillation *(do not forget to dry your setup after the distillation in step 7)*. Collect the fraction that distills between 150° and 156°C. Calculate the percent yield. Note that the equation must be balanced before the limiting reagent can be determined.
11. Perform a gas chromatogram for a water-cyclohexanol mixture, for pure cyclohexanone (provided by the instructor), and for your product. Determine the purity of the distillate.
12. Run and compare the IR spectra of both pure cyclohexanol and the final distillate.

**Questions**

1. In this experiment, what is the purpose of the following:
    a. not allowing the temperature to go above 35°C when sodium dichromate is added to the alcohol
    b. oxalic acid
    c. sodium chloride
    d. anhydrous magnesium sulfate.
2. Provide balanced equations for the oxidation of the following:
    a. 2-propanol to acetone, by means of $Cr_2O_7^{-2}/H^+$
    b. oxalic acid to carbon dioxide, by means of $Cr_2O_7^{-2}/H^+$
    c. benzaldehyde in basic $MnO_4^-$ to yield $MnO_2$ and the benzoate anion.
3. What are the products of the $Cr_2O_7^{-2}/H^+$ (excess) oxidation of the following compounds? If there is no reaction, state so.
    a. 2-methyl-2-butanol
    b. 3-methyl-2-butanol
    c. 3-methyl-1-butanol
    d. 3,3-dimethyl-2-butanol
    e. 3,3-dimethyl-1-butanol
    f. 2,3-dimethyl-1-butanol

## 17.2  REDUCTION OF CYCLOHEXANONE TO CYCLOHEXANOL

Sodium borohydride, $NaBH_4$, is an efficient reagent for reducing aldehydes, ketones, and acid chlorides to the corresponding alcohols.

$$R - \underset{\underset{O}{\|}}{C} - H \quad \xrightarrow[\text{2. } H^+/H_2O]{\text{1. } NaBH_4} \quad R - CH_2OH$$

aldehyde                    1° alcohol

$$R' - \underset{\underset{O}{\|}}{C} - R'' \quad \xrightarrow[\text{2. } H^+/H_2O]{\text{1. } NaBH_4} \quad R' - \underset{\underset{OH}{|}}{\overset{\overset{H}{|}}{C}} - R''$$

ketone                      2° alcohol

$$R - \underset{\underset{O}{\|}}{C} - Cl \quad \xrightarrow[\text{2. } H^+/H_2O]{\text{1. } NaBH_4} \quad R - CH_2OH$$

acyl halide                 1° alcohol

Other functional groups—such as the carbon-carbon double and triple bonds, the carbonyl group of esters, amides, lactones, and carboxylic acids, the nitro group, and the cyano group—are not reduced by $NaBH_4$.

Sodium borohydride is soluble in most protic solvents (such as water, methanol, and ethanol) with little evolution of hydrogen gas. Acidification leads to a much more vigorous reaction, with rapid hydrolysis of the compound. The reaction with the carbonyl group involves a hydride transfer from the borohydride to the carbonyl group, followed by work-up of the borate ester with water.

$$4 \underset{R'}{\overset{R}{>}}C=O + NaBH_4 \longrightarrow \left[ \left( H - \underset{\underset{R'}{|}}{\overset{\overset{R}{|}}{C}} - O \right)_4 B \right]^- Na^+ \xrightarrow{H_2O} 4 \, H - \underset{\underset{R'}{|}}{\overset{\overset{R}{|}}{C}} - OH + Na_3BO_3$$

ketone                          borate ester

This reagent has the stoichiometric advantage of reducing *four* mole equivalents of the carbonyl compound for each mole equivalent of sodium borohydride. Moreover, its low molecular weight (37.8) allows the use of a very small quantity of the reagent for the reduction.

In this experiment, cyclohexanone will be treated with sodium borohydride, and then acidified to yield cyclohexanol, which will be dried and then purified by distillation.

cyclohexanone              cyclohexanol

**Suggested References**     Carey, F. A., *Organic Chemistry.* 3rd ed., 608–11. New York: McGraw-Hill, 1996.

Loudon, G. Marc, *Organic Chemistry.* 3rd ed., 890–94. Redwood City, CA: Benjamin/Cummings, 1995.

McMurry, J., *Organic Chemistry.* 4th ed., 641–42, 723. Pacific Grove, CA: Brooks/Cole, 1996.

March, J., *Advanced Organic Chemistry.* 4th ed., 910–12. New York: McGraw-Hill, 1992.

"Sodium Borohydride." In *Reagents for Organic Synthesis,* edited by L. F. Fieser and M. Fieser, 1049–55. New York: John Wiley, 1967.

Solomons, T. W. G., *Organic Chemistry.* 6th ed., 406, 429–31. New York: John Wiley, 1996.

Vollhardt, K. P. C., and N. E. Schore, *Organic Chemistry.* 2nd ed., 251–53. New York: W. H. Freeman, 1994.

Wade, L. G., *Organic Chemistry.* 3rd ed., 419–21. Englewood Cliffs, NJ: Prentice-Hall, 1995.

 **SAFETY TIPS**

✔ Sodium borohydride is a corrosive, moisture-sensitive compound and should be handled with gloves. When not in use it should be stored in a dessicator.

✔ Ether is a flammable liquid that forms peroxides, which are sensitive to shock, sparks, heat and friction. All flames should be extinguished during the experiment. See also the Safety Tips of Experiment 14.1.

✔ Methanol is also a flammable liquid and is highly toxic if ingested.

✔ Sodium methoxide is a stong base and should be handled with care, preferably with plastic gloves.

**Experimental**

1. Dissolve 10.3 mL (d = 0.95 g/mL; 9.8 g) of cyclohexanone in 25 mL of methanol in a 100-mL round-bottomed flask.

2. Prepare a solution of 0.5 g of dry sodium methoxide and 2.0 g of sodium borohydride in 25 mL of anhydrous methanol. Add this solution to the cyclohexanone solution of step 1 while swirling the reaction flask.

3. Reflux for about 50 minutes, then pour the reaction mixture into a 400-mL beaker that contains 75 mL of ice and water. Add carefully 10 mL of 10% hydrochloric acid. If the solution is not acidic (pH = 4–6) add dropwise more hydrochloric acid until the desired pH is reached.

4. Transfer the mixture to a separatory funnel, and extract with 50 mL of ether. Wash the organic layer (which one is it?) with three 15-mL portions of water.

5. Dry the organic layer with about 1 g of *anhydrous* magnesium sulfate, and let stand while covered with a small watch glass for about 5 minutes.

6. Gravity-filter the solution into a 100-mL round-bottom flask, and carry out a simple distillation. Collect the fraction that distills at temperatures up to 100°C in a 50-mL round-bottom flask immersed in an ice-water bath. This will be mostly ether (BP range 34°–40°C). Transfer the distillate to a container labeled "waste ether." Continue the distillation and collect the cyclohexanol portion that boils in the 155°–165°C range in a preweighed vial. Calculate the percent yield.

7. Run IR spectra for both the reactant and the product and compare.

8. Peform a gas chromatogram for a water-cyclohexanone mixture, for pure cyclohexanol (provided by the instructor) and for your product. Is the prepared cyclohexanol free of any unreacted cyclohexanone?

**Questions**    1. Give the reaction products of sodium borohydride (followed by acidification) with the following compounds. If there is no reaction, state so.
a. acetone
b. 3-phenylpropanal
c. acetonitrile
d. 3-buten-2-one
e. ethyl acetate
f. *m*-nitroacetophenone
g. *p*-benzoquinone
h. ethyl acetoacetate
2. Explain the purpose of the following in this experiment:
a. dilute (and not concentrated) hydrochloric acid
b. water extraction (step 4)
c. anhydrous magnesium sulfate.
3. How could you distinguish between the organic and aqueous layers during the ether extraction?
4. Describe the differences in the IR spectra of the reactant and product in this experiment.
5. Describe the results of the GC experiment. Did it confirm the identity of your product?

## 17.3 ACETAL FORMATION: SYNTHESIS OF 4,5-DIMETHYLDIOXOLANE

When an aldehyde (or ketone) is reacted with excess alcohol in an acidic medium, an **acetal** (or **ketal**) is produced. First, the intermediate hemiacetal (or hemiketal) is formed in a series of *acid-catalyzed* equilibria:

$$
\underset{\text{aldehyde}}{R-\overset{\overset{\displaystyle O}{\|}}{C}-H} + \underset{\text{alcohol}}{R'OH} \underset{H^+}{\rightleftharpoons} \underset{\text{hemiacetal}}{R-\overset{\displaystyle OH}{\underset{\displaystyle H}{C}}-OR'} \underset{R'OH/H^+}{\rightleftharpoons} \underset{\text{acetal}}{R-\overset{\displaystyle OR'}{\underset{\displaystyle H}{C}}-OR'} + H_2O
$$

$$
\underset{\text{ketone}}{R_2C=O} + \underset{\text{alcohol}}{R'OH} \underset{H^+}{\rightleftharpoons} \underset{\text{hemiketal}}{R_2\overset{\displaystyle}{\underset{\displaystyle OH}{C}}-OR'} \underset{R'OH/H^+}{\rightleftharpoons} \underset{\text{ketal}}{R_2\overset{\displaystyle}{\underset{\displaystyle OR'}{C}}-OR'} + H_2O
$$

Acetals (or ketals) are frequently used to protect the carbonyl group during the course of a multistep synthesis. They are stable in neutral or alkaline media, but hydrolyze readily in acidic solution. For example, they are unreactive towards the highly basic Grignard reagent. Hemiketals and ketals are generally more difficult to prepare than their aldehyde counterparts. This is shown by unfavorable equilibrium constants for ketal formation. The major reason for this is that the entropy of the reaction of ketones with alcohols is less favorable than that of the reaction of aldehydes. To overcome this difficulty, ketones are often reacted with 1,2- or 1,3-diols to form cyclic ketals. For instance,

$$
\begin{array}{c} CH_3CH_2 \\ \\ CH_3CH_2 \end{array}\!\!\!C=O \; + \; \begin{array}{c} HO-CH_2 \\ | \\ HO-CH_2 \end{array} \xrightarrow{H^+} \begin{array}{c} CH_3CH_2 \\ \\ CH_3CH_2 \end{array}\!\!\!C\!\!\begin{array}{c} O-CH_2 \\ | \\ O-CH_2 \end{array} \; + \; H_2O
$$

In this experiment, formaldehyde (in the form of paraformaldehyde) will be reacted with 2,3-butanediol in the presence of sulfuric acid. Once it is formed, the acetal will be isolated by azeotropic distillation to prevent its hydrolysis.

$$
\begin{array}{c} CH_3 \\ | \\ (CHOH)_2 \\ | \\ CH_3 \end{array} \; + \; H_2C=O \; \rightleftharpoons \; \begin{array}{c} H_3C \\ | \\ HC-O \\ | \qquad\;\; CH_2 \\ HC-O \\ | \\ H_3C \end{array} \; + \; H_2O
$$

2,3-butanediol    formaldehyde    4,5-dimethyldioxolane

The distillate will then be treated with solid sodium chloride to salt out the product which is partially soluble in water. 4,5-Dimethydioxolane will then be treated with anhydrous sodium carbonate to remove any traces of both acid and water and thus avoid the reversal of the reaction during the final distillation.

**Suggested References**

Carey, F. A., *Organic Chemistry*. 3rd ed., 696–700. New York: McGraw-Hill, 1996.

Loudon, G. Marc, *Organic Chemistry*. 3rd ed., 898–902. Redwood City, CA: Benjamin/Cummings, 1995.

McMurry, J., *Organic Chemistry*. 4th ed., 740–43. Pacific Grove, CA: Brooks/Cole, 1992.

Solomons, T. W. G., *Organic Chemistry*. 6th ed., 724–26. New York: John Wiley, 1996.

Vollhardt, K. P. C., and N. E. Schore, *Organic Chemistry*. 2nd ed., 645–48. New York: W. H. Freeman, 1994.

Wade, L. G., *Organic Chemistry*. 3rd ed., 843–50. Englewood Cliffs, NJ: Prentice-Hall, 1996.

---

⚠️ **SAFETY TIPS**

✔ Paraformaldehyde is a moisture-sensitive irritant and should be handled with gloves.

✔ Sulfuric acid is corrosive and should be handled with care.

---

**Experimental**

1. Place 100 mL water in a 250-mL round bottom flask and then add 6 mL concentrated sulfuric acid while cooling in ice.
2. Add 6 g of paraformaldehyde, 13.6 mL (d = 0.99 g/mL; 13.5 g) of 2,3-butanediol, a few boiling chips and perform a simple distillation.
3. Collect 30–35 mL of the 4,5-dimethyldioxolane/water mixture in a 50-mL round bottom flask immersed in an ice-water bath. The temperature of the distillate will rise to 100°–105°C.
4. Transfer the distillate to a 125-mL separatory funnel and add solid sodium chloride until the water layer is saturated. Separate the water layer (which one is it?) and discard.
5. Pour the organic layer into a *dry,* stoppered 50-mL conical flask containing about 3 g *anhydrous* sodium carbonate and let stand for about 20 minutes.
6. Gravity filter the organic layer into a dry 50-mL round bottom flask and distill into a pre-weighed vial or test tube. Collect the fraction that distills over the range 90°–105°C.
7. Calculate the % yield.
8. Run a gas chromatogram and calculate the purity of the compound.

**Questions**

1. Propose a mechanism for the acid-catalyzed synthesis of 4,5-dimethyldioxolane from formaldehyde and 2,3-butanediol.
2. Formaldehyde (BP –6°C) is commercially available as a 37% aqueous solution. Yet solid paraformaldehyde, a polymer of gaseous formaldehyde, is used in this experiment. Explain.
3. Use the principle of acetal/ketal formation to identify the products of the following reactions.

a. 
$$HO-CH_2$$
$$\quad\quad\;\,|$$
$$HO-CH \;\; + \;\; H_2C{=}O \;\; \xrightarrow{H^+} \;\; C_4H_8O_2 \;+\; H_2O$$
$$\quad\quad\;\,|$$
$$\quad\quad\;\,CH_3$$

b. $3\,CH_3CHO \;\; \xrightarrow{H^+} \;\; C_6H_{12}O_3$

4. Identify the following as hemiacetals, acetals, hemiketals or ketals:

5. Identify the aldehydes or ketones from which the above compounds are prepared.
6. As isolated, 4,5-dimethyldioxolane is optically inactive; yet, it has chiral centers. Explain why the product is optically inactive. How many stereoisomers are present?
7. Compare the ketalization of acetone when reacted with 2 moles of ethyl alcohol, and when reacted with 1 mole of 1,2-ethanediol, with regard to the following:
   a. change in entropy
   b  change in free energy
   c. equilbrium constant.

## 17.4  THE PINACOL REARRANGEMENT: A ¹H NMR STUDY

When treated with strong acids, 1,2-diols yield not only the expected alkenes (or dienes), but also ketones via the so-called **pinacol rearrangement**. This classic reaction involves the formation of a carbocation, which then rearranges.

In this experiment, 2,3-dimethyl-2,3-butanediol (also known as *pinacol*) will be converted to 3,3-dimethyl-2-butanone (also known as *pinacolone* and *t*-butyl methyl ketone).

$$
\underset{\substack{\text{2,3-dimethyl-2,3-butanediol}\\\text{(pinacol)}}}{\begin{array}{c} H_3C \quad CH_3 \\ | \qquad | \\ CH_3-\overset{|}{\underset{|}{C}}-\overset{|}{\underset{|}{C}}-CH_3 \ + \ H^+ \\ HO: \quad OH \end{array}}
\longrightarrow
\begin{array}{c} H_3C \quad CH_3 \\ CH_3-\overset{|}{\underset{|}{C}}-\overset{|}{\underset{|}{C}}-CH_3 \\ HO: \quad OH_2 \\ \qquad \qquad + \end{array}
\xrightarrow[\text{:CH}_3 \text{ shift}]{-H_2O}
\begin{array}{c} CH_3 \\ | \\ CH_3-\overset{|}{\underset{|}{C}}-\overset{|}{\underset{|}{C}}-CH_3 \\ \| \quad | \\ HO: \quad CH_3 \\ + \end{array}
$$

$$
\underset{\substack{\text{3,3-dimethyl-2-butanone}\\\text{(pinacolone)}}}{\begin{array}{c} CH_3 \\ | \\ CH_3-\overset{|}{\underset{\|}{C}}-\overset{|}{\underset{|}{C}}-CH_3 \\ O: \quad CH_3 \end{array}}
\xleftarrow{\ -H^+\ }
$$

The reaction proceeds by a 1,2 shift of a methyl group with its electrons (i.e., a methide shift). In this case, only one rearrangement product is possible. When different groups are present, the migration follows the following sequence:

$$C_6H_5 \text{ (aryl)} > R \text{ (alkyl)}$$

Rearrangement of a hydrogen (as a hydride, :H⁻) is unpredictable; the hydrogen may or may not migrate in preference to an alkyl or aryl group.

**Suggested References**     Ege, S. N., *Organic Chemistry*. 691. Lexington: D. C. Heath, 1989.

Loudon, G. Marc, *Organic Chemistry*. 3rd ed., 904–07, 915–17. Redwood City, CA: Benjamin/Cummings, 1995.

March, J., *Advanced Organic Chemistry*. 4th ed., 1058–59. New York: John Wiley, 1992.

Wade, L. G., *Organic Chemistry*. 3rd ed., 454–55. Englewood Cliffs, NJ: Prentice-Hall, 1995.

**Experimental**

1. In a 100-mL round-bottomed flask, place 5.9 g of anhydrous pinacol, 20 mL of a 20% sulfuric acid solution, and set up the apparatus for a simple distillation.
2. Distill approximately 10–15 mL of the pinacolone-water azeotrope mixture, using a 25-mL round-bottomed flask as the collecting flask immersed in an ice-water bath.
3. Transfer the distillate to a separatory funnel, and separate the organic (top) layer from the aqueous (bottom) layer.
4. Extract the organic layer with 10 mL of 10% sodium bicarbonate to remove any traces of the acid.
5. Transfer the organic layer to a 50-mL conical flask, add 0.5 g anhydrous sodium sulfate, cover with a watch glass, and let dry for about 5 minutes.

6. Gravity-filter into a 50-mL round-bottomed flask, and perform a simple distillation, using *dry* glassware.

7. Collect the fraction that distills between 100° and 106°C in a preweighed vial (or test tube). Record the mass, percent yield, and BP range. Obtain an IR spectrum.

8. If an NMR instrument is available, collect the fraction between 69° and 106°C, and run a $^1$H NMR spectrum. Calculate the relative percentages of pinacolone and 2,3-dimethyl-1,3-butadiene in the product (see NMR section pp. 105–112). Also, perform a gas chromatogram, and determine the relative percentages of pinacolone and 2,3-dimethyl-1,3-butadiene from it.

**Questions**

1. The by-product of this experiment's reaction is 2,3-dimethylbutadiene. The boiling point is 69°–70°C.

   a. Give a detailed mechanism that explains the formation of 2,3-dimethylbutadiene.

   b. If a few drops of the final distillation product are collected at 69°C, what simple chemical test can you employ to identify that fraction as the diene or the pinacolone?

   c. Describe the expected NMR spectra of both pinacolone and 2,3-dimethylbutadiene.

   d. In what way would the IR spectra of pinacolone and 2,3-dimethylbutadiene differ?

2. Write equations to show how one could prepare pivalic acid, $(CH_3)_3C$—COOH, starting with pinacol.

3. The $^1$H NMR spectrum of the pinacolone-2,3-dimethylbutadiene mixture includes signals at 2.2 and 5.0 ppm. Identify these signals.

## 17.5 QUALITATIVE TESTS FOR ALDEHYDES AND KETONES

Both aldehydes ($R-\overset{\overset{\displaystyle O}{\|}}{C}-H$) and ketones ($R-\overset{\overset{\displaystyle O}{\|}}{C}-R$) contain the carbonyl group, and both undergo **nucleophilic addition reactions**, with a nucleophile, Nu⁻:

$$R-\overset{\overset{\displaystyle O}{\|}}{C}-H \;+\; H^+/Nu^-\text{:} \;\rightleftharpoons\; R-\overset{\overset{\displaystyle OH}{|}}{\underset{\underset{\displaystyle Nu}{|}}{C}}-H$$

aldehyde

$$R-\overset{\overset{\displaystyle O}{\|}}{C}-R' \;+\; H^+/Nu^-\text{:} \;\rightleftharpoons\; R-\overset{\overset{\displaystyle OH}{|}}{\underset{\underset{\displaystyle Nu}{|}}{C}}-R'$$

ketone

On the other hand, carboxylic acids ($R-\overset{\overset{\displaystyle O}{\|}}{C}-G$, G =—OH) and their derivatives that also contain a carbonyl group—such as acyl halides (G = halogen), esters (G = —OR′), amides (G = $NH_2$), and anhydrides ($G=-O-\overset{\overset{\displaystyle O}{\|}}{C}-R$)— are characterized by **addition-elimination** reactions when treated with nucleophiles. For instance,

$$R-\overset{\overset{\displaystyle O}{\|}}{C}-G \;+\; H^+/Nu^-\text{:} \;\rightleftharpoons\; R-\overset{\overset{\displaystyle O}{\|}}{C}-Nu \;+\; GH$$

Aldehydes are much more reactive than ketones towards oxidizing agents (such as $AgNO_3$ in ammonia, $Cr_2O_7^{-2}/H^+$, $MnO_2$ or $Cu^{+2}$) and yield the corresponding carboxylic acids under conditions where ketones are unreactive. Thus,

$$R-\overset{\overset{\displaystyle O}{\|}}{C}-H \;\xrightarrow{[O]}\; R-\overset{\overset{\displaystyle O}{\|}}{C}-OH$$

aldehyde          carboxylic acid

while

$$R-\overset{\overset{\displaystyle O}{\|}}{C}-R' \;\xrightarrow{[O]}\; \text{No reaction}$$

ketone

where [O] is one of the oxidizing agents mentioned previously.

Note that aromatic ketones of the form $Ar-\overset{\overset{\displaystyle O}{\|}}{C}-R$ (Ar = aryl group) are also susceptible to strong oxidizing agents (e.g., $Cr_2O_7^{-2}/H^+$, $MnO_4^-/OH^-$), and yield the corresponding benzoic acid (Ar—COOH).

In this experiment, a series of tests will be performed on a known aldehyde, a known ketone, and an unknown compound that will be identified from the compounds listed in the table accompanying this section.

## Tests to Distinguish between an Aldehyde and a Ketone

Aldehydes can be distinguished from ketones by their reaction toward the following mild oxidizing agents:

i.  Silver oxide in ammonia (to yield a silver mirror), otherwise known as **Tollens' test**.

$$R-\overset{\overset{\displaystyle O}{\|}}{C}-H \;+\; Ag_2O/OH^-/NH_3 \longrightarrow R-\overset{\overset{\displaystyle O}{\|}}{C}-O^- \;+\; Ag$$

aldehyde      silver oxide      carboxylate  silver
anion     mirror

ii. Buffered (blue) cupric oxide solution (to yield the red cuprous oxide) otherwise known as **Fehling's** or **Benedict's test**.

$$R-\overset{\overset{\displaystyle O}{\|}}{C}-H \;+\; CuO \longrightarrow R-\overset{\overset{\displaystyle O}{\|}}{C}-O^- \;+\; Cu_2O$$

aldehyde    cupric        carboxylate   cuprous
oxide (blue)     anion   oxide (red ppt)

iii. Orange acidic dichromate solution (to yield green chromic salts), otherwise known as the **Jones oxidation test**.

$$R-\overset{\overset{\displaystyle O}{\|}}{C}-H \;+\; Cr_2O_7^{-2} \longrightarrow R-\overset{\overset{\displaystyle O}{\|}}{C}-OH \;+\; Cr^{+3}$$

aldehyde  dichromate       carboxylic  chromium (III)
anion (orange)     acid   cation (green)

**Suggested References**    Carey, F. A., *Organic Chemistry*. 3rd ed., 709–10. New York: McGraw-Hill, 1996.

Loudon, G. Marc, *Organic Chemistry*. 3rd ed., 882–83. Redwood City, CA: Benjamin/Cummings, 1995.

McMurry, J., *Organic Chemistry*. 4th ed., 723–24. Pacific Grove, CA: Brooks/ Cole, 1996.

Morrison, R. T., and R. N. Boyd, *Organic Chemistry*. 6th ed., 671, 675–77, 1149–50. Englewood Cliffs, NJ: Prentice-Hall, 1992.

Solomons, T. W. G., *Organic Chemistry*. 6th ed., 740–42. New York: John Wiley, 1996.

Vollhardt, K. P. C., and N. E. Schore, *Organic Chemistry*. 2nd ed., 662–63, 651–53, 866–67. New York: W. H. Freeman, 1994.

Wade, L. G., *Organic Chemistry*. 3rd ed., 850–51. Englewood Cliffs, NJ: Prentice-Hall, 1995.

### ⚠ SAFETY TIPS

✔ Silver nitrate is a toxic oxidant and a skin-stainer. Its solutions, therefore, should be handled with gloves.

✔ Sodium dichromate is a toxic compound, and the use of plastic gloves is recommended.

---

**Experimental**

Perform the following tests on a known aldehyde, a known ketone, and an unknown.

## Tollens' Test

Place 2 mL of a 5% $AgNO_3$ solution in a freshly cleaned test tube. Add dropwise 5% NaOH solution until a black precipitate forms. Dissolve the precipitate by adding dropwise with stirring dilute $NH_4OH$ until the solution *just* becomes clear. Avoid adding excess ammonia. Divide the solution into three parts. Add 10–15 drops of the compound to be tested to each of the parts, *cork* and heat for about one minute in a 60°C water bath. *The test is very sensitive to traces of aldehyde vapors, so special care should be taken to avoid contaminating the unknown test tube.* Record your results.

## Fehling's Test

Mix 5 mL of Fehling's solution A (see Appendix IV) with 5 mL of Fehling's solution B (see Appendix IV) in a test tube. Divide the resulting solution in three parts, and add 1 mL of each of the compounds to be tested to each of the parts. Cork the three test tubes, place them in a 60°C water bath and heat for about 10 minutes. A positive test is indicated by a change in the blue color of the solution and the slow formation of a fine brick-red $Cu_2O$ precipitate. Record your results.

## Benedict's Test

Add 1 mL of each of the compounds to be tested to 3 mL of Benedict's solution (see Appendix IV) contained in test tubes. Cork the three test tubes, place them in a 60°C water bath, and heat for about 10 minutes. A positive test is indicated by a change in the blue color of the reagent and the slow formation of a fine brick-red $Cu_2O$ precipitate. Record your results.

## Jones Oxidation

Perform the test as described in Section 14.2, Qualitative Tests for Alcohols. Substitute for the alcohol a known aldehyde, a known ketone and the unknown. Beware that prolonged heating of aryl ketones and cyclic ketones will give positive results also.

## Preparation of derivatives

Organic compounds can generally be identified by making solid derivatives that can be easily purified and that possess a sharp melting point in the range of 80°–240°C. Aldehydes and ketones form crystalline condensation products with ammonia derivatives (see Table 17.1). The 2,4-dinitrophenylhydrazones are generally the easiest compounds to crystallize.

Prepare the 2,4-dinitrophenylhydrazone for the given unknown *only*. If the results of the MP determination are inconclusive, prepare the semicarbazone as described in the experimental section.

**Table 17.1**  Reaction of aldehydes and ketones, $R\!-\!\overset{\overset{\displaystyle O}{\|}}{C}\!-$, with ammonia derivatives

$$R\!-\!\overset{\overset{\displaystyle O}{\|}}{C}\!- \; + \; H_2N\!-\!H \;\xrightarrow{H^+}\; R\!-\!\overset{\overset{\displaystyle N\,-\,H}{\|}}{C}\!- \; + \; H_2O$$

ammonia                                imine

$$R\!-\!\overset{\overset{\displaystyle O}{\|}}{C}\!- \; + \; H_2N\!-\!NH_2 \;\xrightarrow{H^+}\; R\!-\!\overset{\overset{\displaystyle N\,-\,NH_2}{\|}}{C}\!- \; + \; H_2O$$

hydrazine                              hydrazone

$$R\!-\!\overset{\overset{\displaystyle O}{\|}}{C}\!- \; + \; H_2N\!-\!NH\!-\!\bigcirc \;\xrightarrow{H^+}\; R\!-\!\overset{\overset{\displaystyle N\,-\,NH-\bigcirc}{\|}}{C}\!- \; + \; H_2O$$

phenylhydrazine                        phenylhydrazone

$$R\!-\!\overset{\overset{\displaystyle O}{\|}}{C}\!- \; + \; H_2N\!-\!NH\!-\!\bigcirc^{O_2N}_{\;\;\;\;NO_2} \;\xrightarrow{H^+}\; R\!-\!\overset{\overset{\displaystyle N\,-\,NH-\bigcirc^{O_2N}_{\;NO_2}}{\|}}{C}\!- \; + \; H_2O$$

2,4-dinitrophenyl                      2,4-dinitrophenyl hydrazone
hydrazine                              (2,4-DNP derivative)

$$R\!-\!\overset{\overset{\displaystyle O}{\|}}{C}\!- \; + \; H_2N\!-\!OH \;\xrightarrow{H^+}\; R\!-\!\overset{\overset{\displaystyle N\,-\,OH}{\|}}{C}\!- \; + \; H_2O$$

hydroxylamine                          oxime

$$R\!-\!\overset{\overset{\displaystyle O}{\|}}{C}\!- \; + \; H_2N\!-\!NHCNH_2 \;\xrightarrow{H^+}\; R\!-\!\overset{\overset{\displaystyle N\,-\,NHCNH_2}{\|}}{C}\!- \; + \; H_2O$$

semicarbazide                          semicarbazone

**Suggested References**  Carey, F. A., *Organic Chemistry.* 3rd ed., 700–01. New York: McGraw-Hill, 1996.

Horak, V., and R. F. X. Klein, "Microscale Group Test for Carbonyl Compounds." *J. Chem. Education* 62(1985): 806.

McMurry, J., *Organic Chemistry.* 4th ed., 733–36. Pacific Grove, CA: Brooks/Cole, 1996.

Morrison, R. T., and R. N. Boyd, *Organic Chemistry.* 6th ed., 673, 679–80. Englewood Cliffs, NJ: Prentice-Hall, 1992.

Solomons, T. W. G., *Organic Chemistry.* 6th ed., 728–32. New York: John Wiley, 1996.

Wade, L. G., *Organic Chemistry.* 3rd ed., 851–53. Englewood Cliffs, NJ: Prentice-Hall, 1995.

**Experimental** **2,4-Dinitrophenylhydrazone**

Place 4.0 mL of the 2,4-dinitrophenylhydrazine reagent (see Appendix IV) in a test tube, and add 1–2 mL of the unknown compound to be tested. Isolate the orange-yellow precipitate by vacuum filtration, and recrystallize from a minimal amount (approximately 10 mL) of hot ethanol. Determine the MP and compare it with those on the list of compounds in Table 17.2. Write the formulas for each of the compounds listed in Table 17.2 on a separate sheet of paper and attach it to your report.

**Semicarbazone**

In a large test tube place 1.0 g of semicarbazide hydrochloride and 10 mL of distilled water. Add 1.5 g of sodium acetate and 1 mL of the unknown compound. Stopper the test tube with a cork, and shake it vigorously for about one minute. Place the test tube in an ice-water bath, isolate the crystals via vacuum filtration, determine the MP, and compare the MP with those on the list of compounds in Table 17.2.

**Table 17.2**  Physical properties of aldehydes and ketones and their derivatives

| Compound | Formula | BP | 2,4-DNP MP | Semicarbazone MP |
|---|---|---|---|---|
| *Aldehydes* | | | | |
| Formaldehyde | HCHO | −21[a] | 167 | 169[b] |
| Acetaldehyde | | 21 | 168 | 162 |
| Propionaldehyde | | 50 | 154 | 89; 154 |
| *n*-Butyraldehyde | | 74 | 123 | 106 |
| Isobutyraldehyde | | 64 | 187 | 125 |
| *n*-Valeraldehyde | | 103 | 98; 107 | 108 |
| *n*-Caproaldehyde | | 131 | 107 | 106 |
| Benzaldehyde | | 179 | 237 | 222 |
| Salicylaldehyde | | 197 | 252[b] | 231 |
| *p*-Anisaldehyde | | 248 | 254 | 209 |
| *o*-Chlorobenzaldehyde | | 213 | 209 | 146; 229 |
| | | | | |
| *Ketones* | | | | |
| Acetone | $CH_3-\overset{\overset{\displaystyle O}{\|}}{C}-CH_3$ | 56 | 128 | 187 |
| Butanone | | 80 | 115 | 146 |
| 2-Pentanone | | 102 | 143 | 112 |
| 3-Pentanone | | 102 | 156 | 139 |
| 3-Methyl-2-butanone | | 94 | 120 | 113 |
| 3,3-Dimethyl-2-butanone | | 106 | 125 | 157 |
| 4-Methyl-2-pentanone | | 119 | 95 | 133 |
| Cyclopentanone | | 131 | 146 | 210 |
| Cyclohexanone | | 156 | 162 | 166 |
| Acetophenone | | 202 | 237; 250 | 198 |
| Benzophenone | | 48[c] | 239 | 165 |

[a]Gas at room temperature; normally handled as a 37% aqueous solution
[b]Decomposes
[c]Melting point

### The Iodoform test

A positive iodoform test is characteristic not only for alcohols that carry the group

$$CH_3 - \underset{\underset{R}{\displaystyle |}}{\overset{\overset{H}{\displaystyle |}}{C}} - OH \qquad \text{R = alkyl, aryl or hydrogen}$$

but also for carbonyl compounds that are of the form

$$CH_3 - \overset{\overset{O}{\displaystyle \|}}{C} - R \qquad \text{R = alkyl, aryl or hydrogen}$$

Therefore, the test can distinguish between methyl ketones (or acetaldehyde) and non-methyl carbonyl compounds. The yellow iodoform, $CHI_3$, precipitates out of solution with slight warming of the test tube.

$$CH_3 - \underset{\underset{O}{\displaystyle \|}}{C} - R + I_2/KI/OH^- \longrightarrow {}^-O - \underset{\underset{O}{\displaystyle \|}}{C} - R + CHI_3$$

---

**Experimental**

Perform the iodoform test (as described in Chapter 14, pp. 182) on a known compound that gives a positive iodoform reaction (e.g., acetone, ethyl or isopropyl alcohol) and also on your unknown.

**Questions**

1. Give each of the aldehydes in Table 17.2 an IUPAC name.
2. Give each of the ketones in Table 17.2 a common name.
3. Which of the compounds in Table 17.2 gives a positive iodoform test?
4. How many aldehydes give a positive iodoform test? Give the structure and name for each aldehyde.
5. Acetophenone and benzophenone form 2,4-dinitrophenylhydrazones that have similar melting points (237°C and 239°C, respectively). How could someone distinguish between them without preparing a semicarbazone?
6. Formaldehyde and acetaldehyde form 2,4-dinitrophenylhydrazones that have similar melting points (167°C and 168°C, respectively). How could someone distinguish between them without preparing a semicarbazone?
7. Acetaldehyde and cyclohexanone form semicarbazones that have similar melting points (168°C and 162°C, respectively). Suggest two ways that would help distinguish between them without preparing a 2,4-dinitrophenylhydrazone.
8. Suggest two different chemical tests that would help distinguish between the following pairs of compounds. State what you would do and what you would see.
   a. benzaldehyde and benzyl alcohol
   b. 2-pentanone and 3-pentanone
   c. propanal and diethyl ether
   d. acetophenone and phenylacetaldehyde
   e. propanal and propanoic acid
9. Suggest an IR frequency that would help distinguish between each pair of compounds in Question 8.

Name: ______________________________

Unknown #: ___________________________

# DATA SHEET: QUALITATIVE TESTS FOR ALDEHYDES AND KETONES

**Tollens' Test**

*Observations*

Known aldehyde______________________________________________

Known ketone________________________________________________

Unknown____________________________________________________

Conclusion:__________________________________________________

**Fehling's Test**

*Observations*

Known aldehyde______________________________________________

Known ketone________________________________________________

Unknown____________________________________________________

Conclusion:__________________________________________________

**Benedict's Test**

*Observations*

Known aldehyde______________________________________________

Known ketone________________________________________________

Unknown ____________________________________________________

Conclusion:__________________________________________________

**Iodoform Test**

*Observations*

Known compound______________________________________________

Unknown____________________________________________________

Conclusion:__________________________________________________

**2,4-Dinitrophenylhydrazone** MP: _______ °C

***Conclusion:*** Unknown # _______________ is ______________________

# chapter 18

# Carbanions and α,β-Unsaturated Carbonyls

## 18.1 THE ALDOL CONDENSATION: SYNTHESIS OF DIBENZALACETONE

Base-catalyzed condensations involve **carbanion intermediates** that are readily formed via abstraction of an α-hydrogen and stabilized by resonance.

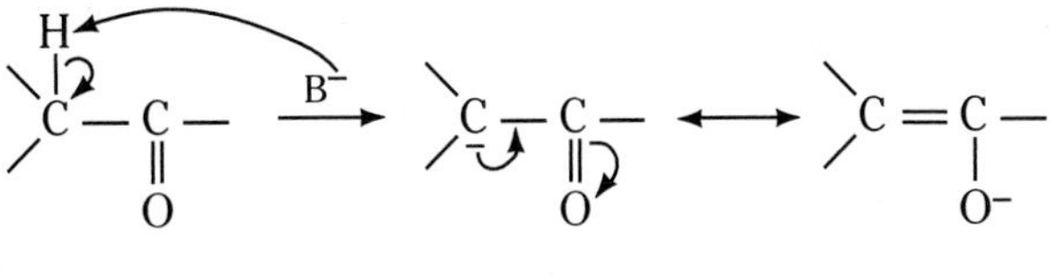

where B⁻ = base (e.g., OH⁻, $CH_3O^-$, $CH_3COO^-$, $C_2H_5O^-$).

The carbanion in turn can react with electropositive carbon atoms, thus creating new carbon-carbon bonds. Examples include the following:

*Aldol condensation*

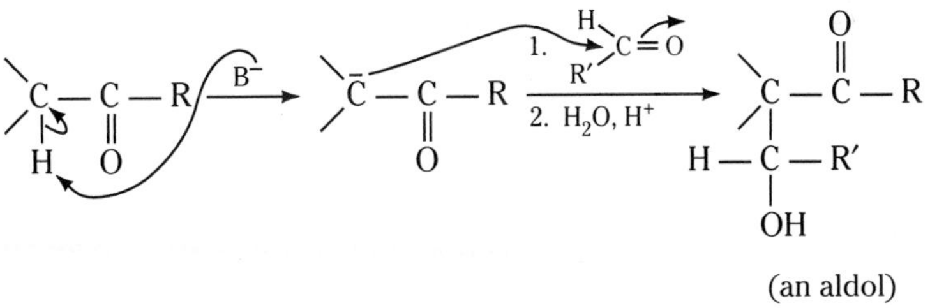

(an aldol)

*Claisen condensation*

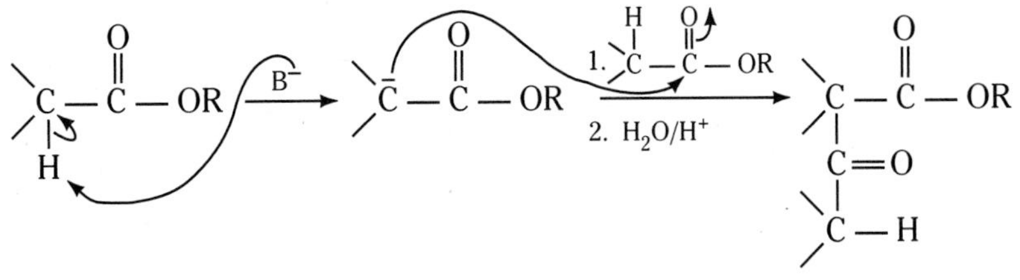

(a β-ketoester)

**243**

*Perkin condensation*

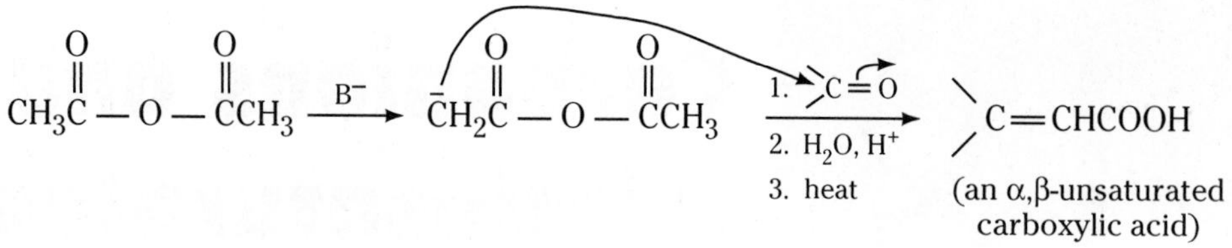

$$CH_3\overset{\overset{\displaystyle O}{\|}}{C}-O-\overset{\overset{\displaystyle O}{\|}}{C}CH_3 \xrightarrow{B^-} \ ^-\overset{}{C}H_2\overset{\overset{\displaystyle O}{\|}}{C}-O-\overset{\overset{\displaystyle O}{\|}}{C}CH_3 \xrightarrow[\text{2. H}_2\text{O, H}^+]{\text{1. } >\!\!C=\!O \\ \text{3. heat}} \ >\!\!C=CHCOOH$$

(an α,β-unsaturated
carboxylic acid)

*Knoevenagel condensation*

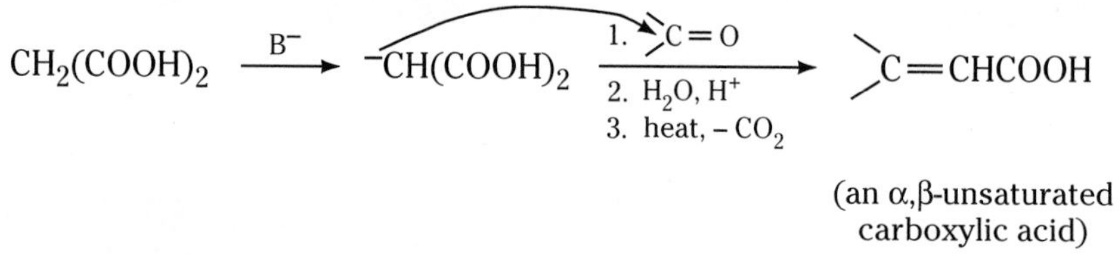

$$CH_2(COOH)_2 \xrightarrow{B^-} \ ^-CH(COOH)_2 \xrightarrow[\substack{\text{2. H}_2\text{O, H}^+ \\ \text{3. heat, } -\text{CO}_2}]{\text{1. } >\!\!C=\!O} \ >\!\!C=CHCOOH$$

(an α,β-unsaturated
carboxylic acid)

In the aldol condensation, a hydrogen and a base (e.g., sodium hydroxide, sodium methoxide, or sodium ethoxide) react to produce a carbanion **(enolate)**, which reacts with the electrophilic carbonyl to yield a β-hydroxyketone (or aldehyde). When the pH is considerably acidic or basic, water is spontaneously eliminated to yield an α,β-unsaturated ketone (or aldehyde), especially if this leads to a conjugated system.

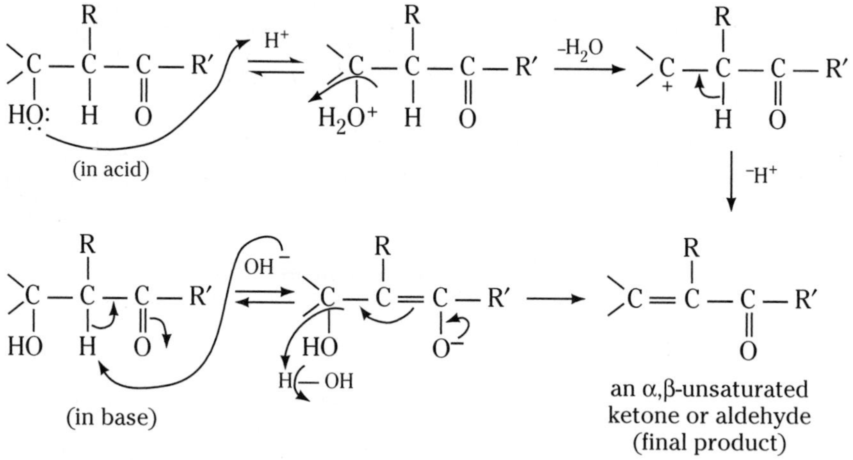

(in acid)

(in base)

an α,β-unsaturated
ketone or aldehyde
(final product)

All steps of the aldol condensation are reversible, especially when higher temperatures are employed. The reversibility of the aldol condensation may seem to be a severe disadvantage when this synthesis is compared with other types of chemical reactions. The backward reaction is known as the **reverse (or retro) aldol condensation.** In basic solution, it begins via a Michael addition to the α,β-unsaturated aldehyde or ketone:

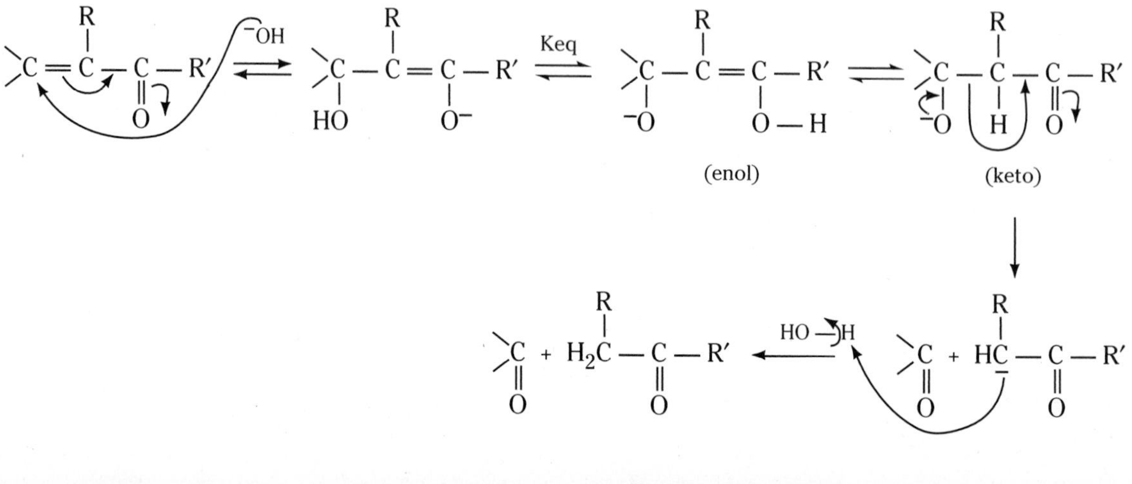

(enol)            (keto)

In some cases, the experimental conditions can minimize the reverse aldol reaction, thus increasing yields. For instance, by choosing a solvent that dissolves both reactants but not the product, the reaction will be driven to the right, thus increasing the percent yield as the product precipitates.

When ketones are used as one component and aldehydes as the other component, the crossed aldol condensations are called **Claisen-Schmidt reactions**, after the German scientists L. Claisen and J. G. Schmidt. The reaction combines a ketone with an aldehyde that does not have an α-hydrogen and, thus, cannot form an enolate anion. The products rarely arise from self-condensation of ketones when aldehydes are present.

In this experiment, acetone will be condensed with benzaldehyde in basic solution to form an α,β-unsaturated ketone, dibenzalacetone. The product is not soluble in the water-ethanol solvent, and precipitates out of solution. The loss of two moles of water is spontaneous as an extensive conjugated system is produced.

| acetone | benzaldehyde | dibenzalacetone | water |

The product is washed with dilute acetic acid to remove traces of the base which would hinder product crystallization, especially, during the recrystallization process.

**Suggested References**

Carey, F. A., *Organic Chemistry.* 3rd ed., 741–48. New York: McGraw-Hill, 1996.

Loudon, G. Marc, *Organic Chemistry.* 3rd ed., 1055–58. Redwood City, CA: Benjamin/Cummings, 1995.

March, J., *Advanced Organic Chemistry.* 4th ed., 175–86, 937–45. New York: John Wiley, 1992.

McMurry, J., *Organic Chemistry.* 4th ed., 898–908. Pacific Grove, CA: Brooks/Cole, 1996.

Morrison, R. T., and R. N. Boyd, *Organic Chemistry.* 6th ed., 805–11. Englewood Cliffs, NJ: Prentice-Hall 1992.

Solomons, T. W. G., *Organic Chemistry.* 6th ed., 762–65. New York: John Wiley, 1996.

Vollhardt, K. P. C., and N. E. Schore, *Organic Chemistry.* 2nd ed., 685–93. New York: W. H. Freeman, 1994.

Wade, L. G., *Organic Chemistry.* 3rd ed., 1055–65. Englewood Cliffs, NJ: Prentice-Hall, 1995.

 **SAFETY TIPS**

✔ The product, dibenzalacetone, is a skin irritant, and working with gloves is advisable. If skin irritation occurs, wash thoroughly with water and a dilute sodium bicarbonate solution.

✔ Acetone and ethanol are highly flammable liquids; no open flames should be used.

**Experimental**

1. Place 2.5 mL (density is 1.05 g/mL; 2.6 g) of benzaldehyde, 0.8 mL (density is 0.87 g/mL; 0.7 g) of acetone, and 20 mL of ethanol in a 125-mL conical flask.

2. Add 25 mL of 10% sodium hydroxide solution, stopper with a cork, and swirl the flask for at least 15 minutes. The solution will initially be clear, but should quickly become cloudy as the yellow crystals of the product precipitate.

3. Collect the product by vacuum filtration, using about 30 mL of water to transfer and thoroughly wash the crystals. Press the solid product against the filter paper with a glass stopper to remove as much water as possible.

4. Turn off the suction and add a solution of 1 mL of glacial acetic acid in 20 mL of 95% ethanol. Allow the crystals to soak for a minute or so, reapply the vacuum, and drain off as much liquid as possible.

5. Recrystallize the product from 20 mL of ethanol, determine the MP (literature value 113°C) and calculate the percent yield.

**Questions**

1. Suggest a mechanism for the preparation of dibenzalacetone as described in this experiment.

2. What would be the effect on this experiment if the student
   a. added half the amount of benzaldehyde?
   b. forgot to add benzaldehyde?
   c. forgot to add acetone, but added a high (50%) concentration of sodium hydroxide?
   d. washed the product with large amounts of hot ethanol?

3. Give the products of the following condensations:

a. $C_6H_5 - \overset{\overset{\displaystyle O}{\|}}{C} - H + CH_3 - \overset{\overset{\displaystyle O}{\|}}{C} - C_6H_5 \xrightarrow[C_2H_5OH]{OH^-} \mathbf{A}\ (C_{15}H_{12}O)$

b. $C_6H_5 - \overset{\overset{\displaystyle O}{\|}}{C} - H + CH_3 - \overset{\overset{\displaystyle O}{\|}}{\overset{+}{N}} - O^- \xrightarrow[C_2H_5OH]{OH^-} \mathbf{B}\ (C_8H_7NO_2)$

c. $CH_3 - O - \langle\!\!\!\bigcirc\!\!\!\rangle - \overset{\overset{\displaystyle O}{\|}}{C} - H \ +\ CH_3 - C \equiv N$

$\qquad\qquad\qquad\qquad\qquad\qquad \Big\downarrow OH^-/C_2H_5OH$

$\mathbf{D}\ (C_{10}H_{10}O_3) \xleftarrow{\text{hydrolysis}} \mathbf{C}\ (C_{10}H_9NO)$

(IR peaks at 3300–2500 (broad), 1690, 1640 cm$^{-1}$)

d. $2\ CH_3COOC_2H_5 \xrightarrow[\text{2. } H_2O/H^+]{\text{1. } NaOC_2H_5\,/\,C_2H_5OH} \mathbf{E}\ (C_6H_{10}O_3)$

(IR peaks at 1710 and 1740 cm$^{-1}$)

e. $H - COOC_2H_5 \ + \ \langle\!\!\!\bigcirc\!\!\!\rangle\!\!=\!\!O \xrightarrow{NaOC_2H_5} \mathbf{F}\ (C_7H_{10}O_2)$

(IR peaks at 1710 (broad), 2750 and 2850 cm$^{-1}$)

f. $H_5C_2OOC(CH_2)_4COOC_2H_5 \xrightarrow{\text{base}} \mathbf{G}\ (C_8H_{12}O_3)$

(IR peaks at 1710 and 1740 cm$^{-1}$)

## 18.2 A CYCLIZATION REACTION: SYNTHESIS OF CYCLOPENTANONE FROM ADIPIC ACID

Pyrolysis of carboxylate salts under strongly basic conditions leads to **decarboxylation** (i.e., loss of carbon dioxide) to form alkanes; for example,

$$CH_3 - \underset{\underset{O}{\|}}{C} - O^-Na^+ \; + \; NaOH \; \xrightarrow{\text{heat}} \; CH_4 \; + \; Na_2CO_3$$

Pyrolysis of the carboxylate salts of divalent metals (e.g., calcium, barium) results in partial decarboxylation (i.e., loss of only one carbon dioxide molecule) to form a ketone. This is especially useful in the synthesis of five- and six-carbon cyclic ketones by cyclization and decarboxylation of $\alpha,\delta$- and $\alpha,\varepsilon$-dicarboxylic acids (also called 1,6- and 1,7-dioic acids, respectively). In this experiment, cyclopentanone is formed by partial decarboxylation of adipic acid (IUPAC name: 1,6-hexanedioic acid).

$$HO - \underset{\underset{O}{\|}}{C}(CH_2)_4\underset{\underset{O}{\|}}{C} - OH \; \xrightarrow[\text{heat}]{Ba(OH)_2} \; \text{}=O \; + \; CO_2 \; + \; H_2O$$

adipic acid        cyclopentanone
(1,6–hexanedioic acid)

The reaction involves basic conditions, high temperatures (up to 300°C), and a distillation. Special care should be taken in handling the glassware because strong bases and high temperatures can fuse the joints. After the experiment is completed, the distillation apparatus must be dismantled and washed thoroughly. Under no circumstances should traces of base be left on the ground-glass joints.

The distillate is then treated with anhydrous potassium carbonate twice: once to salt out the cyclopentanone, and once to dry the product.

---

⚠️ **SAFETY TIPS**

✔ The high temperature (295°C) of the pyrolysis reaction might break the thermometer, especially if it is defective. If this happens, add sulfur powder to the round-bottomed flask (after cooling). The contents should then be disposed of properly by the instructor.

---

**Suggested References**

Loudon, G. Marc, *Organic Chemistry*. 3rd ed., 958. Redwood City, CA: Benjamin/ Cummings, 1995.

March, J., *Advanced Organic Chemistry*. 4th ed., 627–30. New York: John Wiley, 1992.

Thorpe, J. F., and G. A. R. Kon. "Cyclopentanone." *Organic Synthesis*, coll. vol. 1(1941): 192.

Vollhardt, K. P. C., and N. E. Schore, *Organic Chemistry*. 2nd ed., 747–49. New York: W. H. Freeman, 1994.

**Experimental**

1. Grind 21.8 g of adipic acid and 1.5 g barium hydroxide with a mortar and pestle until the mixture is a fine powder.
2. Pour the mixture into a 50-mL round-bottomed flask, wipe out its joint with a paper towel, and set up the apparatus for a simple distillation. Lubricate all joints (except those connecting the vacuum adaptor and receiving flask) with an ample amount of grease.
3. Insert a 400°C thermometer, and adjust it so that its bulb is within 0.5 cm (1/5 inch) of the bottom of the flask.
4. Place the receiving flask in an ice-water bath.
5. Heat the flask with a Bunsen burner carefully until the solid has melted. Heat more rapidly until the temperature reaches 285°C. At this temperature, decarboxylation occurs, and cyclopentanone accompanied by water distills slowly. As soon as the distillate starts accumulating, lift the thermometer slowly to the level appropriate for a simple distillation. The temperature should then read approximately 140°C. This will prevent the breaking of the thermometer. Collect the distillate until only a small amount of dry residue is left in the distilling flask.
6. Carefully disconnect the joints connecting the round-bottomed flask, connecting adaptor, adaptor outlet, and distilling adaptor, *but do not dismantle the apparatus until the glassware is cold.* Leave the thermometer in place until the apparatus cools to room temperature. Any sudden changes in temperature might break the thermometer. When the setup has cooled down, clean all grease and base from the joints.
7. Transfer the two-phase distillate to a separatory funnel, and add about 0.5 g of anhydrous potassium carbonate. Separate the (bottom) water layer, and then transfer the organic layer to a dry Erlenmeyer flask. Add another 0.5 g of anhydrous potassium carbonate and swirl the flask.
8. Gravity-filter the suspension into a *dry* 50-mL round-bottomed flask. Distill and collect the cyclopentanone (literature BP 131°C).
9. Calculate the percent yield, and run infrared spectra for both the adipic acid and cyclopentanone. Also run a gas chromatograph and compare it with that of an authentic sample of cyclopentanone.

**Questions**

1. What is the purpose of the following in this experiment?
   a. removing traces of barium hydroxide left on the joints
   b. using barium hydroxide
   c. adding potassium carbonate to the distillate.
2. Give the product of the following reactions:
   a. $HOOC(CH_2)_5COOH \xrightarrow{Ba(OH)_2/heat}$

   b. $HOOCCH_2COOH \xrightarrow{Ba(OH)_2/heat}$
3. If 9.0 g of water is isolated in the partial decarboxylation of adipic acid, what is the maximum weight of cyclopentanone that can be obtained?

## 18.3 MICHAEL ADDITION: REACTION OF ANILINE WITH BENZALACETOPHENONE

Benzalacetophenone (also called chalcone) is an α,β-unsaturated ketone that undergoes reactions typical of both the ketone and the alkene functional group. Thus, 2,4-dinitrophenylhydrazine reacts with it to give the corresponding 2,4-dinitrophenylhydrazone.

Normally nucleophiles react with α,β-unsaturated carbonyl compounds by a Michael-type addition to yield β-substituted carbonyl compounds.

3-(N-phenylamino)-1,3-diphenyl-1-propanone

**Suggested References**   Carey, F. A., *Organic Chemistry.* 3rd ed., 749–51. New York: McGraw-Hill, 1996.

Loudon, G. Marc, *Organic Chemistry.* 3rd ed., 1090–92. Redwood City, CA: Benjamin/Cummings, 1995.

McMurry, J., *Organic Chemistry.* 4th ed., 748–52. Pacific Grove, CA: Brooks/Cole, 1996.

Morrison, R. T., and R. N. Boyd, *Organic Chemistry.* 6th ed., 979–82. Englewood Cliffs, NJ: Prentice-Hall, 1992.

Solomons, T. W. G., *Organic Chemistry.* 6th ed., 833–35, 910–11. New York: John Wiley, 1996.

Vollhardt, K. P. C., and N. E. Schore, *Organic Chemistry.* 2nd ed., 699–704. New York: W. H. Freeman, 1994.

Wade, L. G., *Organic Chemistry.* 3rd ed., 1085–88. Englewood Cliffs, NJ: Prentice-Hall, 1995.

---

⚠️ **SAFETY TIPS**

✔ Both aniline and chalcone are skin irritants and skin contact should be avoided. Use of plastic gloves is recommended—especially, when handling solutions.

---

**Experimental**

1. Dissolve 1.0 g of *dry trans*-benzalacetophenone (chalcone) in 5 mL of dry ethanol in a 25-mL round bottomed flask. Add 1.0 mL (density = 1.02 g/mL, 1.0 g) of (preferably) freshly distilled aniline, and reflux for 1 hour. Freshly distilled aniline can be conveniently prepared by distillation over zinc dust.

2. Cool the mixture in ice, and isolate the crystals via vacuum filtration. If the crystals do not appear, let the solution evaporate to dryness by placing it in your drawer until the next lab period. Do not attempt to concentrate your solution by heating it.

3. Determine the MP (literature value 175°C) and percent yield.

**Questions**

1. Describe the infrared spectrum differences between the starting material, benzalacetophenone, and the product, 3-(N-phenylamino)-1,3-diphenyl-1-propanone, in this experiment.

2. Suggest a synthesis of benzalacetophenone using a crossed aldol condensation (Claisen-Schmidt reaction).

3. Identify the products in the following reactions:

a. =O   +   $CH_3NH_2$   $\xrightarrow{H^+}$

b. =O   +   HCl   $\longrightarrow$

c. $(CH_3)_2CHCH = CH - \overset{\overset{\displaystyle O}{\|}}{C}H$   +   $NH_2CH_2NH_2$   $\xrightarrow{H^+}$   $C_7H_{14}N_2$

d. =O   +   ⬡—$NH_2$   $\longrightarrow$

4. Why is 3-(N-phenylamino)-1,3-diphenyl-1-propanone white, while the reactant (chalcone) is yellow? (*Hint:* Use your knowledge about conjugation and UV/visible spectrophotometry.)

5. If the product does not precipitate, it is recommended that the solution *not* be boiled down. Why?

# chapter 19

# Amines

## 19.1 REDUCTION OF A NITRO COMPOUND TO AN AMINE: SYNTHESIS OF *m*-AMINOACETOPHENONE FROM *m*-NITROACETOPHENONE

One of the major problems facing synthetic chemists is the question of **chemoselectivity**. Simply put, this question asks: of two (or more) functional groups present in a molecule, which one will react under a given set of conditions? A great deal of research is directed toward the development of synthetic reagents that will selectively react with only one type of functional group. Remember that enzymes are also highly chemoselective, often so selective that only one specific molecule (rather than a whole class of molecules containing a given functional group) will react—not so useful, from a synthetic viewpoint!

As an example, consider the nitro group ($R - NO_2$) and the carbonyl group ($RR'C = O$). Both functional groups are susceptible to reduction, the former to an amino group and the latter to an alcohol. Suppose we had a molecule containing both groups. How, then, could we selectively reduce *only one* of them? Fortunately, this problem has been solved, as this experiment will demonstrate. The carbonyl group can be reduced to a secondary alcohol using a mild source of hydride, such as $NaBH_4$, while the nitro group can be reduced to the amino group using a metal (such as Fe or Sn) in hydrochloric acid. In both cases, the other functional group does not react.

In this experiment, the reduction of a nitro group in the presence of a carbonyl group will be accomplished. Treatment of *m*-nitroacetophenone with granular tin metal in refluxing hydrochloric acid cleanly reduces the $-NO_2$ to $-NH_2$, without touching the carbonyl group.

| *m*-nitroacetophenone | tin | | *m*-acetylanilinium chloride (water-soluble salt) | | *m*-aminoacetophenone (water-insoluble) |

Since the reaction is performed in acidic medium, the amino group in the product, *m*-aminoacetophenone, will be protonated, yielding the water-soluble *m*-acetylanilinium chloride salt. Addition of excess NaOH

will deprotonate the salt, thus causing the precipitation of *m*-aminoace-
tophenone, which will be purified by recrystallization from ethanol-water.

The mechanism of the reduction of the nitro group by tin (or other
metal) is rather complicated and involves the stepwise transfer of electrons and protons. The reduction proceeds through two intermediates,
each of which can be isolated under other conditions.

First, two-electron reduction of the nitro group yields a nitroso compound;
then, a second two-electron reduction of the nitroso group produces a
hydroxylamine. Finally, a third two-electron reduction gives the amine.

One important implication of this reaction is that it simplifies the introduction of amino groups onto aromatic rings. Thus, one can easily make
aniline by electrophilic aromatic nitration ($HNO_3/H_2SO_4$), followed by
reduction (Sn/HCl). The regioselectivity of the reaction is governed by the
usual directing influences. Also, notice that by reducing the nitro group to
an amino group, the aromatic ring is changed from electron-poor to
electron-rich; as a result, both the reactivity and regioselectivity of subsequent reactions will be dramatically changed. After nitration, further electrophilic aromatic substitutions yield *m*-substitution products, while after
reduction of the nitro group, further electrophilic aromatic substitutions
yield *o*-/*p*-substitution products.

For your pre-lab, be sure to balance the overall equation. You can use
the half-reaction method; start by determining how many electrons it
takes to reduce the nitro group to the amino group (*Hint:* re-read the above
paragraph), then, knowing that the Sn(0) is oxidized to Sn(IV), calculate
how many equivalents of tin are required.

**Suggested References**    Carey, F. A. *Organic Chemistry.* 3rd ed., 916. New York: McGraw-Hill, 1996.

Jones, A. G., "Selective reduction of meta- (and para-)nitroacetophenone." *J. Chem. Educ.* (1975): 52, 668–69.

Gibson, M. S. In *The Chemistry of the Amino Group,* ed. S. Patai, 66. New York: Interscience, 1968.

Hudlicky, M., *Reductions in Organic Chemistry.* 13. Chichester: Ellis Horwood, 1984.

Fessenden, R. J., and J. S. Fessenden, *Organic Chemistry.* 5th ed., 758. Pacific Grove, CA: Brooks/Cole, 1994.

Loudon, G. M., *Organic Chemistry.* 3rd ed., 1147–48. Redwood City, CA: Benjamin/Cummings, 1995.

McMurry, J., *Organic Chemistry.* 4th ed., 977–78. Pacific Grove, CA: Brooks/Cole, 1996.

Vollhardt, K. P. C., and N. E. Schore, *Organic Chemistry.* 2nd ed., 610–11. New York: W. H. Freeman, 1994.

Wade, L. G., Jr., *Organic Chemistry.* 3rd ed., 868. Englewood Cliffs, NJ: Prentice-Hall, 1995.

**Safety Tips**

✔ Hydrochloric acid and sodium hydroxide are corrosive; in case of skin contact, wash with lots of water.

✔ *m*-Aminoacetophenone is an irritant; in case of skin contact, wash with lots of water.

**Experimental**

1. Place 1.00 g of m-nitroacetophenone, 1.20 g of granular tin metal, and 20 mL of 3*M* HCl in a 100 mL round-bottomed flask.

2. Attach a reflux condenser, and reflux gently until all of the tin has dissolved (approximately 30 minutes).

3. While the mixture is refluxing, place 20 mL of 6*M* NaOH in a small beaker and cool it in an ice-water bath.

4. After 30 minutes, or after all of the tin has dissolved, discontinue heating and let the contents of the reaction flask cool to room temperature. Place the reaction flask in the ice-water bath and let it cool thoroughly for about 5 minutes.

5. Carefully add the sodium hydroxide (from step 3) dropwise to the reaction mixture with continued stirring. (Record your observations—what just happened? Why?)

6. Allow the resulting suspension to cool a little longer in the ice-water bath. Collect the yellow precipitate by suction filtration using a Büchner funnel. Use 1 or 2 mL of the cold filtrate to transfer the last bits of product from the reaction flask to the filter paper and wash the product with 5–10 mL of ice-cold water. Put the aqueous filtrates in the appropriate waste container in the hood, *not* down the sink.

7. Allow the product to air-dry with suction for a few minutes.

8. Transfer the crude product to a 50 mL conical flask and recrystallize from approximately 10–15 mL of a 3:1 water-ethanol mixture using a hot-plate in a hood. If any undissolved solids remain, perform a hot filtration.

9. Cool the hot solution to room temperature slowly and then add water until cloudiness ensues. Cool the flask in an ice-water bath to complete the crystallization.

10. Collect the crystals via vacuum filtration, rise with a small amount of ice water, and allow to air-dry (using suction) for at least 30 minutes.

11. While waiting for the product to dry completely, dissolve a few crystals of it in acetone in a small test tube. Spot this solution, plus a solution of *m*-nitroacetophenone (in the hood) on a silica gel-coated TLC plate, and develop it in dichloromethane (see Experiment 5.2 for a review of thin-layer chromatography). Visualize the developed plate with a UV lamp or an iodine chamber. Calculate the $R_f$ values for the starting material and product, label the TLC plate clearly, and tape it to your lab report. Discuss the results of your TLC analysis in your report.

12. When your product is dry, weigh it in a pre-weighed vial or test tube, and calculate the percent yield. Measure and record the melting point (lit. 96–98°C).

13. If time permits, measure and compare the infrared spectra of the starting material and product.

**Questions**

1. If you obtained IR spectra of the starting material and product, assign as many of the absorptions as possible.

2. Based on the TLC analysis of the starting material and product, which compound is more polar? If the reduction product was isolated before neutralization with NaOH (step 5), would you expect its $R_f$ to be higher or lower than that of *m*-aminoacetophenone? Explain.

3. Write a balanced equation for the reduction of nitrobenzene to aniline using iron in acidic medium. (*Hint:* the Fe(0) is oxidized to Fe(III).)

4. Suggest syntheses of *m*-bromoaniline and *p*-bromoaniline from benzene. More than one step may be required for each molecule.

5. Balance the following equation:

## 19.2 SYNTHESIS OF ACETANILIDE

Amines are the organic analogs of ammonia, and as such, they are both basic and nucleophilic. They are classified as **primary (1°), secondary (2°), and tertiary (3°)** amines, depending on the number of hydrogens attached on the nitrogen.

$$C_2H_5NH_2 \qquad (C_2H_5)_2NH \qquad (n\text{-}C_3H_7)_2NCH_3$$

ethylamine      diethylamine      di-*n*-propylmethylamine
(1° amine)      (2° amine)      (3° amine)

Primary amines react with acylating agents to form amides by nucleophilic attack at the carbonyl group. Acyl chlorides, anhydrides, and (less frequently) acids are usually employed. Acyl chlorides react most rapidly; acids require high temperatures and long reaction times for conversion of the initially formed carboxylic acid salt to the amide.

$$R-NH_2 \; + \; R'-\overset{\overset{\displaystyle O}{\|}}{C}-Cl \longrightarrow R-\underset{\underset{\displaystyle H}{|}}{N}-\overset{\overset{\displaystyle O}{\|}}{C}-R' \; + \; HCl$$

1° amine     acyl chloride      N-substituted amide

$$R-NH_2 \; + \; R'\underset{\underset{\displaystyle O}{\|}}{C}-O-\underset{\underset{\displaystyle O}{\|}}{C}R' \longrightarrow R-\underset{\underset{\displaystyle H}{|}}{N}-\overset{\overset{\displaystyle O}{\|}}{C}-R' \; + \; R'COOH$$

1° amine     anhydride      N-substituted amide

$$R-NH_2 \; + \; R'COOH \longrightarrow R'COO^-H_3NR^+ \xrightarrow[-H_2O]{heat} R-\underset{\underset{\displaystyle H}{|}}{N}-\overset{\overset{\displaystyle O}{\|}}{C}-R'$$

1° amine     carboxylic     carboxylic acid      N-substituted amide
(a base)     acid      salt

In this experiment, aniline (a 1° aromatic amine) will be converted to acetanilide (an amide) by reacting it with acetic anhydride.

$$\text{C}_6\text{H}_5-NH_2 \; + \; (CH_3-\overset{\overset{\displaystyle O}{\|}}{C})_2O \longrightarrow \text{C}_6\text{H}_5-NH\overset{\overset{\displaystyle O}{\|}}{C}CH_3 \; + \; CH_3COOH$$

aniline     acetic      acetanilide     acetic
    anhydride      acid

Acetanilide, which in the past was used as an analgesic and antipyretic, can serve as the starting material for the synthesis of two analgesics, ***acetaminophen*** (Tylenol) and ***phenacetin,*** as well as for drugs in the sulfanilamide family (see Experiment 14.3).

acetaminophen                                              phenacetin

sulfanilamide

---

⚠️ **SAFETY TIPS**

✔ Aniline is a skin irritant and skin contact should be avoided.
✔ Acetic anhydride is a lachrymator and should be used with adequate ventilation.

---

**Suggested References**     "Acetic Anhydride." In *Reagents for Organic Synthesis,* edited by L. F. Fieser and M. Fieser, vol. 1, 3. New York: John Wiley, 1967.

Carey, F. A., *Organic Chemistry.* 3rd ed., 833–35. New York: McGraw-Hill, 1996.

Loudon, G. Marc, *Organic Chemistry.* 3rd ed., 1004–05. Redwood City, CA: Benjamin/Cummings, 1995.

McMurry, J., *Organic Chemistry.* 4th ed., 822–24. Pacific Grove, CA: Brooks/Cole, 1996.

Morrison, R. T., and R. N. Boyd, *Organic Chemistry.* 6th ed., 857–60. Englewood Cliffs, NJ: Prentice-Hall, 1992.

Solomons, T. W. G., *Organic Chemistry.* 6th ed., 821–22, 864. New York: John Wiley, 1996.

Vollhardt, K. P. C., and N. E. Schore, *Organic Chemistry.* 2nd ed., 613–14. New York: W. H. Freeman, 1994.

Wade, L. G., *Organic Chemistry.* 3rd ed., 997–1006. Englewood Cliffs, NJ: Prentice-Hall, 1995.

---

**Experimental**

1. Place 10 mL of glacial acetic acid and 2.0 mL (d = 1.08 g/mL; 2.2 g) of acetic anhydride in a 50-mL round-bottomed flask. Add dropwise 2.2 mL (d = 1.02 g/mL; 2.2 g) of aniline, and reflux for 1 hour.

2. Pour the hot contents of the reaction flask into a beaker containing 50 mL of cold water, and stir vigorously.

3. Filter with suction, and wash the crystals with about 5 mL of ice-cold water.

4. Transfer the crystals to another beaker containing 60 mL of water, and heat to dissolve the acetanilide.

5. If the solution is colored, add a pinch of charcoal and heat to boiling. Filter with suction *while still hot,* and cool the filtrate to room temperature.

6. Immerse the mixture in an ice-water bath, and let stand for 5–10 minutes.

7. Collect the crystals by vacuum filtration through a Büchner funnel, and air dry.

8. Determine the MP (literature value 114°C) and percent yield.

**Questions**

1. Explain the purpose of the following in this experiment:
   a. acetic acid
   b. water (in step 4)
   c. charcoal.
2. Why is acetic anhydride a better acylating agent than acetic acid?
3. Acylations are often carried out in the presence of a base, such as a 3° amine or pyridine. What is the function of the base in these reactions?
4. Monoacylation of a primary amine is far easier than diacylation. Explain.
5. What are the products of the reaction of acetic anhydride with:
   a. diphenylamine
   b. water (with heat)
   c. benzene in the presence of a Lewis acid
   d. toluene in the presence of a Lewis acid
   e. ethanol (with heat)?

## 19.3  THE COUPLING OF AROMATIC DIAZONIUM COMPOUNDS: AZO DYE FORMATION

Aromatic diazonium salts (formed by the reaction of nitrites with 1° aromatic amines) couple with electron-rich aromatic rings to form **azo dyes**. The coupling reaction is carried out in an alkaline, neutral, or weakly acidic medium.

In this experiment, a dye will be formed and applied to cloth. The experimental quantities can be increased for better results. The preparation involves the reaction of *p*-nitrobenzene diazonium chloride with a basic solution of 2-naphthol to yield Para Red, a bright-red azo dye.

$$O_2N-\!\!\langle\bigcirc\rangle\!\!-NH_2 \;+\; NaNO_2 \xrightarrow{\text{HCl}} O_2N-\!\!\langle\bigcirc\rangle\!\!-\overset{+}{N}\equiv N \;\; Cl^-$$

*p*-nitroaniline          sodium nitrite          *p*-nitrobenzene diazonium chloride

2-naphthol $\xrightarrow{\text{NaOH}}$ naphthol-$O^-Na^+$

Para Red
(bright red dye)

**Suggested References**

Carey, F. A., *Organic Chemistry*. 3rd ed., 934–36. New York: McGraw-Hill, 1996.

Loudon, G. Marc, *Organic Chemistry*. 3rd ed., 1143–44. Redwood City, CA: Benjamin/Cummings, 1995.

McMurry, J., *Organic Chemistry*. 4th ed., 984–85. Pacific Grove, CA: Brooks/Cole, 1996.

Morrison, R. T., and R. N. Boyd, *Organic Chemistry*. 6th ed., 873–75. Englewood Cliffs, NJ: Prentice-Hall, 1992.

"Sodium Nitrite." In *Reagents for Organic Synthesis,* edited by L. F. Fieser and M. Fieser, vol. 1, 1097–1101. New York: John Wiley, 1967.

Solomons, T. W. G., *Organic Chemistry*. 6th ed., 930–32. New York: John Wiley, 1996.

Vollhardt, K. P. C., and N. E. Schore, *Organic Chemistry*. 2nd ed., 896–97. New York: W. H. Freeman, 1994.

Wade, L. G., *Organic Chemistry*. 3rd ed., 906–07. Englewood Cliffs, NJ: Prentice-Hall, 1995.

Wishnok, J. S., and M. C. Archer, "The Nitrous Acid Test for Amines—A Potentially Hazardous Reaction." *J. Chem. Educ.* 53 (1976): 551.

⚠ **SAFETY TIPS**

✔ It is important that extreme care be taken when handling the solutions. Diazonium salts are explosive, unless prepared and kept at low temperatures.

✔ Although the toxicity of 2-naphthol is of low order, skin contact may produce peeling of the skin and pigmentation.

> ✔ The use of plastic gloves in this experiment is required.
> ✔ Concentrated hydrochloric acid is a corrosive irritant and should be poured with adequate ventilation.

**Experimental**

*Note:* The quantities given for this procedure are enough to dye a 6″ × 6″ piece of cloth.

1. In a 50-mL beaker, place 1.0 g of *p*-nitroaniline, 25 mL of water, and 1 mL of concentrated hydrochloric acid. Place the beaker in an ice bath, and cool to 0°C.
2. Add slowly, and with stirring, a cold (0°–5°C) solution of 1.0 g of sodium nitrite in 5 mL of water.
3. In a separate 250-mL beaker, prepare a solution of 1.0 g of 2-naphthol in 50 mL of 10% sodium hydroxide.
4. While wearing plastic gloves, immerse a piece of clean cotton cloth in the alkaline 2-naphthol solution (from step 3), squeeze out as much solution as possible (avoiding direct contact with your skin), and hang the cloth up to dry for a few minutes.
5. Immerse the dried cloth (from step 4) in the solution (from step 2). Wear plastic gloves and puddle the cloth with a glass rod. Remove the cloth, squeeze as much solution as possible (avoiding direct contact with your skin), and again air dry.

**Questions**

1. Suggest structures for the following reactions:

   a. 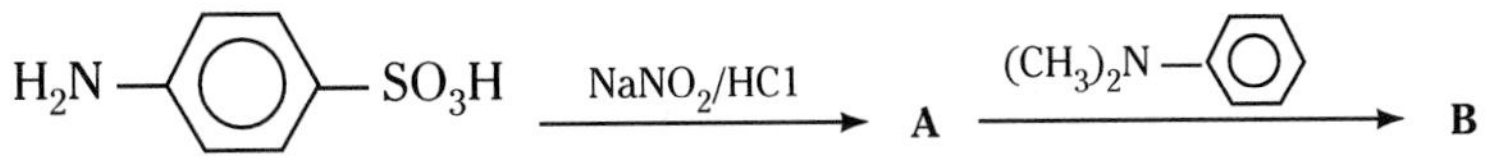 $H_2N$—⟨○⟩—$SO_3H$ $\xrightarrow{NaNO_2/HCl}$ **A** $\xrightarrow{(CH_3)_2N-⟨○⟩}$ **B**

   b. 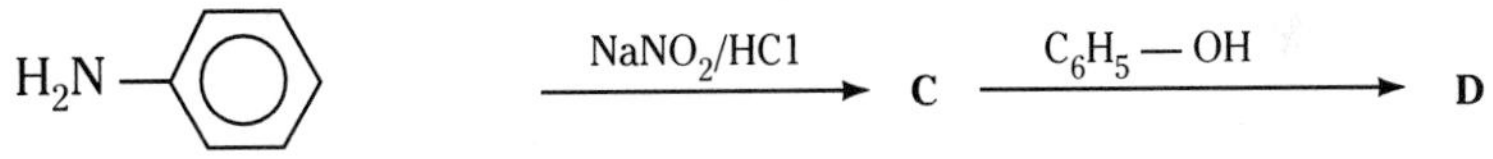 $H_2N$—⟨○⟩ $\xrightarrow{NaNO_2/HCl}$ **C** $\xrightarrow{C_6H_5-OH}$ **D**

   c. $^+N_2$—⟨○⟩—⟨○⟩—$N_2^+$ + ⟨○○⟩—$O^-Na^+$ $\longrightarrow$ **E**

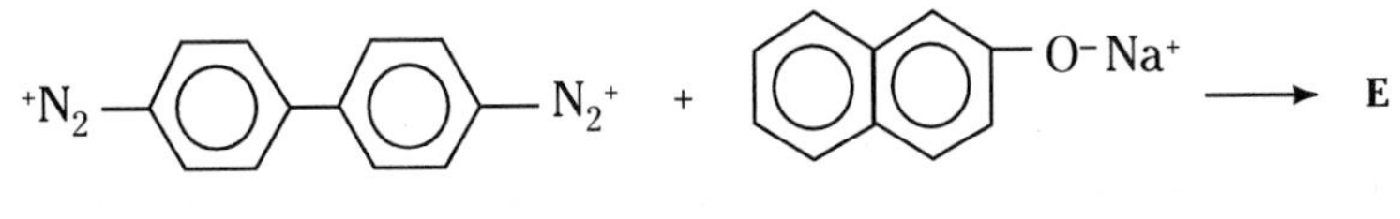

2. During the coupling reaction with 2-naphthol, why does the coupling (electrophilic substitution) occur only in the 1-position? (*Hint:* Write resonance structures for 2-naphthol in basic solution to justify electrophilic attack at the 1-position).
3. Provide the mechanism for the preparation of *p*-nitrobenzene diazonium cation from *p*-nitroaniline and nitrous acid, $HNO_2$.

# chapter 20
# Polynuclear Aromatics and Heterocycles

## 20.1  OXIDATION OF 2-METHYLNAPHTHALENE

**Condensed benzenoid compounds** are characterized by two or more benzene rings that are fused together at *ortho* positions in such a way that each pair of rings shares two carbons. Such compounds are also called **polynuclear aromatics**. The simplest members of this group are naphthalene, anthracene, and phenanthrene.

naphthalene
R.E. = 61 kcal/mole

anthracene
R.E. = 83 cal/mole

phenanthrene
R.E. = 91 kcal/mole

As a general rule, the more benzene rings in the system the smaller the amount of resonance stabilization energy (R.E.) *per ring*. The resonance energy of benzene is 36 kcal/mole, while that of *each* naphthalene ring is only 30.5 kcal/mole (i.e., 61 kcal/mole/2 rings). As a result, naphthalene can be oxidized to products that retain much of their aromaticity, under conditions where benzene is inert.

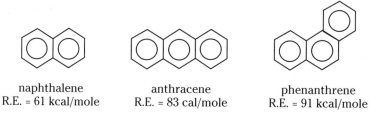

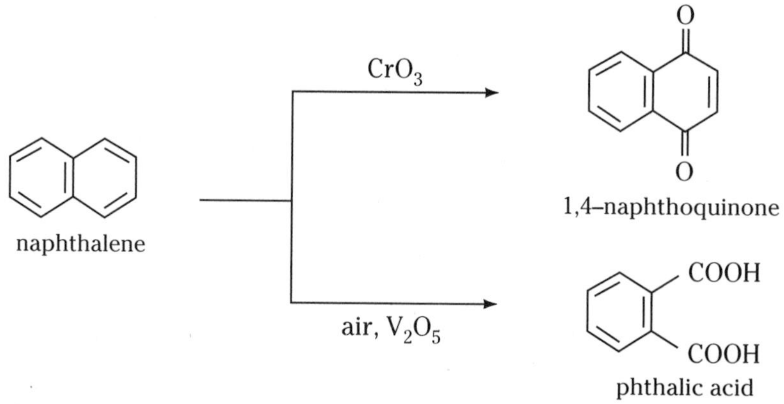

naphthalene

CrO₃ → 1,4–naphthoquinone

air, V₂O₅ → phthalic acid

In this experiment, 2-methylnaphthalene will be oxidized with chromium (VI) oxide to yield 2-methyl-1,4-naphthoquinone. The reaction temperature must be controlled in the early stage of the oxidation to prevent ring opening and further oxidation to phthalic acid.

2-methylnaphthalene    chromium
                       (VI) oxide                          2-methyl 1,4-naphthoquinone

**Suggested References**    "Chromic Anhydride." In *Reagents for Organic Synthesis,* edited by L. F. Fieser and
M. Fieser, vol. 1, 144–47. New York: John Wiley, 1967.

Loudon, G. Marc, *Organic Chemistry.* 3rd ed., 941–43. Redwood City, CA:
Benjamin/Cummings, 1995.

Morrison, R. T., and R. N. Boyd, *Organic Chemistry.* 6th ed., 510–13, 545–46.
Englewood Cliffs, NJ: Prentice-Hall, 1992.

Wade, L. G., *Organic Chemistry.* 3rd ed., 734–36. Englewood Cliffs, NJ: Prentice-Hall,
1995.

---

⚠ **SAFETY TIPS**

✔ Chromium (VI) oxide forms a corrosive dust which is particularly
dangerous when inhaled. It is strongly recommended that the
compound be handled with gloves at all times.

✔ Methanol is a flammable liquid and is highly toxic if ingested.

---

**Experimental**

1. In a 125-mL Erlenmeyer flask, dissolve 5 g of chromium (VI) oxide in
5 mL of water.

2. Add 5 mL of acetic acid, and then in small proportions (while stirring),
add a solution of 1.4 g of 2-methylnaphthalene in 15 mL of acetic acid,
keeping the temperature below 60°C.

3. When all the solution has been added and the temperature starts
dropping, place the Erlenmeyer flask in a hot-water bath, and heat for
one hour at 90°C.

4. Add 40 mL of water and isolate the product by vacuum filtration.

5. Isolate the product as yellow plates by recrystallization from approximately 10–15 mL of methanol. Determine the MP (literature value
106°C) and the percent yield. The compound is light-sensitive and
should be stored in dark, non-transparent vials wrapped with aluminum foil.

**Questions**

1. Explain the purpose of the following in the above experiment:
   a. acetic acid
   b. not allowing the temperature to go above 60°C.

2. Balance the oxidation reaction of naphthalene in acid medium using
the following:
   a. $CrO_3$ (to yield $Cr^{+3}$ and 1,4-naphthoquinone)
   b. $Cr_2O_7^{-2}$ (to yield $Cr^{+3}$ and phthalic acid)
   c. $MnO_4^-$ (to yield $Mn^{+2}$ and phthalic acid).

3. Give all resonance forms for the following:
   a. naphthalene (3 structures)
   b. anthracene (4 structures)
   c. phenanthrene (5 structures)

4. Nitration of naphthalene yields 1-nitro (and not 2-nitro) naphthalene.
Explain. (*Hint:* Write resonance structures for the intermediate formed
upon reaction of naphthalene with the nitronium, $NO_2^+$ ion.)

## 20.2  SYNTHESIS OF BENZIMIDAZOLE

There are many cyclic structures in which oxygen, nitrogen, or sulfur atoms form part of the cyclic π-system. If the nonbonding electrons of the non-carbon atom satisfy Hückel's rule, the structure can be aromatic. Common examples of such heterocyclic aromatics are furan, pyrrole, and thiophene.

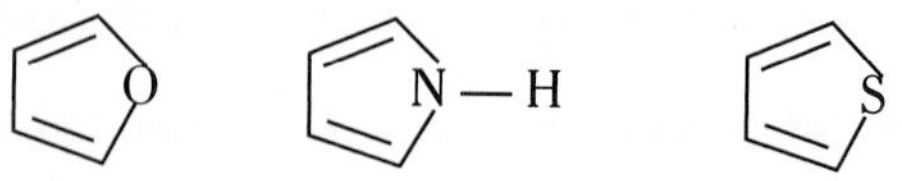

furan        pyrrole        thiophene

Often heterocyclic compounds contain two or more heteroatoms. Some examples are thiazole, imidazole, and pyrimidine.

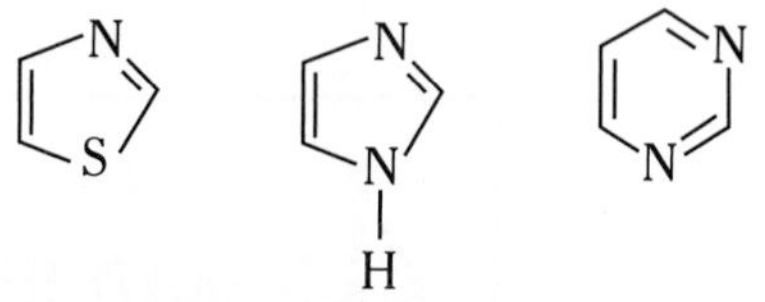

thiazole        imidazole        pyrimidine

Some heterocyclic compounds can be conveniently prepared by cyclization with concomitant elimination of water or another small molecule. Benzimidazole and its derivatives can be synthesized by the condensation of o-phenylenediamine (1,2-diaminobenzene) and a carboxylic acid at high temperatures (at room temperature, these reactions normally result in the formation of a carboxylate salt). In this experiment o-phenylenediamine will be condensed with formic acid in a neat (no solvent) reaction.

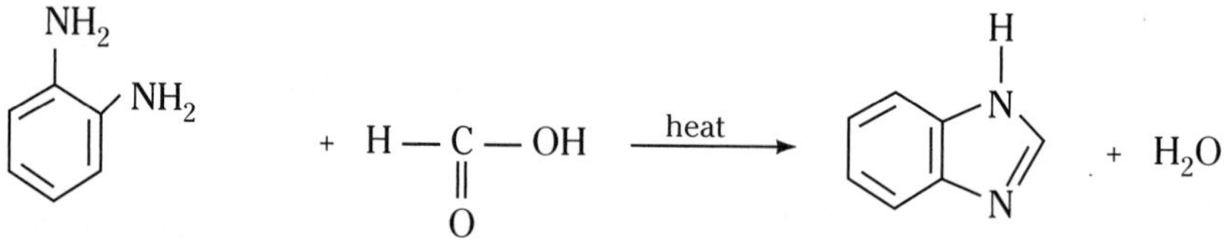

o–phenylenediamine        formic acid        benzimidazole

The reaction mixture is further treated with a weak base (ammonia) to neutralize the excess formic acid and to precipitate the free benzimidazole. Because of the easy air-oxidation of o-phenylenediamine, the crude product must be treated with activated charcoal to remove the colored impurities.

---

⚠ **SAFETY TIPS**

✔  o-Phenylenediamine is a toxic compound and should be handled with gloves.

✔  Formic acid is corrosive to the skin. Exposure to formic acid vapors may produce irritation of the eyes, skin and mucous membranes. Adequate ventilation is recommended.

**Suggested References**

Loudon, G. Marc, *Organic Chemistry*. 3rd ed., 1291–98. Redwood City, CA: Benjamin/Cummings, 1995.

McMurry, J., *Organic Chemistry*. 4th ed., 553–55, 1128–36, 1107–17. Pacific Grove, CA: Brooks/Cole, 1996.

"*o*-Phenylenediamine." In *Reagents for Organic Synthesis,* edited by L. F. Fieser and M. Fieser, vol. 1, 415–24. New York: John Wiley, 1967.

Patnaik, P., *A Comprehensive Guide to the Hazardous Properties of Chemical Substances*. 42. New York: Van Nostrand Reinhold, 1992.

Vollhardt, K. P. C., and N. E. Schore, *Organic Chemistry*. 2nd ed., 744. New York: W. H. Freeman, 1994.

**Experimental**

1. Place 2.1 g of *o*-phenylenediamine and 1.5 mL (d = 1.22 g/mL; 1.8 g) of formic acid in a test tube, and heat in a water bath for one hour.
2. Cool the test tube under tap water and add dropwise 6*M* ammonium hydroxide while stirring the dark paste with a glass rod until the mixture is just alkaline to litmus.
3. Add 2 mL of water, stir, vacuum-filter, and wash with 1 mL of ice-cold water.
4. Place the isolated crude benzimidazole in 30 mL of boiling water, and add 0.2 to 0.3 g of charcoal. Heat for about 5 minutes and vacuum-filter while still hot. Isolate the filtrate in a large test tube that has been previously placed inside the filter flask (see page 63).
5. Allow the filtrate to cool to room temperature, then cool in ice and isolate the crystals via vacuum filtration.
6. Determine the MP (literature value 171°C) and the percent yield.
7. If in this experiment, acetic acid is used in place of formic acid, the product is 2-methylbenzimidazole (MP literature value 176°–177°C).

**Questions**

1. In this experiment, what is the purpose of the following:
   a. ammonium hydroxide?
   b. charcoal?
2. Suggest a plausible mechanism for the synthesis of benzimidazole as prepared in this experiment.
3. Complete the following reactions:

a. [structure: benzene ring with two NH$_2$ groups ortho] + CH$_3$COOH/heat ⟶

b. [structure: benzene ring with two NH$_2$ groups ortho] + S=C=S/heat ⟶

c. [structure: benzene ring with NH$_2$] + CH$_3$COOH $\xrightarrow{\text{room temperature}}$

d. H$_2$N—[benzene ring]—NH$_2$ + CH$_3$COOH (excess) $\xrightarrow{\text{heat}}$

# chapter
# 21

# Carbohydrates

## 21.1 THE ACID CATALYZED HYDROLYSIS OF SUCROSE: A KINETIC STUDY

The disaccharide sucrose is hydrolyzed in acidic solution to the monosaccharides (+)-glucose and (–)-fructose.

(+)-sucrose        (+)-glucose      (–)-fructose

This bimolecular **second-order** reaction becomes **pseudo-first-order** in the presence of excess water, and can conveniently be followed by monitoring the change in optical rotation of the solution over time. Initially the solution is dextrorotary, but it becomes levorotary because highly levorotary fructose is formed.

### Kinetics

The decrease in [A], the concentration of sucrose, follows the first-order rate equation:

$$-d[A]/dt = k[A] \quad \text{(equation 21.1)}$$

When this expression is integrated between limits $[A_0] \longrightarrow [A]$ and from $t = 0 \longrightarrow t$ it becomes

$$\ln \{[A]/[A_0]\} = -kt$$

or

$$\ln[A] = -kt + \ln[A_o] \quad \text{(equation 21.2)}$$

where [A] is the concentration of sucrose at any time $t$, and $[A_0]$ is the initial sucrose concentration.

A plot of ln[A] versus $t$ is linear, with a slope of $(-k)$. As with any first-order reaction, the half-life, $\tau$, is given by the expression

$$\tau = 0.693/k \quad \text{(equation 21.3)}$$

The rate will be determined by recording the change in optical rotation with time with a polarimeter. The initial reading, $\alpha_o$, is the optical rotation of the original sucrose solution. It is assumed that little or no hydrolysis occurs until concentrated HCl is added. Subsequent readings, $\alpha_t$, at time $t$, record the formation of fructose and glucose and the disappearance of sucrose. The last reading, $\alpha_\infty$, will be taken to represent the starting concentration of sucrose, so that the concentration of sucrose at any time $t$, can be represented as $(a_\infty - a_t)$. Due to the nature of the kinetics of the reaction, the more significant changes occur at the beginning of the reaction.

**Suggested References**

Loudon, G. Marc, *Organic Chemistry.* 3rd ed., 1337–40. Redwood City, CA: Benjamin/Cummings, 1995.

McMurry, J. *Organic Chemistry.* 4th ed., 1041–42. Pacific Grove, CA: Brooks/Cole, 1996.

Morrison, R. T., and R. N. Boyd. *Organic Chemistry.* 6th ed., 1191–92. Englewood Cliffs, NJ: Prentice-Hall, 1992.

Solomons, T. W. G. *Organic Chemistry.* 6th ed., 1046–47, 1075–76. New York: John Wiley, 1996.

"Sucrose." In *McGraw-Hill Encyclopedia of Science and Technology,* 7th ed., vol. 17, 571–72. New York: McGraw-Hill, 1992.

Vollhardt, K. P. C., and N. E. Schore, *Organic Chemistry.* 2nd ed., 968–70. New York: W. H. Freeman, 1994.

Wade, L. G. *Organic Chemistry.* 3rd ed., 1139–40. Englewood Cliffs, NJ: Prentice-Hall, 1995.

  **Safety Tip**

✔ Concentrated hydrochloric acid is extremely corrosive; it should be poured carefully and with adequate ventilation.

**Experimental**

1. Measure and record the length of the polarimeter tube in decimeters (dm).
2. Prepare a solution of 10 g of sucrose in 25 mL of water in a 50-mL beaker.
3. Fill a polarimeter tube with the sucrose solution and obtain a reading on the polarimeter. Record as $\alpha_o$ on the data sheet. Use this reading to calculate the value of the specific rotation of sucrose. It is assumed that very little hydrolysis has occurred at this point before the acid has been added to the solution.
4. Empty the polarimeter tube into the original 50-mL beaker, add 2 mL of concentrated hydrochloric acid (12 M), stir vigorously with a stirring rod for about 5 seconds, and record the time as $t = 0$.
5. Refill the polarimeter tube and obtain readings at $t = 1, 2, 3, 4, 5, 10, 15$, and 20 minutes, and record them as $\alpha_t$ on the data sheet.
6. Take a final reading at $t = 30$ minutes and record as $\alpha_\infty$.
7. Tabulate your data in five columns labeled $t$, $\alpha_t$, $\alpha_\infty - \alpha_t$, $\ln(\alpha_\infty - \alpha_t)$ and $1/(\alpha_\infty - \alpha_t)$.

8. Plot ln $(\alpha_\infty - \alpha_t)$ versus $t$ on the graph paper provided. Measure the slope and calculate $\tau_{1/2}$.

9. If the reaction had been second-order, a plot of $\{1/(\alpha_\infty - \alpha_t)\}$ versus $t$ would have been linear. Make this plot on the graph paper provided and compare it with the first-order plot.

**Questions**

1. Which fits the data better, the first- or second-order plot?

2. From the length of the polarimeter tube ($l$), the concentration, $c$ in g/mL, and the initial value, $\alpha_o$, calculate the specific rotation $[\alpha]_D$ of sucrose using the equation:

$$[\alpha]_D = \alpha_o/lc \quad \text{(equation 21.4)}$$

3. From the integrated first-order rate law (equation 21.2), derive the expression for the half-life (equation 21.3).

4. The hydrolysis of sucrose is acid-catalyzed, and the reaction is pseudo-first-order. The rate can be more rigorously expressed by the second-order rate law, which includes the acid catalyst:

$$-d[A]/dt = k[H^+][A] \quad \text{(equation 21.5)}$$

But at any *constant* concentration of $H^+$ the equation simplifies to the first order expression

$$-d[A]/dt = k_{obs}[A] \quad \text{(equation 21.6)}$$

where $k_{obs} = k[H^+]$ (i.e., the $H^+$ concentration becomes part of the observed rate constant). Would $t_{1/2}$ become a larger or smaller value if the concentration of H+ doubled? Explain in terms of the previous rate equations.

5. a. Calculate $k_{obs}$ from the slope of the first-order rate plot.

     b. Calculate the value of $k$ from equation 21.5. (*Hint:* Use $[H^+]$ concentration, item d, from your data sheet.)

# DATA SHEET: *FIRST* ORDER KINETICS PLOT
## FOR THE ACID-CATALYZED HYDROLYSIS OF SUCROSE

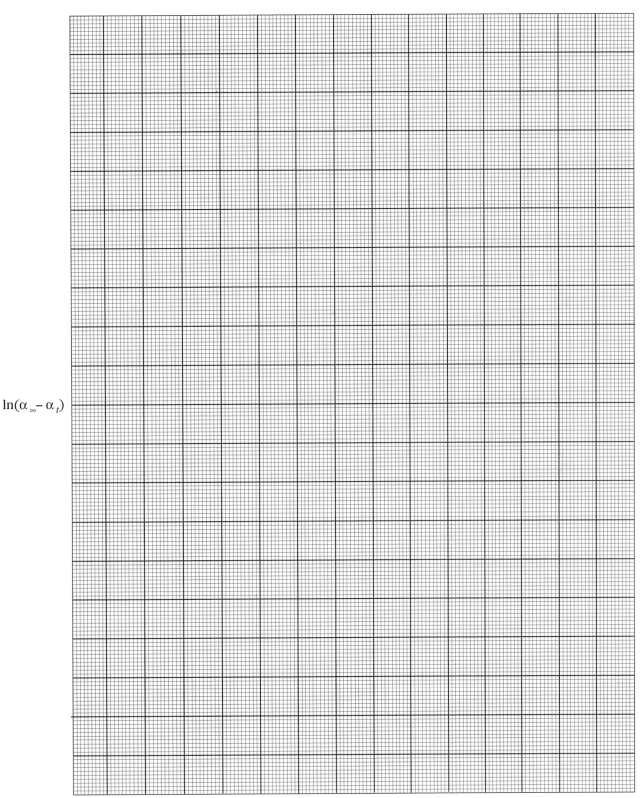

$\ln(\alpha_{\infty} - \alpha_t)$

time (min)

## DATA SHEET: *SECOND* ORDER KINETICS PLOT
## FOR THE ACID-CATALYZED HYDROLYSIS OF SUCROSE

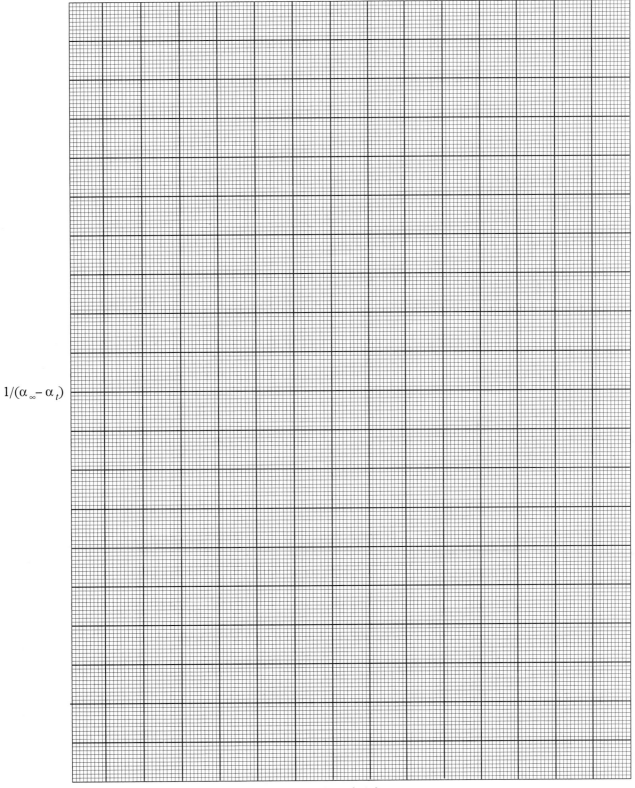

$1/(\alpha_{\infty} - \alpha_t)$

time (min)

## DATA SHEET: THE ACID-CATALYZED HYDROLYSIS OF SUCROSE: A KINETIC STUDY

a. Molarity of HCl used: 12 M

b. Volume of HCl added: 2 mL

c. Final volume of the solution (including the 2 mL of 12 M HCl): _____ mL

d. $[H^+]$ concentration: _____ M

e. Temperature _____ °C

f. Tube length _____ dm

g. $\alpha_o$ = _____ °

h. $[\alpha]_D$ = _____ °

i. $\alpha_\infty$ = _____ °

| $t$ (min) | $\alpha_t$ | $\alpha_\infty - \alpha_t$ | $\ln(\alpha_\infty - \alpha_t)$ | $1/(\alpha_\infty - \alpha_t)$ |
|---|---|---|---|---|
| 1 | | | | |
| 2 | | | | |
| 3 | | | | |
| 4 | | | | |
| 5 | | | | |
| 10 | | | | |
| 15 | | | | |
| 20 | | | | |

From attached graph:

$k_{obs}$ = _____ (show units)

## 21.2 QUALITATIVE TESTS FOR CARBOHYDRATES

Simple **carbohydrates**, also known as **sugars** or **saccharides**, are polyhydroxy aldehydes or ketones generally containing an oxygen atom attached on each carbon atom. In solution, they exist as intramolecular hemiacetals or hemiketals which are in equilibrium with an open form.

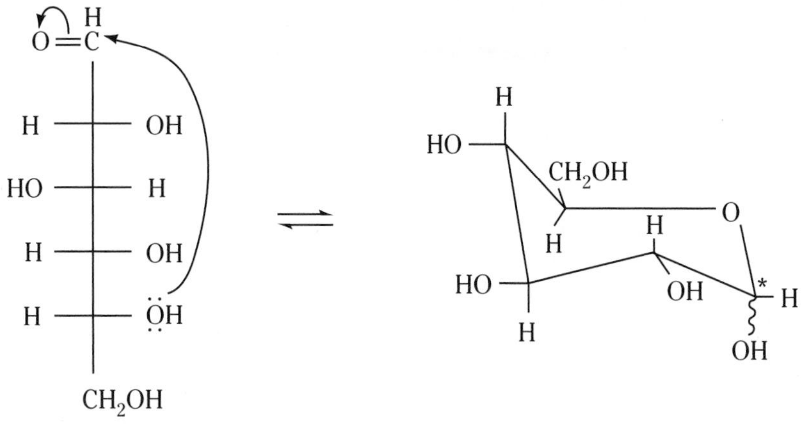

**D** (+)–glucose
(open form)

**D** (+)–glucopyranose
(cyclic hemiacetal)

The cyclic form of a sugar is more stable and commonly exists as either a six-membered ring, called a **pyranose**, or a five-membered ring, called a **furanose**. The open-chain aldehyde form is an **aldose**, and if it contains six carbon atoms, it is an **aldohexose**. (+) -Glucose is an aldohexose, while (–) -arabinose, with five carbons, is an **aldopentose**. An open-chain ketosugar is a **ketose**. Fructose is a **ketohexose** that exists primarily in the furanose form.

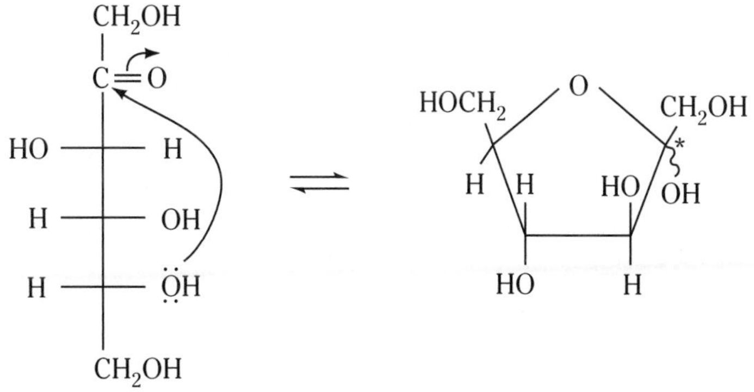

**D** (–)–fructose
(a ketohexose)

fructofuranose

The cyclic forms of the sugar have one more chiral carbon than the open forms. This extra chiral carbon, labelled with *, is generated at the carbonyl carbon during ring formation. The two diastereomeric forms of the sugar which thus arise are called **anomers**. Using glucose as an example, the two anomers are in the pyranose (six-membered ring) form.

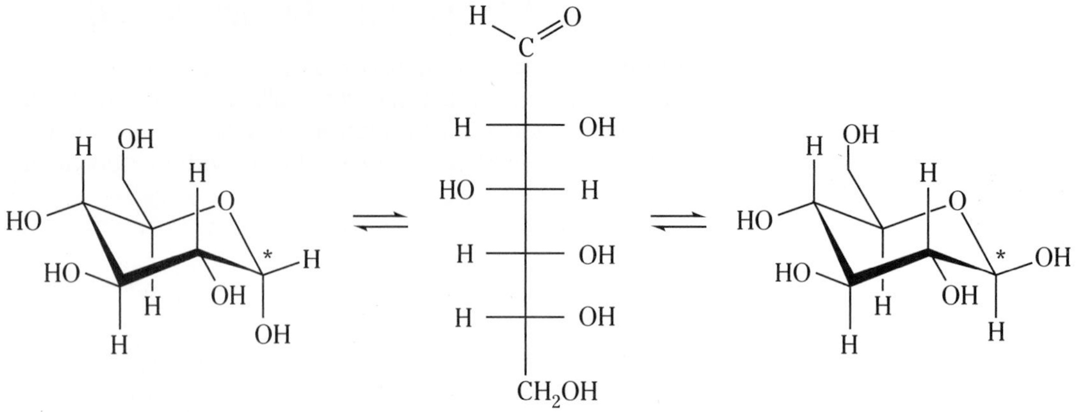

α–D–(+)–glucopyranose
$[\alpha]_D = +112°$

open–chain form
(Fischer Projection)

β–D–(+)–glucopyranose
$[\alpha]_D = +18.7°$

The anomers, designated α and β, are in equilibrium with each other and also with the open form, which is present in very small amounts. When either pure α-D-(+)-glucose, or pure β-D-(+)-glucose, is placed in solution, a drift of optical rotation to an equilibrium value of +52.7° is observed. Such a drift in rotation is termed **mutarotation** and is the result of the formation of an equilibrium mixture of anomers, in this case, 36.4% of α-D-glucopyranose and 63.6% of β-D-glucopyranose.

The absolute configuration of sugars is still identified by the older D and L designations that are based on the two enantiomeric forms of glyceraldehyde (an aldotriose). As with the *R,S* system, the sign of rotation is not related to the D,L designations. Thus, D-glucose is dextrorotary (i.e., it rotates plane-polarized light clockwise, hence the name *dextrose*) while D-fructose is levorotary (and is called *levulose*). The formal names for the two anomeric forms of glucose are α-D-(+)-glucopyranose and β-D-(+)-glucopyranose. From the Fisher projection formula, one can place the sugar in the D- or L- family. In this flat open-chain representation, the carbonyl group is placed as near as possible to the top. Looking at the first chiral carbon from the bottom (usually the penultimate carbon), if the — OH group appears on the right side of the carbon-carbon backbone, the sugar is in the D family, if on the left it is in the L family.

## Fischer Projection Formulas

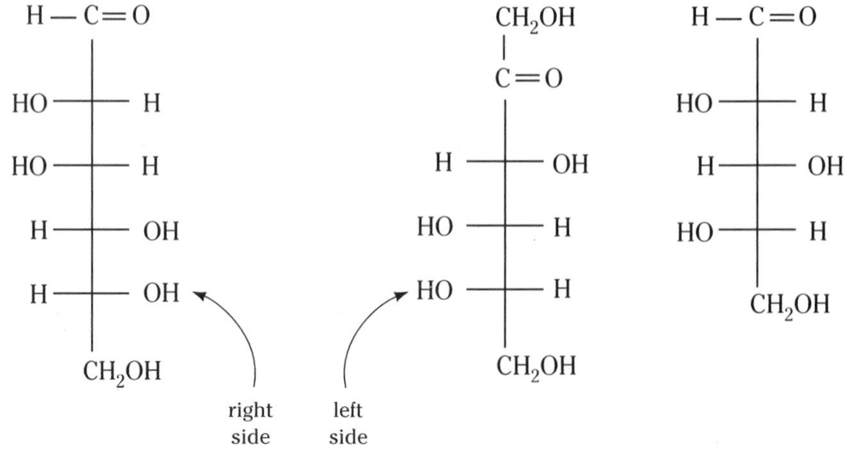

D - (+) - mannose

L - (+) - fructose

an L - aldopentose

Carbohydrates are designated as monosaccharides, disaccharides, trisaccharides, tetrasaccharides, etc., to polysaccharides. *Monosaccharides* are those sugars that cannot be broken down into simpler sugars by hydrolysis. *Disaccharides, trisaccharides, tetrasaccharides,* etc., are dimers, trimers, and tetramers. Maltose (a disaccharide found in malt, used in making beer) is a dimer of α-D-glucose. The bond holding the two glucose units together is called an α-**linkage**. Another disaccharide, cellobiose (a disaccharide obtained from cellulose) is a dimer of β-D-glucose, and is held together by a β-**linkage**. Such a difference in stereochemistry also occurs in starch and cellulose, which are polymers of α-D-glucose and β-D-glucose, respectively. Sucrose (cane sugar) is a cross-dimer (co-dimer) of glucose and fructose, while lactose (found in mammalian milk to an extent of about 5%) is a co-dimer of β-galactose (an aldohexose) and α-D-glucose.

Sugars can be classified as **reducing** or **nonreducing**. All sugars that exist in the hemiacetal or hemiketal form, and can equilibrate in solution with a small but significant amount of the open aldehyde or α-hydroxyketone form, will act as mild reducing agents and will therefore give positive tests with either the Tollens' reagent or Benedict's solution. All reducing sugars mutarotate for the same reason; i.e., the anomers and the open aldehyde or ketone forms equilibrate in solution.

In this experiment, several qualitative tests for sugars will be carried out, and an unknown sugar will be identified.

A generalized flow chart for the identification of an unknown sugar involves a series of simple, quick test-tube experiments. Perform these experiments on the given known solutions as well as on your unknown solution. Record your data and conclusions on the data sheet.

**Suggested References**

"Carbohydrates." In *McGraw-Hill Encyclopedia of Science and Technology,* 7th ed., vol. 3, 206–14. New York: McGraw-Hill, 1992.

Carey, F. A., *Organic Chemistry.* 3rd ed., 1011–1051. New York: McGraw-Hill, 1996.

Furniss, R. S., A. J. Hannaford, P. W. G. Smith, and A. R. Tatchell. *Vogel's Textbook of Practical Organic Chemistry.* 5th ed., Essex: Longman, 1989.

Loudon, G. Marc, *Organic Chemistry.* 3rd ed., 1325–64. Redwood City, CA: Benjamin/Cummings, 1995.

McMurry, J. *Organic Chemistry.* 4th ed., 1010–50. Pacific Grove, CA: Brooks/Cole, 1996.

Morrison, R. T., and R. N. Boyd. *Organic Chemistry.* 6th ed., 1143–1201. Englewood Cliffs, NJ: Prentice-Hall, 1992.

Solomons, T. W. G. *Organic Chemistry.* 6th ed., 1046–89. New York: John Wiley, 1996.

"Sucrose." In *McGraw-Hill Encyclopedia of Science and Technology,* 7th ed., vol. 17, 571–72. New York: McGraw-Hill, 1992.

Vollhardt, K. P. C., and N. E. Schore, *Organic Chemistry.* 2nd ed., 942–83. New York: W. H. Freeman, 1994.

Wade, L. G. *Organic Chemistry.* 3rd ed., 1103–63. Englewood Cliffs, NJ: Prentice-Hall, 1995.

## Individual Tests for Saccharides

In this experiment, the Molish test, Barfoed's test, Benedict's test, Bial's test, Seliwanoff's test, and the Mucic acid test will be performed on the following 1% carbohydrate solutions:

a. glucose (an aldohexose)
b. fructose (a ketohexose)
c. lactose (a reducing disaccharide)
d. sucrose (a non-reducing disaccharide)
e. galactose (an aldohexose)
f. arabinose (an aldopentose) and
g. an unknown.

The identification of the unknown may be effected by the following flow chart.

**Flow Chart For Qualitative Analysis of Sugars**

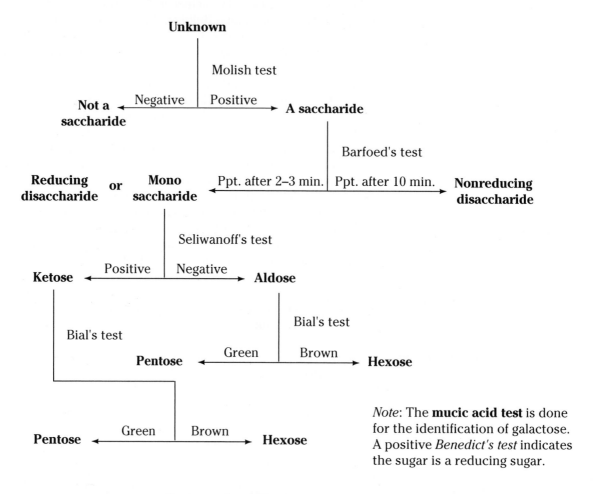

*Note*: The **mucic acid test** is done for the identification of galactose. A positive *Benedict's test* indicates the sugar is a reducing sugar.

Perform the following tests on an aldohexose, a ketohexose, an aldopentose, a reducing disaccharide, a non-reducing disaccharide and an unknown. The following sugars may be used as 1% aqueous solutions: glucose, fructose, arabinose, lactose, sucrose and galactose.

***The Molish Test*** The **Molish test** is a universal test for carbohydrates. Polysaccharides undergo at least partial hydrolysis under the experimental conditions, and the resulting monosaccharides can all be dehydrated to furfural or 5-hydroxymethylfurfural.

$$\begin{array}{l}\text{ketopentose}\\ \text{ketohexose}\\ \text{aldopentos}\\ \text{aldohexose}\end{array} \xrightarrow{\text{H}^+} \quad$$

The furfural will undergo rapid condensation with two moles of α-naphthol under acidic conditions.

The reaction scheme shows:

$$2 \text{ (1-naphthol)} + \text{HC}(=\text{O})\text{-furan-R} \xrightarrow{\text{H}^+} \text{deep red product}$$

deep red color

The test is positive for all saccharides and electron-activated aromatic aldehydes (such as furfural). Comparison of the rates of color formation can be used as evidence to distinguish aldoses and ketoses from polysaccharides.

---

**Experimental**

Place 1 mL of the 1% carbohydrate solution in a test tube, and add 2 drops of a 5% solution of 1-naphthol in ethanol. Add dropwise 10 drops of concentrated sulfuric acid. A positive test is indicated by the formation of a red-violet color at the interface of the two layers after 2–3 minutes.

***Barfoed's Test*** This test distinguishes nonreducing di- or polysaccharides from monosaccharides. It depends on a sugar's ability to reduce a specific reagent, i.e., on its ability to be oxidized.

$$\text{Aldoses} \longrightarrow \text{fast oxidation to RCOO}^-$$
$$\text{Ketoses} \longrightarrow \text{aldoses} \longrightarrow \text{RCOO}^- \text{ (slower)}$$

The oxidizing agent is the blue cupric ($Cu^{+2}$) ion, which is converted to the red cuprous oxide, $Cu_2O$, precipitate.

$$\underset{\text{blue}}{R\overset{\overset{\displaystyle O}{\|}}{C}-H + Cu^{+2}} \longrightarrow \underset{\text{red ppt}}{R\overset{\overset{\displaystyle O}{\|}}{C}-O^- + Cu_2^+O}$$

Nonreducing sugars like sucrose undergo slow hydrolysis under the test conditions and give a positive test after an extended time. Since nonreducing disaccharides hydrolyze very slowly, the absence of any color in 2–3 minutes, indicates that the given sugar is not a monosaccharide.

---

**Experimental**

Place 3 mL of the Barfoed's reagent and 1 mL of 1% carbohydrate solution in a test tube. Place the test tube in boiling water for 2–3 minutes, then cool under running water. A red precipitate of cupric oxide indicates a positive test. A change in the color of the solution from blue to green suggests the presence of a reducing sugar.

***Seliwanoff's Test*** Seliwanoff's test distinguishes between ketoses and aldoses, and it is based on their ability to form furfurals. Ketoses form furanoses more rapidly than do aldoses and aldoses form pyranoses more rapidly than do ketoses. In addition, furanoses dehydrate more rapidly to furfurals than do pyranoses. This is the basis for the test. The dehydrating medium is 6*N* hydrochloric acid, and the reaction is carried out at an elevated temperature. These are milder reaction conditions than those encountered in the Molish test, and allows a greater kinetic selectivity.

Furanose $\xrightarrow{\text{H}^+}$ R—furan—CHO $\xrightarrow{\text{HO} \quad \text{OH}}$ colored products

Ketoses (like fructose) give color in about 2 minutes. Sucrose (which hydrolyzes to fructose) requires a longer time. Ketohexoses give a red color, while ketopentoses give a blue to green color.

**Experimental**     Place 0.5 mL of 1% carbohydrate solution, 1.5 mL of distilled water, and 9 mL of the Seliwanoff's reagent in a test tube. Heat in a hot-water bath for 2–3 minutes, and then cool. Typical rates of color appearance: fructose, ~2 minutes; sucrose, ~5 minutes; glucose and galactose, slow; maltose, very slow.

***Benedict's Test***     Benedict's test is a characteristic test for reducing sugars and α-hydroxyaldehydes. It is similar to Barfoed's test, but is performed under basic conditions. The test depends on the aldehyde's ability to reduce $Cu^{+2}$ and be oxidized to the carboxylic acid. A red precipitate after 2–3 minutes indicates a positive test. Its formation depends on the amount of oxidizing agent present. Prolonged heating (10 minutes) indicates a non-reducing disaccharide, due to partial hydrolysis to monosaccharides.

**Experimental**     In a test tube, place 5 mL of Benedict's reagent and 1 mL of the carbohydrate solution. Heat the solution in a hot-water bath and record the time needed for the red precipitate to form.

***Mucic Acid Test***     Treatment of the monosaccharide with 8*N* nitric acid yields saccharic acids $HOOC(CHOH)_x COOH$. **Mucic acid** (otherwise known as *tetrahydroxyadipic or galactaric acid*) is the acid formed on galactose oxidation.

CHO
H——OH
HO——H
HO——H
H——OH
CH$_2$OH

$\xrightarrow{\text{HNO}_3}$

COOH
H——OH
HO——H
HO——H
H——OH
COOH

$\xleftarrow{\text{HNO}_3}$

CHO
HO——H
H——OH
H——OH
HO——H
CH$_2$OH

D (+) -galactose            Mucic acid            L (−) -galactose

Mucic acid has a high molecular symmetry, and is much less soluble in acidic medium than other dicarboxylic acids. Hence the appearance of a fine white precipitate is considered to be a positive test for D- or L-galactose.

**SAFETY TIPS**

✔    Concentrated nitric acid causes severe burns on contact with skin.
     If a spill occurs, wash immediately with cold water.

---

**Experimental**

In a test tube, place 1 mL of 1% carbohydrate solution and 1 mL of concentrated nitric acid. Heat in a boiling-water bath for one hour. Cool the test tube in running water. A white precipitate is indicative of galactose.

***Bial's Test***   Bial's test can distinguish between pentoses and hexoses. It is based on the nature of the color that develops when the monosaccharide is dehydrated in the presence of orcinol (5-methylresorcinol) and ferric chloride.

Monosaccharide   +   (orcinol structure) $\xrightarrow{FeCl_3}$   colored complex

The initial step involves the condensation of a pentose to furfural, and a hexose to 5-hydroxymethylfurfural. Further reaction of the condensation products with orcinol yields a bluish-green color for furfural (pentoses) and muddy-brown to gray color for hydroxyfurfural (hexoses). The colors also appear with disaccharides and polysaccharides; however, in this case sufficient time is needed for hydrolysis to occur. Pentoses and hexoses provide a positive test in shorter time than disaccharides or polysaccharides because of their more rapid dehydration.

The nature of the colored products is not known. The possibility of an electrophilic aromatic substitution is suspected because of orcinol's ring electron activation.

---

**Experimental**

To 1 mL of a 1% carbohydrate solution, add 2 mL of Bial's reagent. Tilt the tube and heat to boil using a Bunsen burner. If no color is observed, add 2 mL of water and 1 mL of 1-pentanol and mix. Record any color changes.

**Questions**

1. An equilibrium mixture of glucose consists of 36.4% of the α-anomer and 63.6% of the β-anomer. Show how to calculate the equilibrium value (52.7%) for the specific rotation of the mixture, given the specific rotations $[\alpha]_\alpha$ = +112° and $[\alpha]_\beta$ = +18.7°.

2. Calculate the free energy change ($\Delta G°$) for the equilibrium at 25°C, using the relative ratios of the α and the β anomers.

   α-D-(+)-glucopyranose ⇌ β-D-(+)-glucopyranose

3. The β-anomer is somewhat more thermodynamically stable than the α-anomer. Look at the two structures to explain the small difference in stability.

Name: _____

Unknown #: _____

# DATA SHEET: QUALITATIVE TESTS FOR CARBOHYDRATES

| Test | Glucose (aldohexose) | Fructose (ketohexose) | Arabinose (aldopentose) | Lactose (reducing disaccharide) | Sucrose (non-reducing disaccharide) | Galactose (aldohexose) | Unknown (?) |
|------|------|------|------|------|------|------|------|
| Molish | | | | | | | |
| Barfoed's | | | | | | | |
| Benedict's | | | | | | | |
| Bial's | | | | | | | |
| Seliwanoff's | | | | | | | |
| Mucic acid | | | | | | | |

Based on the above data, unknown # _____ is a _____

281

# chapter 22

# Amino Acids and Proteins

## 22.1 SYNTHESIS OF A PEPTIDE

**Amino acids** are the building blocks of peptides and proteins. Two amino acids can be joined by an amide bond to form a **peptide.** The amide bond between two amino acids is called a **peptide bond.** Depending on the number of amino acids involved, the joining of amino acids results in the formation of **dipeptides, tripeptides, . . . , polypeptides.** Polypeptides of molecular weight greater than 10,000 are called **proteins.**

We can represent the formation of a dipeptide by the reaction of two amino acids with the elimination of water. In practice, the carboxyl group is activated to allow the reaction to be carried out under mild conditions.

Since amino acids are difunctional, two amino acids can join to form two different peptides:

$$H_2N - \underset{\underset{R^1}{|}}{CH} - \underset{\underset{O}{\|}}{C} - OH \; + \; H - \underset{\underset{H}{|}}{N} - \underset{\underset{R^2}{|}}{CH} - \underset{\underset{O}{\|}}{C} - OH \longrightarrow H_2N - \underset{\underset{R^1}{|}}{CH} - \underset{\underset{O}{\|}}{C} - \underset{\underset{H}{|}}{N} - \underset{\underset{R^2}{|}}{CH} - \underset{\underset{O}{\|}}{C} - OH \; + \; H_2O$$

Amino acid 1            Amino acid 2                    Peptide 1

$$H_2N - \underset{\underset{R^2}{|}}{CH} - \underset{\underset{O}{\|}}{C} - OH \; + \; H - \underset{\underset{H}{|}}{N} - \underset{\underset{R^1}{|}}{CH} - \underset{\underset{O}{\|}}{C} - OH \longrightarrow H_2N - \underset{\underset{R^2}{|}}{CH} - \underset{\underset{O}{\|}}{C} - \underset{\underset{H}{|}}{N} - \underset{\underset{R^1}{|}}{CH} - \underset{\underset{O}{\|}}{C} - OH \; + \; H_2O$$

Amino acid 2            Amino acid 1                    Peptide 2

In writing structures of polypeptides, the convention is to place the N-terminal amino acid on the left, and the C-terminal amino acid on the right. Then the amino acids are named from left to right using the standardized three-letter codes for their common names (e.g., alanine as **ala,** serine as **ser,** tyrosine as **tyr,** glycine as **gly**). For example,

$$H_2N - CH_2 - \overset{\overset{O}{\|}}{C} - NH - \underset{\underset{CH_3}{|}}{CH} - \overset{\overset{O}{\|}}{C} - OH$$

Gly-ala
(glycylalanine)

$$H_2N - \underset{\underset{CH_2}{|}}{CH} - \overset{\overset{O}{\|}}{C} - NH - \underset{\underset{CH_2OH}{|}}{CH} - \overset{\overset{O}{\|}}{C} - OH$$

Tyr-ser
(tyrosylserine)

A list of the most common amino acids and their structures is given in Table 22.1. Included are also the corresponding *isoelectric points,* as well as pK$_a$s of the α-COOH and α-NH$_3^+$ groups (whether an amino acid or a

**Table 22.1** The twenty common amino acids found in proteins

| Name | Abbreviations | Molecular Weight | Structure | Isoelectric Point | pK$_a$ $\alpha$-COOH | pK$_a$ $\alpha$-NH$_3$ |
|---|---|---|---|---|---|---|
| **Neutral amino acids** | | | | | | |
| Alanine | Ala (A) | 89 | | 6.0 | 2.3 | 9.8 |
| Asparagine | Asn (N) | 132 | | 5.4 | 2.0 | 8.8 |
| Cysteine | Cys (C) | 121 | | 5.0 | 1.8 | 10.2 |
| Glutamine | Gln (Q) | 146 | | 5.7 | 2.2 | 9.1 |
| Glycine | Gly (G) | 75 | | 6.0 | 2.3 | 9.7 |
| Isoleucine* | Ile (I) | 131 | | 6.0 | 2.3 | 9.7 |
| Leucine* | Leu (L) | 131 | | 6.0 | 2.3 | 9.7 |
| Methionine* | Met (M) | 149 | | 5.7 | 2.3 | 9.2 |
| Phenylalanine* | Phe (F) | 165 | | 5.9 | 2.6 | 9.2 |
| Proline | Pro (P) | 115 | | 6.3 | 2.0 | 10.6 |
| Serine | Ser (S) | 105 | | 5.7 | 2.2 | 9.2 |

**Table 22.1**   The twenty common amino acids found in proteins—cont'd

| Name | Abbreviations | Molecular Weight | Structure | Isoelectric Point | pK$_a$ α-COOH | pK$_a$ α-NH$_3$ |
|------|---------------|------------------|-----------|-------------------|---------------|-----------------|
| Threonine* | Thr  (T) | 119 | | 5.6 | 2.1 | 9.1 |
| Tryptophan* | Trp  (W) | 204 | | 5.9 | 2.4 | 9.4 |
| Tyrosine | Tyr  (Y) | 181 | | 5.7 | 2.2 | 9.1 |
| Valine* | Val  (V) | 117 | | 6.0 | 2.3 | 9.7 |
| **Acidic amino acids** | | | | | | |
| Aspartic acid | Asp  (D) | 133 | | 3.0 | 2.1 | 9.8 |
| Glutamic acid | Glu  (E) | 147 | | 3.2 | 2.1 | 9.5 |
| **Basic amino acids** | | | | | | |
| Arginine* | Arg  (R) | 174 | | 10.8 | 2.0 | 9.0 |
| Histidine* | His  (H) | 155 | | 7.6 | 1.8 | 9.1 |
| Lysine* | Lys  (K) | 146 | | 9.7 | 2.2 | 8.9 |

*Essential amino acid

protein). The isoelectric point of a molecule is the specific pH at which the compound will be electrically neutral and will not move in an electric field.

One standard experimental procedure used in the formation of a peptide bond is to activate the carboxyl group by converting it to the more reactive acyl halide (e.g., via thionyl chloride, $SOCl_2$) before reacting it with the amino group of the second amino acid.

The main problem with this approach arises from the possibility of a reaction of the acyl halide functional group with an amino group of the *same* amino acid. Therefore, the amino group of amino acid 1 must be protected so that it does not interfere. The protection of the amino group in amino acid 1 is carried out by adding a **blocking group.** In the peptide synthesis described here, glycylglycine will be prepared using phthalic anhydride to block the amino group of the first glycine molecule. After the peptide bond is formed, the phthaloyl group can be removed by reacting it with hydrazine ($NH_2NH_2$). However, this last step is not included in the experiment because of the toxicity of hydrazine.

**Suggested References**

Carey, F. A., *Organic Chemistry.* 3rd ed., 1117–1127. New York: McGraw-Hill, 1996.

Loudon, G. Marc, *Organic Chemistry.* 3rd ed., 1264. Redwood City, CA: Benjamin/Cummings, 1995.

McMurry, J., *Organic Chemistry.* 4th ed., 1076–1082. Pacific Grove, CA: Brooks/Cole, 1996.

Morrison, R. T., and R. N. Boyd., *Organic Chemistry.* 6th ed., 1221–25. Englewood Cliffs, NJ: Prentice-Hall, 1992.

Solomons, T. W. G., *Organic Chemistry.* 6th ed., 1153, 1170. New York: John Wiley, 1996.

Vollhardt, K. P. C., and N. E. Schore, *Organic Chemistry.* 2nd ed., 1035–67. New York: W. H. Freeman, 1994.

Wade, L. G., *Organic Chemistry.* 3rd ed., 1164–1210. Englewood Cliffs, NJ: Prentice-Hall, 1995.

⚠ **SAFETY TIP**

✔  Phthalic anhydride (step 1) is a corrosive eye and skin irritant and should be handled with gloves. Thionyl chloride (step 2) is a low-boiling (BP 79°C), fuming, corrosive liquid and reacts violently with water. It reacts with water to form hydrochloric acid and sulfur dioxide. It is irritating to the skin, eyes and respiratory system. If skin contact occurs, wash thoroughly with cold water. Tetrahydrofuran (step 3) is a low-boiling (BP 66°C), extremely flammable liquid. It is also irritating to the eyes and respiratory system. More importantly, it forms explosive peroxides on prolonged exposure to light and air which must be decomposed before distilling to small volumes.

## Synthetic Scheme for the Synthesis of Glycylglycine

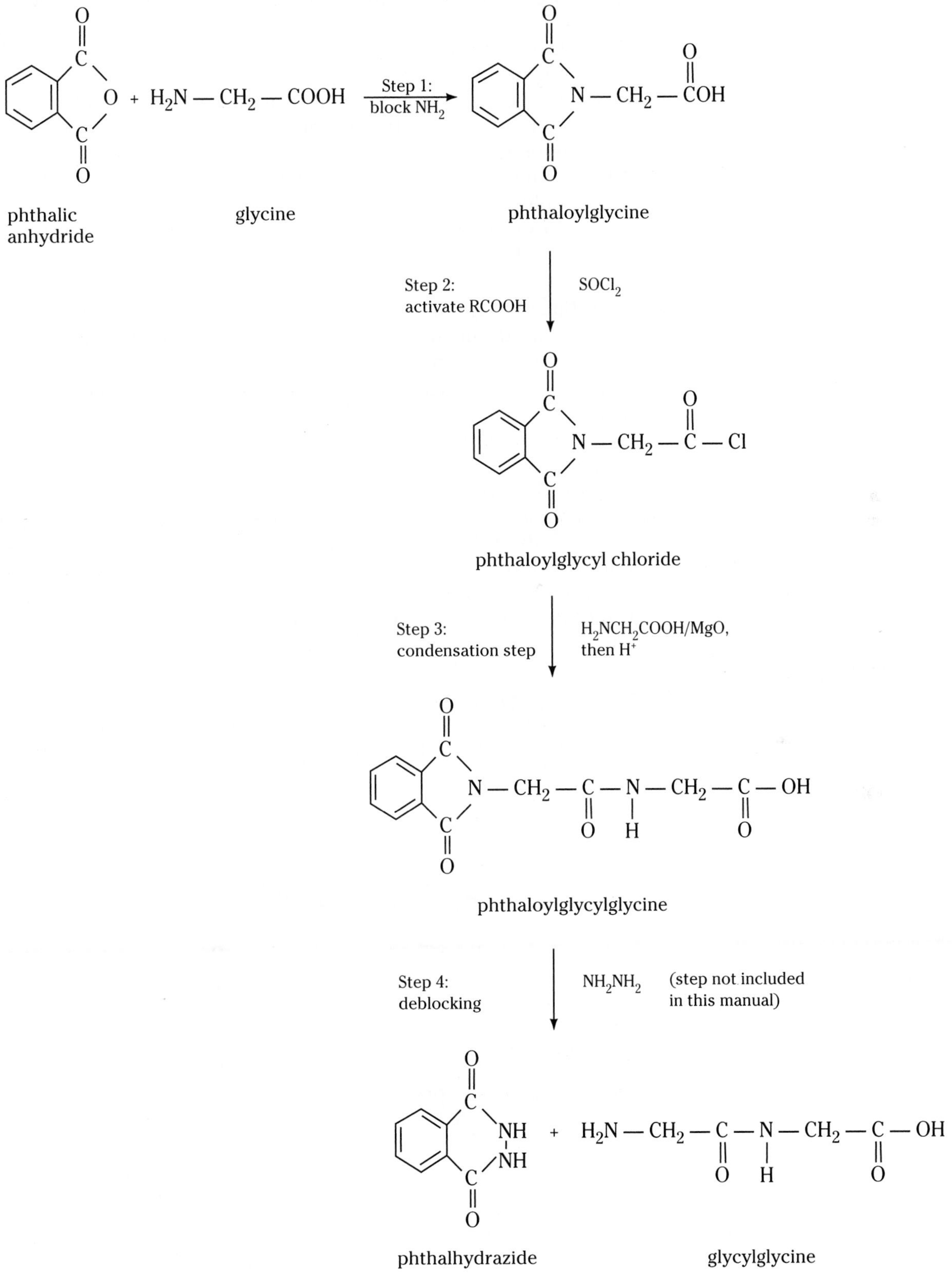

phthalic
anhydride

glycine

$\xrightarrow[\text{block NH}_2]{\text{Step 1:}}$

phthaloylglycine

Step 2:
activate RCOOH      $SOCl_2$

phthaloylglycyl chloride

Step 3:
condensation step      $H_2NCH_2COOH/MgO$,
then $H^+$

phthaloylglycylglycine

Step 4:
deblocking      $NH_2NH_2$      (step not included
in this manual)

phthalhydrazide

glycylglycine

**Experimental**

## Step 1: Phthaloylglycine

1. Place a well-pulverized mixture of 6.0 g of phthalic anhydride and 3.0 g of glycine in a large test tube.
2. Clamp the tube and insert a thermometer so that the solid mixture covers the mercury bulb.
3. Heat the tube gently in an oil bath until the solids melt, and then heat strongly until the temperature reaches 150°–170°C. Continue heating for 30 minutes.
4. Allow the mixture to cool, add 100 mL boiling water, and gravity-filter into a conical flask.
5. Let cool to room temperature, collect the crystals via vacuum filtration, and determine the MP (literature value 196°–198°C).

## Step 2: Phthaloylglycyl chloride

1. Place in a 100-mL round-bottomed flask 4 g of the recrystallized phthaloylglycine (from **step 1**), 20 mL of thionyl chloride *(Caution: Thionyl chloride's fumes are irritating, and the solution should be poured in a well-ventilated hood!)*, 10 drops of dimethylformamide, and a couple of boiling chips. Set up the reflux apparatus as seen in Figure 22.1 and reflux for one hour.
2. Remove the condenser, add 1–2 fresh boiling chips, and distill the excess thionyl chloride using a hot water bath. Collect the distillate in a round-bottomed flask *immersed in ice. (Caution:* avoid any contact of thionyl chloride with water.) Place the distillate *(Use the hood!)* in a container labelled "waste thionyl chloride," which should be properly disposed of by your instructor.
3. Use the solidified residue without further purification in **step 3**.

## Step 3: Phthaloylglycylglycine

1. Add 1.4 g of glycine and 1.0 g of magnesium oxide to 60 mL of water in a 125-mL flask. Place the flask in an ice bath, and cool the mixture to 5°C.

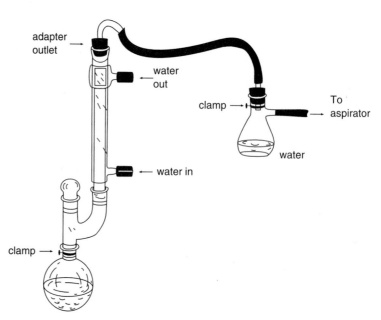

**Figure 22.1**    Reflux set-up for the synthesis of phthaloylglycyl chloride

2. Add dropwise to the cooled suspension, a solution of 3.5 g of phthaloylglycyl chloride (from **step 2**) in 20 mL tetrahydrofuran, maintaining the temperature at 5°C during the addition.
3. Stir the mixture for 10–15 minutes at room temperature to complete the reaction.
4. Acidify the mixture with concentrated hydrochloric acid (check acidity with litmus paper).
5. Cool the reaction mixture in an ice bath for 20 minutes, and collect the crystalline product via vacuum filtration.
6. Recrystallize the product from water (75 mL) and determine the MP (literature value 224°–26°C).

**Questions**
1. Outline the potential hazards encountered in this experiment.
2. Write the structures of all the possible tripeptides that can be made from glycine, alanine, and serine.
3. For each of the possibilities in question 1, write its name using the standardized three-letter codes, e.g., gly- ala- ser.
4. Outline the synthesis of any one of the possibilities described in question 2.
5. What is the reaction of phthaloylglycyl chloride with:
   a. water?
   b. tyrosine?
   c. serine?
   d. ethanol?
6. How many different combinations are possible for:
   a. four different amino acids?
   b. five different amino acids?
   Assume each amino acid appears only once.

## 22.2 QUALITATIVE TESTS FOR AMINO ACIDS AND PROTEINS

Amino acids are the building blocks of proteins. They are organic compounds that contain both the amino ($-NH_2$) and carboxyl ($-COOH$) groups, and as a result, possess both basic and acidic properties (i.e., they are **amphoteric**). Their relatively high melting points and solubility properties can be accounted for by a **dipolar ion structure,** in which the amino and carboxyl groups are converted to ammonium and carboxylate ions. These internal salt structures are also called **zwitterions.**

$$\underset{\text{un-ionized molecule}}{H_2N - \overset{\overset{\textstyle R}{|}}{CH} - COOH} \qquad\qquad \underset{\text{zwitterion}}{\overset{+}{H_3N} - \overset{\overset{\textstyle R}{|}}{CH} - COO^-}$$

The amino acids found in proteins are all α-**amino acids.** Table 22.1 lists the twenty most common amino acids. Marked with an asterisk are the **essential amino acids,** which must be included in the human diet.

Amino acids are joined together via the **peptide bond** (see Experiment 22.1). The bond is formed by condensation of the carboxyl group of one amino acid with the amino group of a second amino acid, and is accompanied by a loss of a molecule of water. Proteins are complex molecules described by the main sequence of amino acids (***primary structure,*** 1°), the three-dimensional shape of the peptides (***secondary structure,*** 2°), the arrangement in space of the molecular threads (***tertiary structure,*** 3°) and the bonding of very large molecules with multiple chains (***quaternary structure,*** 4°).

**Denaturation** is the process that modifies the spatial arrangement of proteins without breaking any covalent bonds. This modification occurs as a result of the breaking of hydrogen bonds and/or hydrophobic and hydrophilic interactions. The change in conformation can affect the molecule's solubility and its biological function, such as its enzyme activity. Denaturation is generally an irreversible process, and the resulting compounds tend to form aggregates and precipitate out of solution.

Denaturation can result from

A **physical** process, such as heating, freezing, irradiation (ultraviolet, X-rays), and ultrasound.

A **chemical** process, such as pH changes and the actions of organic solvents and reagents.

A **biological** process, such as the action of proteolytic enzymes before hydrolysis.

In this experiment, denaturation will be studied through the effects of heat, an organic solvent (ethanol), various inorganic salt solutions, and acids and bases. Color reactions will include the Biuret test, the ninhydrin test, the lead acetate test, Millon's reagent test, the xanthoproteic test, the Hopkins-Cole reagent test, and the Sakaguchi reaction test. A 5% aqueous egg-albumin solution and a 5% aqueous casein solution will be tested, both of which will be referred to as "stock solutions."

**Suggested References**     Carey, F. A., *Organic Chemistry.* 3rd ed., 1092–1133. New York: McGraw-Hill, 1996.

Furniss. R. S., A. J. Hannaford, P. W. G. Smith, and A. R. Tatchell., *Vogel's Textbook of Practical Organic Chemistry.* 5th ed., 1230–31. Essex: Longman, 1989.

McMurry, J. *Organic Chemistry.* 4th ed., 1055–93. Pacific Grove, CA: Brooks/Cole, 1996.

Morrison, R. T., and R. N. Boyd., *Organic Chemistry.* 6th ed., 1205–35. Englewood Cliffs, NJ: Prentice-Hall, 1992.

"Protein." In *McGraw-Hill Encyclopedia of Science and Technology.* 7th ed., vol. 14, 443–50. New York: McGraw-Hill, 1992.

Solomons, T. W. G., *Organic Chemistry.* 6th ed., 1143–75. New York: John Wiley, 1996.

Vollhardt, K. P. C., and N. E. Schore, *Organic Chemistry.* 2nd ed., 1024–40. New York: W. H. Freeman, 1994.

Wade, L. G., *Organic Chemistry.* 3rd ed., 1164–1210. Englewood Cliffs, NJ: Prentice-Hall, 1995.

## i. DENATURATION

### a. Effect of Heat

**Experimental**

Place 2 mL of the stock solution in a test tube, and place the test tube in a boiling water bath for 5 minutes. Record your observations.

### b. Effect of an Organic Solvent

**Experimental**

Place 2 mL of the stock solution in a test tube. Add 2 mL of 95% ethanol and shake the test tube. Record your observations.

### c. Effect of Inorganic Salts

  **SAFETY TIPS**

✔ Lead and mercuric salts are toxic pollutants. All solutions should be properly disposed of by the instructor and should *never* be poured in the sink.

**Experimental**

Place 2 mL of the stock solution in a test tube. Add 2 mL of 5% cupric sulfate and record your observations. Repeat the same test using 5% solutions of lead acetate, of sodium chloride, and of mercuric chloride in place of cupric sulfate. Record your observations.

### d. Effects of Acids and Bases

**Experimental**

Place 2 mL of the stock solution in a test tube. Add 20 drops of concentrated sulfuric acid and record your observations. Repeat using 20 drops of 6N sodium hydroxide in place of sulfuric acid. Record your observations.

## ii. COLOR REACTIONS

### a. Biuret Test
The Biuret test involves complexation of the cupric ion with two peptide bonds to yield a purple color. The test does not occur with free amino acids.

**Experimental**

Place 2 mL of the stock solution in a test tube. Add 2 mL of 3*N* sodium hydroxide solution and 10 drops of 0.5% cupric sulfate solution. Record your observations. Repeat using a 5% solution of glycine, proline and cysteine in place of the stock solution.

***b. Ninhydrin Test***   A positive ninhydrin test is characteristic of free amino acids. Ninhydrin is a mild oxidizing agent that reacts with amino acids in the pH range of 4 to 8 to give a purple color. Only proline and hydroxyproline produce a yellow color.

**Experimental**

Place 5 mL of the stock solution in a test tube. Add 1 mL of a 0.1% aqueous solution of ninhydrin, and observe any color changes. Place the test tube in a boiling water bath for five minutes and observe again. Repeat using 5% aqueous solutions of proline, glycine, and cysteine in place of the stock solution.

***c. Lead (II) Acetate Test***   The lead ( II ) acetate test involves the heavy $Pb^{+2}$ ion, which precipitates some proteins. It also yields the black precipitate of lead (II) sulfide (PbS) when some amino acids, such as cysteine, are present in a basic solution. Hydrolysis of cysteine-containing proteins should give a positive lead (II) acetate test.

 **SAFETY TIPS**

✔   Lead salts are toxic pollutants that should be disposed of properly by the instructor. They should *never* be poured down the sink.

**Experimental**

Place 2 mL of the stock solution in a test tube. Add 15 drops of 6*N* sodium hydroxide and 10 drops of 2% lead ( II ) acetate solution. Record your observations. Heat for 5 minutes in a boiling-water bath, and again record your observations. Repeat using 5% aqueous solutions of glycine and of cysteine and using a hair from your head in place of the stock solution.

***d. Millon's Reagent Test***   Millon's reagent test is characteristic of phenolic amino acids, in particular, tyrosine. It involves complexation of the electron-activated aromatic ring with a mixture of $Hg_2^{+2}$, $Hg^{+2}$, $HNO_3$, and $HNO_2$ to yield a red color.

 **SAFETY TIPS**

✔   Mercury salts are toxic pollutants and should be disposed of properly by the instructor. They should *never* be poured down the sink.

**Experimental**  Place 5 mL of the stock solution in a test tube, add 5 mL of the Millon's reagent, and heat for about 5 minutes. Record your observations before and after heating. Repeat using 5% aqueous solutions of glycine, cysteine, phenol and tyrosine in place of the stock solution. Finally repeat using a hair.

*e. Xanthoproteic Test*   The xanthoproteic test is characteristic of amino acids with an aromatic ring. It involves an electrophilic aromatic substitution with nitric acid to yield a yellow color.

**Experimental**  Place 5 mL of the stock solution in a test tube, and add 10 drops of concentrated nitric acid. Heat in a boiling-water bath for about 3 minutes. Allow the contents of the test tube to cool, and add 4 mL of 6N NaOH solution. The yellow color is intensified by making the solution basic. Record your observations before and after heating, and after adding the NaOH solution. Repeat using aqueous 5% solutions of phenol, of tyrosine, and of glycine in place of the stock solution.

*f. Hopkins-Cole Reagent Test*   In the Hopkins-Cole reagent test, the indole group of tryptophan reacts with glyoxylic acid (OHC—COOH) in the presence of concentrated sulfuric acid to yield a purple complex at the interface of the two solutions.

**Experimental**  Place 3 mL of the stock solution in a test tube, together with 3 mL of the Hopkins-Cole reagent. Incline the test tube and add dropwise 5 mL of concentrated sulfuric acid. *Do not mix.* Record your observations. Repeat using 5% aqueous solutions of glycine, proline, phenylalanine and of tryptophan in place of the stock solution.

*g. Sakaguchi Reaction*   The Sakaguchi reaction test is positive when arginine is present. A positive test yields a red or pink color that fades on standing.

**Experimental**  Add 5 drops of 6N NaOH solution to 1 mL of the stock solution. Next add 3 drops of a solution of 0.05% α-naphthol in 95% ethanol, and then 1 mL of a 0.5% sodium hypochlorite solution. Repeat the test using 5% solutions of arginine and glycine in place of stock solution.

**Questions**
1. Define denaturation and list the various ways it can occur.
2. Give three examples of protein denaturation that occur in everyday life.
3. Suggest a test that would help distinguish between the following pairs of compounds. State what you would do and what you would see.
   a. tryptophan and alanine
   b. tyrosine and alanine
   c. cysteine and tryptophan
   d. insulin (a protein) and glycine
   e. phenylalanine and alanine
   f. methionine and cysteine
   g. proline and isoleucine
   h. insulin and leucine
   i. proline and alanine
   j. phenol and glycine
4. Why did phenol, which is not an amino acid, give a positive Millon's test?

# DATA SHEET PAGE 1: QUALITATIVE TESTS FOR AMINO ACIDS AND PROTEINS

**Denaturation Tests**

Record your observations for each of the following conditions:

| Condition | Egg Albumin (5%) | Casein (5%) |
|---|---|---|
| a. Heat | | |
| b. Organic solvent (ethanol) | | |
| c. Inorganic salts<br>　i. $CuSO_4$ | | |
| 　ii. $Pb(OOCCH_3)_2$ | | |
| 　iii. NaCl | | |
| 　iv. $HgCl_2$ | | |
| d. Acids and bases<br>　i. conc. $H_2SO_4$ | | |
| 　ii. 6 $N$ NaOH | | |

# DATA SHEET PAGE 2: QUALITATIVE TESTS FOR AMINO ACIDS AND PROTEINS

**Color Tests**

Record your observations for the tests which were performed.

| Test Solution | Biuret | Ninhydrin | Lead (II) Acetate | Millon's Reagent | Xantho-proteic | Hopkins-Cole | Saka-guchi |
|---|---|---|---|---|---|---|---|
| Egg albumin | | | | | | | |
| Casein | | | | | | | |
| Glycine | | | | | | | |
| Proline | | | XXX | XXX | XXX | | XXX |
| Cysteine | | | | | | XXX | XXX |
| Hair | XXX | XXX | | | XXX | XXX | XXX |
| Phenol | XXX | XXX | XXX | | | XXX | XXX |
| Tyrosine | XXX | XXX | XXX | | | XXXX | XXX |
| Phenylalanine | XXX | XXX | XXX | XXX | XXX | | XXX |
| Tryptophan | XXX | XXX | XXX | XXX | XXX | | XXX |
| Arginine | XXX | XXX | XXX | XXX | XXX | XXX | |

## 22.3 DECOMPOSITION OF HYDROGEN PEROXIDE WITH CATALASE

It is estimated that the human body has more than 100,000 different proteins. They account, on average, for about one-half of the total dry weight of living organisms. For this reason, appropriately, the name "protein" is derived from the Greek word "protos," meaning primary or first. Many proteins serve as important components of extracellular fluids, but their most important function is as biological catalysts or enzymes. In fact, more than 85 percent of cellular protein is enzymatic protein.

Proteins are polyamides of α-amino acids and occur in various sizes and shapes. The smaller members are called **peptides** and are classified as **dipeptides, tripeptides,** and so on, depending on the number of amino acids present. The peptide (amide) bond is the primary covalent link between amino acids, but disulfide bridges between cysteine residues are also important. The disulfide bridge can link two separate peptide chains or form a loop within a single chain. The three-demensional shape (tertiary structure) of a protein arises from the folding of the polyamide chain into a particularly stable and biologically active structure. Globular proteins are highly folded with hydrophilic polar groups on the surface of the molecule and hydrophobic nonpolar groups enclosed in the interior of the structure. In addition to disulfide bridges, several noncovalent interactions between amino acids that are situated either close to each other or far apart largely determine the protein's tertiary structure. These interactions can be classified as: a) salt bridges; b) hydrogen bonding; and c) nonpolar, hydrophobic forces.

Salt bridges result from the interaction between the negatively charged carboxyl group of one amino acid and the positively charged ammonium group of another. They can be intra- or intermolecular and are ionic in character, much stronger than hydrogen bonds but easily weakened by the action of polar solvents (such as alcohol) and especially by acids and bases.

Hydrogen bonds arise from attractions between hydroxyl and amino groups of the amide functionality and are much weaker (5 kcal/mole) than a typical covalent bond (80 kcal/mole). They are of primary importance especially in maintaining a protein's secondary structure (α-helix and β-pleated sheet). Like salt bridges, they are disrupted by polar solvents and acids and bases.

Proteins can be classified based on their water solubility and overall shape as either globular or fibrous. Both types are susceptible to hydrolysis and denaturation. Hydroysis, either by enzymatic action or by heating in strong acid, results in the breakdown of the protein into its component amino acids. Denaturation, on the other hand, does not break down the protein chain but does destroy its characteristic shape (known as the native state), which is so essential to its function. This is especially true for proteins that function as enzymes. In general globular proteins are much more sensitive to denaturation than the fibrous proteins. Denaturation can be brought about by the use of polar chemicals, thermal or radiant energy, and mechanical action. Polar chemicals such as alcohol, detergents, acids, and bases destroy hydrogen bonds and thereby disrupt the protein's shape. For this reason, some of these compounds can serve as cleansers and disinfectants since they can denature bacteria protein. Heavy metals, such as lead and mercury, disrupt disulfide linkages and precipitate proteins. These metals pose a danger, especially to children, since, when accidentally ingested, they can deactivate key enzymes. Heat energy also breaks hydrogen bonds and coagulates proteins, often causing its precipitation from solution. With the exception of a few RNA molecules, all enzymes are proteins. In general, enzymes have an optimum operation temperature near 37°C, the normal body temperature. Abnormally high temperatures can seriously interfere with their performance. Smaller proteins are less susceptible to heat denaturation than larger proteins. Moreover, in general, the greater the number of disulfide linkages, the more stable the enzyme, since the S—S bond is not subject to thermal disruption. All living cells undergo a degree of protein denaturation and replacement. For those cells that exist at the higher ranges of biologically acceptable temperatures, a great amount of the available energy of the cell must be expended to synthesize new protein and replace the rapidly denatured cellular components. Irradiation with different forms of electromagnetic radiation (X-rays, gamma rays, ultraviolet light) has a similar, if not more drastic, effect. Denaturation can also occur by mechanical action. Beating an egg, which leads to the formation of a frothy mixture filled with air spaces, is a common example.

An enzyme reacts stereospecifically with a particular substrate. The two initially combine to form an enzyme-substrate complex. Complex formation often induces specific conformational changes in the enzyme, which increases the stability of the complex. The conformational requirement for maximum stability is so explicit that only a particular substrate will fit onto a particular enzyme, much as only one key will fit into only one particular lock. The position on the enzyme at which the reaction takes place is called the **active site**. Denaturation of the enzyme (disruption of its conformation) consequently destroys or minimizes its effectiveness. This result is usually irreversible.

**Suggested References**     Doty, P. "Proteins." *Scientific American* 197 (1957):173–84.

---

 **Safety Tips**

  Hydrochloric acid is corrosive and should be poured carefully; use with adequate ventilation.

  Silver nitrate is a toxic oxidant and a skin stainer. Its solutions, therefore, should be handled with gloves.

---

**Experimental**

*Note:* All pieces of glassware must be cleaned and washed with deionized water before the actual experiment. This includes the mortar and pestle as well as the side-arm test tube. All pipettes must be rinsed at least three times with deionized water and should not be placed on a bench or be in contact with any contaminated material (such as dirty glassware, etc.) at any time. Students may work in pairs or groups of three.

## Procedures

1. Peel a small potato and grind it quickly and thoroughly with a pestle in a mortar. Add about 5 mL deionized water to the suspension and grind again.
2. While wearing gloves collect the juice from step 1 in a flask by wrapping the contents of the motar in a cheesecloth or paper towel and squeezing it. This will be your catalase solution.

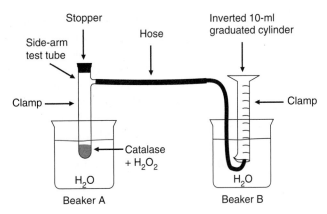

**Figure 22.1** Set up for the oxygen isolation in the decomposition of hydrogen peroxide by catalase

3. Fill a 400-mL beaker (referred to as Beaker A) to about two-thirds its volume with water. Immerse a side-arm test tube equipped with a tight stopper inside the beaker and clamp it as shown in Figure 22.1.

4. Fill two-thirds of an 800- or 1000-mL beaker with water (referred to as Beaker B). Fill a 10-mL graduated cylinder to the brim with water, place your thumb securely over its mouth, and quickly invert it inside the beaker and beneath the water level. Remove your finger and clamp *vertically* the graduated cylinder on a ring stand. At this stage the graduated cylinder must be filled with water. If it is not, then repeat step 4.

5. Attach one side of a 10–20 inch piece of clean rubber tubing to the side-arm of the test tube in Beaker A. Place the other end exactly under the graduated cylinder.

6. Using a graduated pipette, place exactly 2.0 mL of your catalase solution inside the side-arm test tube and let it stabilize for about 2 minutes. Measure the temperature of the water. This will now be referred to as the *ambient temperature*.

7. Add about 10 mL of 3% hydrogen peroxide solution to a test tube. Place the test tube in the beaker for 2–3 minutes to make its temperature identical to the ambient temperature of the catalase solution in step 6.

8. Using a second clean volumetric pipette, transfer exactly 2.0 mL of the hydrogen peroxide solution from step 7 into the catalase solution from step 6; cork quickly, shake the mixture for about 5 seconds, immerse in the water of Beaker A, and then record the time as *time of mixing*.

9. Record the time needed for 4.0 mL of oxygen to accumulate in the inverted graduated cylinder from the time of mixing. Wash both the pipettes and the side-arm test tube with deionized water, and prepare the set-up again as in step 4.

10. Repeat steps 5 through 9, but add 5 drops of HCl together with the hydrogen peroxide solution (step 8). The temperature should again be equal to the one recorded in step 6. Measure the time needed to collect 4.0 mL of oxygen. Is this time shorter, equal to, or longer than the time recorded in step 9?

11. Repeat steps 5 through 9, but this time add 5 drops of 1% $AgNO_3$ together with the hydrogen peroxide solution (step 8). The temperature should again be equal to the one recorded in step 6. Measure the time needed to collect 4.0 mL of oxygen. Is this time shorter, equal to, or longer than the time recorded in step 9?

12. Repeat steps 5 through 9, but this time add 5 drops of ethanol together with the hydrogen peroxide solution (step 8). The temperature should

again be equal to the one recorded in step 6. Measure the time needed to collect 4.0 mL of oxygen. Is this time shorter, equal to, or longer than the time recorded in step 9?

13. Place exactly 2.0 mL of your catalase solution that has been previously heated in a 50°C bath for about 2 minutes inside the side-arm test tube, and repeat steps 6, 7, and 8. Record the time needed to collect 4.0 mL of oxygen. Is this time shorter, equal to, or longer than the time recorded in step 9?

**Questions**

1. Explain the effect of:
   a. hydrochloric acid
   b. silver nitrate
   c. ethanol
   d. heat
   on the catalase reaction with hydrogen peroxide.

2. Describe two ways in which the addition of an acid or a base may affect an enzyme's effectiveness.

3. What effect does a catalyst such as catalase have on:
   a. the speed
   b. the equilibrium constant
   c. the energy of activation
   d. the enthalpy of reaction?

4. Sketch the energy profile diagram for a one-step reaction that is catalyzed and uncatalyzed. Indicate on your diagram the free energy of activation and the overall free energy change of the reaction.

5. Catalase promotes the reaction: $2 H_2O_2 \longrightarrow 2 H_2 + O_2$.
   If 20.0 mL of oxygen is collected at STP conditions, how many millimoles of peroxide was originally present?

6. a. Write out the structure of a heptapeptide having the following primary structure:

$$\overline{\text{Gly-Val-Cys-Ala-Gly-Leu-Cys}}$$

   b. Treatment with certain reducing agents (such as Na/liquid ammonia) causes a change in the protein structure. What has happened?

   c. The reaction in part b can be reversed by treatment with an oxidizing agent (such as hydrogen peroxide). Explain.

## DATA SHEET

a. Ambient room temperature (for steps 5–11): _____ °C

b. Time needed to collect 4.0 mL of oxygen (step 9): _____ min.

c. Time needed to collect 4.0 mL of oxygen when HCl has been added (step 10): _____ min.

d. Time needed to collect 4.0 mL of oxygen when AgNO$_3$ has been added (step 11): _____ min.

e. Time needed to collect 4.0 mL of oxygen when ethanol has been added (step 12): _____ min.

f. Time needed to collect 4.0 mL of oxygen at 50°C (step 13): _____ min.

# chapter 23

# Lipids: Fats, Oils, and Steroids

## 23.1  PREPARATION AND PROPERTIES OF A SOAP

**Lipids** constitute a structurally diverse group of biomolecules that are soluble in one or more nonpolar solvents, such as ether, acetone, chloroform, and benzene (which are sometimes called **fat solvents**). **Fats** and **oils** constitute a major class of lipids. Structurally they are esters of glycerol and fatty acids, and they can be represented by the general formula:

$$
\begin{array}{l}
\text{H}_2\text{CO} - \overset{\displaystyle O}{\overset{\displaystyle \|}{\text{C}}} - \text{R} \\[6pt]
\text{HCO} - \overset{\displaystyle O}{\overset{\displaystyle \|}{\text{C}}} - \text{R} \\[6pt]
\text{H}_2\text{CO} - \overset{\displaystyle O}{\overset{\displaystyle \|}{\text{C}}} - \text{R}
\end{array}
$$

where R represents a chain of carbon atoms (usually from 11 to 17 carbons)

The alkaline hydrolysis of fats is called **saponification,** (see Experiment 16.6), a process that yields glycerol and the salt of the fatty acid. These salts are **soaps**. In this experiment, a soap will be prepared and some of its properties will be examined.

$$
\begin{array}{l}
\text{H}_2\text{CO} - \overset{\displaystyle O}{\overset{\displaystyle \|}{\text{C}}} - \text{R} \\[6pt]
\text{HCO} - \overset{\displaystyle O}{\overset{\displaystyle \|}{\text{C}}} - \text{R} \quad + \ 3\text{NaOH} \longrightarrow \\[6pt]
\text{H}_2\text{CO} - \overset{\displaystyle O}{\overset{\displaystyle \|}{\text{C}}} - \text{R}
\end{array}
\qquad
\begin{array}{l}
\text{H}_2\text{C} - \text{OH} \\[6pt]
\text{HC} - \text{OH} \quad + \ 3\,\text{R} - \overset{\displaystyle O}{\overset{\displaystyle \|}{\text{C}}} - \text{O}^-\text{Na}^+ \\[6pt]
\text{H}_2\text{C} - \text{OH}
\end{array}
$$

fat             sodium           glycerol          soap (salt of
                hydroxide                          a fatty acid)

The ability of soap to dissolve dirt is a direct result of the presence of a hydrophobic portion (water-shy, i.e., nonpolar) *and* a hydrophilic (water-loving, i.e., polar) portion of the molecule. The hydrophobic part is the long-chain R group, and the hydrophilic part is the carboxylate group. Dirt and oils, which are practically insoluble in water, are soluble in soap solutions because of the formation of **micelles.** Each micelle (Figure 23.1)

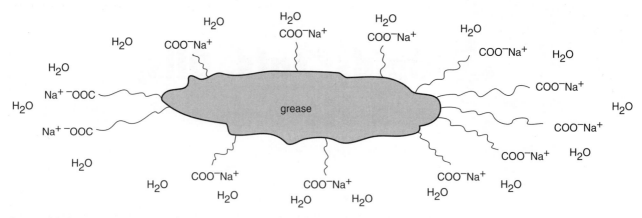

**Figure 23.1**    A simple picture of a soap micelle solubilizing (emulsifying) a grease molecule

is a spherical structure that is made from the dirt or oil unit (located at the center of the sphere) and the soap species (that surrounds it). The hydrophobic regions of the soap are directed *inward* and are associated with the dirt or oil, while the hydrophilic regions of the soap are directed *outward* and toward the water solution. As a result, the dirt or oil molecules are engulfed in the soap species, and can thus be extracted from the cloth or surface to which they were originally attached. This is possible only if the soap is water soluble—such as a sodium or potassium salt.

The fatty acids may be *saturated* or *unsaturated*. Fats, which are solids at room temperature, are generally more saturated than the oils, which are liquids at room temperature. Exceptions are the tropical oils that are highly saturated, such as coconut oil and palm oil. Table 23.1 lists some of the important fatty acids.

**Table 23.1**    Important fatty acids

| Acid Name | Structure | MP (°C) |
|-----------|-----------|---------|
| Lauric | $CH_3(CH_2)_{10}COOH$ | 44 |
| Myristic | $CH_3(CH_2)_{12}COOH$ | 58 |
| Palmitic | $CH_3(CH_2)_{14}COOH$ | 63 |
| Stearic | $CH_3(CH_2)_{16}COOH$ | 70 |
| Oleic | $(Z)-CH_3(CH_2)_7CH = CH(CH_2)_7COOH$ | 4 |
| Linoleic | $(Z,Z)-CH_3(CH_2)_4CH = CH - CH_2 - CH = CH(CH_2)_7COOH$ | −5 |
| Arachidonic | $all-(Z)-CH_3(CH_2)_4(CH = CH - CH_2)_4CH_2CH_2COOH$ | −50 |

**Suggested References**    Carey, F. A., *Organic Chemistry.* 3rd ed., 774–75, 826–31. New York: McGraw-Hill, 1996.

McMurry, J., *Organic Chemistry.* 4th ed., 1103–05. Pacific Grove, CA: Brooks/Cole, 1996.

Morrison, R. T., and R. N. Boyd, *Organic Chemistry.* 6th ed., 1119–28. Englewood Cliffs, NJ: Prentice-Hall, 1992.

Solomons, T. W. G., *Organic Chemistry.* 6th ed., 815–18, 1099–1101. New York: John Wiley, 1996.

Vollhardt, K. P. C., and N. E. Schore, *Organic Chemistry.* 2nd ed., 729. New York: W. H. Freeman, 1994.

Wade, L. G., *Organic Chemistry.* 3rd ed., 1215–17. Englewood Cliffs, NJ: Prentice-Hall, 1995.

"Lipid." In *McGraw-Hill Encyclopedia of Science and Technology,* 7th ed., vol. 10, 100–101. New York: McGraw-Hill, 1992.

**SAFETY TIPS**

✔ Concentrated sodium hydroxide is very caustic. Therefore, care should be taken to avoid skin contact. If this occurs, the affected area should be washed immediately with large amounts of water. To prevent accidents, a funnel should be used when pouring sodium hydroxide solutions from a bottle to a flask.

✔ The use of *several* boiling chips is advisable to prevent overheating and bumping of the solution. The use of a 200- or 250-mL (instead of a 100-mL) round-bottomed flask will also reduce the risk of severe bumping.

**Experimental**

1. Place 10 mL of cottonseed oil, 10 mL of ethanol, 10 mL of 6*M* sodium hydroxide solution, and several boiling chips in a 250-mL round-bottomed flask and reflux for 40 minutes.

2. Carefully pour the mixture into a 400-mL beaker that contains 100 mL of a saturated NaCl solution. Stir vigorously.

3. Isolate the soap by vacuum filtration, and wash the precipitate in the Büchner funnel with 40 mL of water while the vacuum suction is on full force.

4. Place about 1 g of the soap (enough to cover the tip of a spatula) in a 100-mL beaker that contains 50 mL of 50% ethanol-water solution. Test the solution with red and blue litmus paper. Is the solution acidic or basic? Use this solution to perform the following tests in steps 5 through 8. Record your results on the data sheet.

5. Place 5 mL of the soap solution in a test tube and shake thoroughly. Do suds form? Add 1 mL of a 1% calcium chloride solution and shake. Do suds form?

6. Repeat step 5 using a 1% ferrous chloride (instead of calcium chloride) solution. Do suds form? Explain.

7. Repeat step 5 using a 1% magnesium chloride (instead of calcium chloride) solution. Do suds form? Explain.

8. Place 5 mL of the soap solution from step 4 in a test tube. Add 2 drops of mineral oil. Shake the test tube. Does the oil dissolve? Add 1 mL of the calcium chloride solution and shake the test tube again. What happens to the oil? For comparison, add 2 drops of the mineral oil to 5 mL of a 50% ethanol-water solution without the soap. Does the oil dissolve?

**Questions**

1. Consult reference books and provide a paragraph for each of the following terms:
   a. hard water
   b. saponification number
   c. emulsion and emulsifying agent
   d. triglyceride

2. Provide structures for the following:
   a. ethyl palmitate
   b. glycerol trioleate
   c. cholesterol-3-acetate
   d. methyl stearate

3. What are the saponification products of the compounds in question 2?

# DATA SHEET: PREPARATION AND PROPERTIES OF SOAP

| Process | Effect on the Soap Solution |
|---|---|

Step 4:  Red litmus paper: _____

Step 4:  Blue litmus paper: _____

      *Conclusion:* the soap solution is (acidic/basic) _____

Step 5:  Shaking the test tube: _____

Step 5:  Addition of $CaCl_2$: _____

Step 6:  Addition of $FeCl_2$: _____

Step 7:  Addition of $MgCl_2$: _____

      *Conclusion:* _____

Step 8:  Addition of mineral oil to soap solution: _____

Step 8:  Addition of mineral oil *and* $CaCl_2$ to soap solution: _____

Step 8:  Addition of mineral oil to ethanol-water solution: _____

      *Conclusion:* _____

## 23.2 QUALITATIVE DETERMINATION OF UNSATURATION IN LIPIDS: THE HANUS TEST

All fats and oils contain some unsaturation. The degree of unsaturation can be measured by determining the iodine number, which is defined as the number of grams of iodine absorbed by 100 grams of the fat or oil. Because iodine is relatively unreactive to double bonds, the actual reagent is often iodine monobromide (IBr) and not molecular iodine ($I_2$). However, the calculations are performed as if iodine were added. Table 23.2 lists some representative values. The larger the iodine number, the greater the degree of unsaturation. In this experiment, several oils will be tested qualitatively using a solution of iodine in glacial acetic acid to which a small amount of bromine has been added *(Hanus reagent)*. Alternatively, these same oils can be tested qualitatively using a solution of bromine in methylene chloride. The brown color will disappear as the bromine adds to the double bond present in the oil molecule. The greater the unsaturation, the more bromine will be consumed.

Table 23.3 provides a comparison of dietary fats in cholesterol, saturated, monounsaturated and polyunsaturated fats.

**Table 23.2**   Representative iodine numbers

| Fat | Iodine Number |
|---|---|
| *Animal Fats* | |
| Butter | 36 |
| Beef Tallow | 50 |
| Lard | 59 |
| | |
| *Vegetable Oils* | |
| Coconut | 10 |
| Olive | 81 |
| Cottonseed | 106 |
| Corn | 123 |
| Soybean | 130 |
| Linseed | 195 |

**Table 23.3**   Comparison of dietary fats

| Dietary Fat | Cholesterol (mg/tablespoon) | Saturated Fat % | Mono-unsaturated Fat % | Poly-unsaturated Fat % |
|---|---|---|---|---|
| Canola oil | 0 | 5 | 65 | 30 |
| Saflower oil | 0 | 8 | 10 | 82 |
| Sunflower oil | 0 | 10 | 15 | 75 |
| Corn oil | 0 | 10 | 20 | 70 |
| Peanut oil | 0 | 10 | 50 | 30 |
| Olive oil | 0 | 10 | 80 | 10 |
| Soybean oil | 0 | 15 | 25 | 60 |
| Margarine | 0 | 15 | 45 | 35 |
| Cottonseed oil | 0 | 30 | 15 | 55 |
| Chicken fat | 11 | 25 | 50 | 25 |
| Lard | 12 | 45 | 40 | 15 |
| Beef fat | 14 | 50 | 45 | 5 |
| Palm oil | 0 | 50 | 40 | 10 |
| Butter | 33 | 55 | 25 | 5 |
| Coconut oil | 0 | 75 | 5 | 5 |

**Suggested References**

Carey, F. A., *Organic Chemistry*. 3rd ed., 1058–63. New York: McGraw-Hill, 1996.

Mancott, A., and J. Tietjen, "Polyunsaturation in Food Products." *Chemistry 47*, 1974: 29–30.

McMurry, J. *Organic Chemistry*. 4th ed., 1058–63. Pacific Grove, Ca: Brooks/Cole, 1996.

Morrison, R. T., and R. N. Boyd, *Organic Chemistry*. 6th ed., 1119–28. Englewood Cliffs, NJ: Prentice-Hall, 1992.

Solomons, T. W. G., *Organic Chemistry*. 6th ed., 1099–1101. New York: John Wiley, 1996.

Vollhardt, K. P. C., and N. E. Schore, *Organic Chemistry*. 2nd ed., 779–81. New York: W. H. Freeman, 1994.

Wade, L. G., *Organic Chemistry*. 3rd ed., 1211–30. Englewood Cliffs, NJ: Prentice-Hall, 1991.

"Lipid." In *McGraw-Hill Encyclopedia of Science and Technology,* 7th ed., vol. 10, 100–101. New York: McGraw-Hill, 1992.

 **SAFETY TIPS**

 The Hanus reagent contains a small amount of bromine, and therefore, skin contact should be avoided. If contact occurs, the skin should be washed thoroughly with large amounts of water. The experiment should be performed in a well-ventillated area.

 Methylene chloride (BP 40°C) is harmful when ingested or when its vapors are inhaled. It should, therefore, be handled with adequate ventilation.

**Experimental**

1. Arrange four test tubes in a test tube rack, and label them 1 through 4. To the first one, add 10 drops of coconut oil; to the second add 10 drops of cottonseed oil; to the third, 10 drops of olive oil; and to the fourth, 10 drops of linseed oil.

2. Add 5 mL of methylene chloride and 4 drops of the Hanus solution to each test tube, cork the test tube, and shake each sample. Place the test tubes in a cool, dark place for about 30 minutes, and then rank the oils by their ability to absorb the iodine color. Which oil has the highest degree of unsaturation, and which has the lowest?

3. Alternatively add 5 mL of methylene chloride to each test tube to dissolve the oil. Using a medicine dropper, add dropwise to each test tube a 5% solution of bromine in methylene chloride. Shake the test tube after each addition. Continue adding the bromine solution until the brown color of the bromine no longer disappears. Record the number of drops required for each test tube.

**Questions**

1. On the basis of your results with the Hanus solution, and/or the $Br_2$/methylene chloride solution, arrange the vegetable oils in order of *least* saturated to *most* saturated. Which oil would have the highest iodine number and which the lowest? How do these results compare with the values listed in Table 23.2?

2. Rank the following fatty acids in order of *lowest* to *highest* amount of unsaturation: arachidonic, palmitic, oleic, linoleic. (*Hint:* Consult Table 23.1 for structures.)

## DATA SHEET: QUALITATIVE DETERMINATION OF UNSATURATION IN LIPIDS: THE HANUS TEST

| Test Tube # | Oil | Effect on the Hanus Solution | Number of Drops of Bromine Solution Added until Saturation |
|---|---|---|---|
| 1 | Coconut | _____ | _____ |
| 2 | Corn | _____ | _____ |
| 3 | Olive | _____ | _____ |
| 4 | Linseed | _____ | _____ |

*Conclusion:* The most unsaturated oil is _____;

the least unsaturated oil is _____

## 23.3 QUANTITATIVE DETERMINATION OF THE IODINE NUMBER OF AN UNKNOWN FAT OR OIL

The degree of unsaturation in fats and oils is measured by determining the *iodine number*, defined as the number of grams of iodine that can be added to 100 grams of the fat or oil. A highly unsaturated oil will have a high iodine number, while a largely saturated fat will have a very small iodine number. The addition of iodine to a double bond is a rather slow process, and often other halogens are preferred. However, the calculations are performed as if iodine was added. A common reagent for the halogenating process is pyridinium hydrogen sulfate dibromide, which brominates all double bonds in the fat or oil.

$$HSO_4^-. \ Br_2$$

pyridinium hydrogen
sulfate dibromide

Since the degree of unsaturation of a certain fat or oil whose iodine number is to be determined is unknown, an excess of pyridinium hydrogen sulfate dibromide is added. The unreacted bromine is then treated with excess potassium iodide. The free iodine produced is then titrated with a sodium thiosulfate solution of known concentration, using starch as the indicator. The overall sequence of reactions is as follows:

$$lipid + Br_2 \ (excess) \longrightarrow brominated \ lipid + unreacted \ Br_2$$
$$(orange)$$

$$unreacted \ Br_2 + 2KI \ (added \ in \ excess) \longrightarrow 2KBr + \ \ I_2$$
$$(orange) \hspace{8cm} (brown)$$

$$I_2 \ \ + 2Na_2S_2O_3 \longrightarrow \ \ 2NaI \ \ + Na_2S_4O_6$$
$$(brown) \hspace{4cm} (clear)$$

Thus, the amount of $Br_2$ absorbed by the fat or oil can be calculated by determining the difference between the amount of unreacted bromine and the amount initially added. The calculation is done as though iodine were added, and according to the equation, 2 moles of $Na_2S_2O_3$ are equivalent to 1 mole of iodine (which is equivalent to 1 mole of bromine). Thus, one formula unit of $Na_2S_2O_3$ is equivalent to one atom of iodine.

The mass of iodine taken up by the sample can be calculated from the equation:

$$(Normality)_{Na_2S_2O_3} \times (Volume)_{net} = (mass/at.wt.)_{I_2}$$

where

$$(Volume)_{net} = (Volume \ of \ Na_2S_2O_3 \ for \ titration \ of \ blank)$$
$$-(Volume \ of \ Na_2S_2O_3 \ for \ titration \ of \ sample)$$

The iodine number of the fat can be calculated from the equation:

$$\text{Iodine Number} = \frac{\text{Mass of iodine in sample}}{\text{mass of sample}} \times 100$$

**Suggested References**

Mancott, A., and J. Tietjen, "Polyunsaturation in Food Products." *Chemistry 47* (1974): 29–30.

Morrison, R. T., and R. N. Boyd, *Organic Chemistry*. 6th ed., 1119–28. Englewood Cliffs, NJ: Prentice-Hall, 1992.

Solomons, T. W. G., *Organic Chemistry*. 6th ed., 1099–1101. New York: John Wiley, 1996.

Wade, L. G. *Organic Chemistry*. 3rd ed., 1211-30. Englewood Cliffs, NJ: Prentice-Hall, 1995.

"Lipid." In *McGraw-Hill Encyclopedia of Science and Technology*, 7th ed., vol. 10, 100-101. New York: McGraw-Hill, 1992.

 **SAFETY TIPS**

 Methylene chloride (BP 40°C) is harmful when ingested or when its vapors are inhaled. It should, therefore, be handled with adequate ventilation.

 Pyridinium hydrogen sulfate dibromide is corrosive and should be handled with gloves.

**Experimental**

1. Obtain four *dry*, glass-stoppered conical flasks, weigh them accurately, and label them "Blank-1," "Blank-2," "Sample-1," and "Sample-2." Record their masses in the data sheet.

2. Place accurately weighed samples (0.1 to 0.3 g) of the unknown fat or oil in the flasks labeled "Sample-1" and "Sample-2." The masses need not be identical. It is preferable to use an electronic balance.

3. Add 10 mL of methylene chloride to each of the four flasks.

4. Using a pipette, add exactly 25.0 mL of pyridinium hydrogen sulfate dibromide to each of the four flasks. Stopper, shake, and place the flasks in your cabinet. Swirl them in the dark every 5 minutes.

5. At the end of the 30-minute period, add 10 mL of 15% KI solution using your 25-mL or 50-mL graduated cylinder. Stopper and swirl the flasks. The solutions should turn dark brown.

6. Add 75 mL of water to each of the four flasks and mix well.

7. Titrate the contents of each flask with *the same* standardized 0.1$N$ $Na_2S_2O_3$ solution. It is preferable to start with both blanks first. Titrate to a light yellow (straw) color, and then add 5 mL of starch indicator. The solution will turn blue. Continue the titration while swirling the flask. At the end point, the solution becomes colorless. The volumes of $Na_2S_2O_3$ should be almost identical for both blank solutions.

8. Record all four volumes in the data sheet, and calculate the iodine number of the unknown oil or fat.

**Questions**
1. Explain the purpose of the following in this experiment:
   a. KI addition (step 5)
   b. using *dry* flasks at the beginning of the experiment (*Hint:* What is the result of the reaction of $Br_2$ and water?)
   c. starch (step 7)
   d. continuous swirling during the $Na_2S_2O_3$ titration.
2. A triglyceride sample weighing 0.450 g required 13.6 mL of 0.100$N$ $Na_2S_2O_3$. A blank sample of the pyridinium hydrogen sulfate dibromide solution used 23.5 mL.
   a. Calculate the iodine number of the triglyceride.
   b. Is the triglyceride more likely to be a fat or an oil?
3. Calculate the iodine number for pure samples of the following:
   a. ethyl oleate
   b. glycerol trilinoleate
   c. ethyl palmitate.
4. What are the bromination products of the following:
   a. ethyl oleate
   b. ethyl palmitate.

## DATA SHEET: QUANTITATIVE DETERMINATION OF THE IODINE NUMBER OF AN UNKNOWN FAT OR OIL

| | Blank-1 | Blank-2 | Sample-1 | Sample-2 |
|---|---|---|---|---|
| a. Mass of empty flask | XXXXXXX | XXXXXXX | _____ g | _____ g |
| b. Mass of flask + fat/oil | XXXXXXX | XXXXXXX | _____ g | _____ g |
| c. Mass of fat/oil | XXXXXXX | XXXXXXX | _____ g | _____ g |
| d. Normality of $Na_2S_2O_3$ | $0.1N$ | $0.1N$ | $0.1N$ | $0.1N$ |
| e. Volume $Na_2S_2O_3$ | _____ mL | _____ mL | _____ mL | _____ mL |
| f. Average volume for blank | _____ mL | | | |
| g. $(Volume)_{net} = f - e$ (volume of $I_2$ absorbed) | XXXXXXX | XXXXXXX | _____ mL | _____ mL |
| h. Mass of iodine taken by sample $= d \times g \times 126.9$ | | | _____ g | _____ g |
| i. Iodine Number $= (h/c) \times 100$ | | | _____ | _____ |

**Conclusion:** The average iodine number for fat/oil # _____ is _____

## 23.4 THE LIEBERMANN-BURCHARD TEST FOR CHOLESTEROL

**Steroids** are another major class of lipids. They are structural derivatives of the perhydrocyclopentanophenanthrene ring system. In naturally occuring steroids, the BC and CD ring fusions are almost always *trans,* but the AB ring fusion may be *cis* or *trans.* **Cholesterol,** a steroid alcohol, is the biological

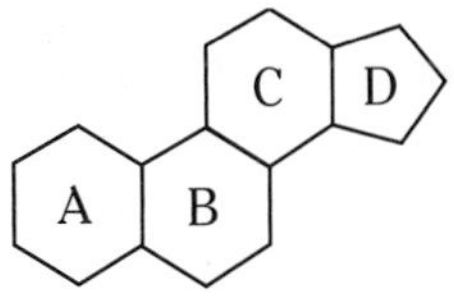

perhydrocyclopentanophenanthrene ring system

precursor of other important steroids. In this part of the experiment, cholesterol will be identified by its reaction with acetic anhydride in the presence of sulfuric acid (the Liebermann-Burchard test). Acetic anhydride condenses with the C-3 hydroxyl group of cholesterol and related sterols to yield esters. If, as in the case of cholesterol, the sterol also has a C-5 unsaturation, epimerization, followed by dehydration occurs which leads to the formation of a characteristic color.

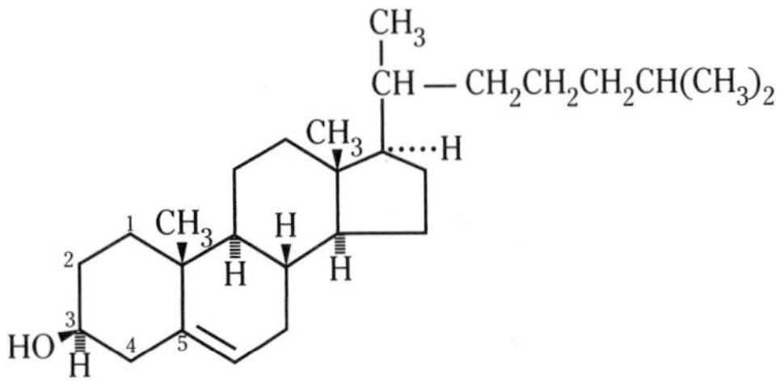

**Suggested References**

McMurry, J., *Organic Chemistry.* 4th ed., 1113–20. Pacific Grove, CA: Brooks/Cole, 1996.

Morrison, R. T., and R. N. Boyd, *Organic Chemistry.* 6th ed., 1134–36, 1140–41. Englewood Cliffs, NJ: Prentice-Hall, 1992.

Solomons, T. W. G., *Organic Chemistry.* 5th ed., 1064–67. New York: John Wiley, 1992.

Wade, L. G. *Organic Chemistry.* 3rd ed., 1220–23. Englewood Cliffs, NJ: Prentice-Hall, 1995.

---

⚠️ **SAFETY TIPS**

✔ Methylene chloride (BP 40°C) is harmful when ingested or when its vapors are inhaled. It should, therefore, be handled with adequate ventilation.

✔ Concentrated sulfuric acid is extremely corrosive and should be handled with care.

✔ Acetic anhydride is a lachrymator and should also be used with adequate ventilation.

**Experimental**

1. Place a few crystals (about 0.2 g) of cholesterol in a test tube containing 3 mL of methylene chloride and dissolve. Add 10 drops of acetic anhydride and 3 drops of concentrated sulfuric acid. Shake the contents of the test tube and record your observations.

2. Repeat the experiment using lanosterol, ergosterol, linseed oil, and coconut oil.

**Questions**

1. Draw the formulas for cholesterol, cholestanol, and ergosterol.
   a. Which of these would react with acetic anhydride? Draw the reaction products.
   b. Which of these steroids would give a positive Liebermann-Burchard test?

## DATA SHEET: THE LIEBERMANN-BURCHARD TEST FOR CHOLESTEROL

| Lipid | Observation |
| --- | --- |
| a. Cholesterol | _____ |
| b. Ergosterol | _____ |
| c. Lanosterol | _____ |
| d. Linseed oil | _____ |
| e. Coconut oil | _____ |

*Conclusion:* Compounds _____ have a hydroxyl *and* a C-5 unsaturation

# chapter 24

# Qualitative Analysis

The identification of a general unknown is one of the experiments that students like best because of the challenge it offers. In this experiment the unknown will contain only *one* functional group, and this group will be limited to amines, carboxylic acids, phenols, alcohols, aldehydes, and ketones. In addition, it will be chosen from one of the tables that appear in this chapter. Unknown aldehydes and ketones are chosen from Table 17.2.

The investigation of the unknown begins with the characterization of the functional group. For this purpose, the infrared spectrum of the unknown, if available, is extremely helpful. Refer to Chapter 9 for a description of the procedure used to obtain a spectrum and for the interpretation of the significant peaks. Some instructors prefer the student to run the infrared spectrum *after* the chemical tests have been performed, because they believe that the spectrum should complement and verify the experimental observations. A complete flow chart appears in Figure 24.1.

The solubility properties of the unknown also provide valuable information. Of the groups that appear in this chapter, many low-molecular-weight compounds with up to four carbon atoms are water soluble. However, if the compound is **not** water soluble, its solubility in acid or base is significant in identifying the functional groups present. Thus, for water-insoluble compounds use the format of Figure 24.1 as described in the sequence of the steps below:

1. If the compound is soluble in dilute (5–10%) HCl solution, it is likely an amine. Moreover, amines have a fishy, unpleasant odor.

$$\underset{\text{amine}}{-\overset{\diagdown}{\underset{\diagup}{N}}{:}} \; + \; HCl \longrightarrow \underset{\substack{\text{amine hydrochloride salt} \\ \text{(water soluble)}}}{-\overset{\diagdown}{\underset{\diagup}{N}}H^{+}Cl^{-}}$$

The results of the Hinsberg test determine whether the amine is primary (1°), secondary (2°), or tertiary (3°). This test involves a reaction of the amine with benzenesulfonyl chloride, $C_6H_5SO_2Cl$ in the presence of sodium hydroxide. Primary amines produce sulfonamides that are soluble in basic solution, while the sulfonamides of secondary amines are insoluble in base. Tertiary amines fail to react with benzenesulfonyl chloride.

$$\underset{\text{1° amine}}{RNH_2} + C_6H_5SO_2Cl \xrightarrow{-HCl} \underset{\substack{\text{benzenesulfonamide} \\ \text{(water insoluble)}}}{C_6H_5SO_2NHR} \xrightarrow[\text{excess}]{NaOH} \underset{\substack{\text{benzenesulfonamide} \\ \text{salt (water soluble)}}}{C_6H_5SO_2NR^-Na^+}$$

$$R_2NH + C_6H_5SO_2Cl \xrightarrow{-HCl} C_6H_5SO_2NR_2 \xrightarrow[\text{excess}]{NaOH} \text{no reaction}$$

2° amine                benzenesulfonamide       (remains water
                         (water insoluble)           insoluble)

$$R_3N + C_6H_5SO_2Cl \longrightarrow \text{no reaction} \xrightarrow[\text{excess}]{NaOH} \text{no reaction} \xrightarrow{HCl} R_3NH^+Cl^-$$

                                        (remains water       (water
                                          insoluble)        soluble)

Primary aromatic amines can be distinguished from primary aliphatic amines by their ability to undergo diazotization and coupling with phenol to form a dye (see Experiment 19.3).

2. If the unknown is soluble in 5–10% sodium bicarbonate solution and produces an effervescence ($CO_2$ evolution), it is a carboxylic acid.

$$R - COOH + NaHCO_3 \longrightarrow R - COO^-Na^+ + H_2O + CO_2$$

carboxylic acid                  sodium carboxylate
                            (water soluble)

The acid can be identified by determining its equivalent weight (see Experiment 16.1), and the MP of its amide or anilide derivative (see Experiment 19.2), according to the flowchart below:

$$R - \overset{\overset{\displaystyle O}{\|}}{C} - OH + SOCl_2 \longrightarrow R - \overset{\overset{\displaystyle O}{\|}}{C} - Cl + SO_2 + HCl$$

$NH_3$ (excess)                                               $C_6H_5 - NH_2$

$$HCl + R - \overset{\overset{\displaystyle O}{\|}}{C} - NH_2 \qquad\qquad R - \overset{\overset{\displaystyle O}{\|}}{C} - NHC_6H_5 + HCl$$

                  amide                             anilide

3. If the compound is soluble in NaOH, but insoluble in $NaHCO_3$, it is a phenol, ArOH.

$$Ar - OH + NaOH \longrightarrow Ar - O^-Na^+ + H_2O$$

Most phenols and enols produce characteristic colors when reacted with a dilute aqueous $FeCl_3$ solution. Although a phenol derivative can be made, such as a dinitrobenzoate ester or a urethane, for the purpose of this experiment, the phenol may be sufficiently identified by its characteristic color with a 5% $FeCl_3$ solution and by its melting point. For this purpose, a series of known phenols will be reacted with $FeCl_3$ solution, and the colors of the resulting solutions recorded. The unknown phenol will be treated in the same manner and its color matched with the colors of the known phenols. This comparison along with the melting point of the unknown, should allow for an accurate identification. The use of the IR aromatic substitution pattern (Table 9.2) should be particularly useful.

4. Alcohols, aldehydes, and ketones that are *water insoluble* will also be insoluble in HCl, NaOH, and $NaHCO_3$ solutions. The presence of the alcohol — OH group, can be identified by the red color the alcohol

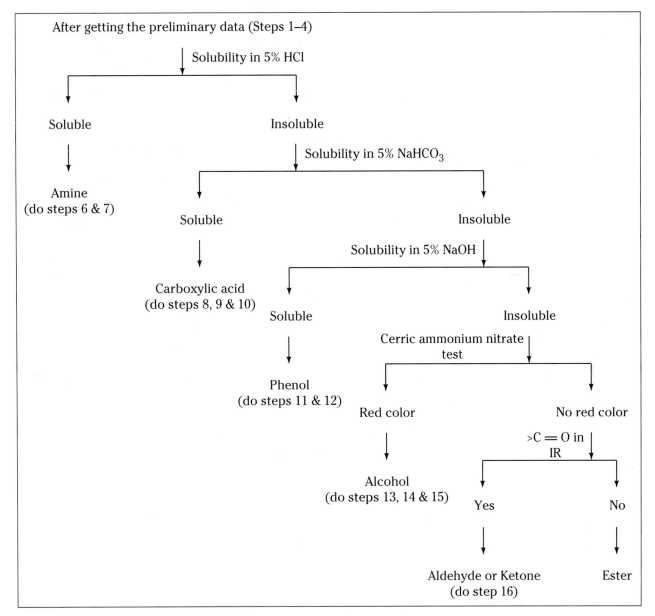

**Figure 24.1** Flowchart for the identification of an organic compound

forms when reacted with cerric aqueous ammonium nitrate. Alcohols can be further analyzed by performing the Jones, Lucas, and iodoform tests (see Experiment 14.2). Reaction of an alcohol with 3,5-dinitrobenzoyl chloride in pyridine to form an ester, or with phenyl isocyanate to form a urethane, yields a crystalline precipitate with a well-defined melting point. These compounds are excellent derivatives for characterizing alcohols.

R'OH  +  —N=C=O  ⟶  —NH—C—OR'

alcohol          phenyl isocyanate                    phenyl urethane

5.  Aldehydes and ketones are distinguished from all other compounds by their ability to form orange or yellow precipitates with 2,4-dinitrophenylhydrazine (2,4-DNP). Only aldehydes give positive Tollens' and Fehling's tests. A positive iodoform test establishes the unknown compound as a methyl ketone or acetaldehyde (see Experiment 14.2).

6.  Finally, the presence of a benzene ring in the compound can be established if the compound gives a yellow precipitate upon nitration (see Experiment 15.1).

   Once the functional group has been identified, the compound's structure may be proven via the preparation of one or more derivatives. A matching of the melting points of the derivative(s) with those listed in the appropriate table(s) is considered as sufficient evidence in the complete identification of the unknown.

**Suggested References**    Carey, F. A., *Organic Chemistry.* 3rd ed., New York: McGraw-Hill, 1996.

Furniss, B. S., A. J. Hannaford, P. W. G. Smith, and A. R. Tatchell, *Vogel's Textbook of Practical Organic Chemistry.* 5th ed., London: Longman Scientific, 1989.

McMurry, J., *Organic Chemistry.* 4th ed., Pacific Grove, CA: Brooks/Cole, 1996.

Morrison, R. T., and R. N. Boyd, *Organic Chemistry.* 6th ed., Englewood Cliffs, NJ: Prentice-Hall, 1992.

Solomons, T. W. G., *Organic Chemistry.* 6th ed., New York: John Wiley, 1996.

Vollhardt, K. P. C., and N. E. Schore, *Organic Chemistry.* 2nd ed. New York: W. H. Freeman, 1994.

Wade, L. G., *Organic Chemistry.* 3rd ed., Englewood Cliffs, NJ: Prentice-Hall, 1995.

⚠ **SAFETY TIPS**

✔  The diazotization process (step 7) involves the reaction of an amine with nitrous acid. Secondary amines yield N-nitrosoamines, which are carcinogenic compounds. Therefore, the use of plastic gloves is highly recommended. Phenol is a skin, eye and respiratory system irritant and should be used in the hood.

✔  Benzenesulfonyl chloride (step 6) is a skin, eye and respiratory system irritant. It should therefore be dealt with in a well-ventilated area.

✔  Contact with or inhalation of thionyl chloride vapors and concentrated ammonium hydroxide (step 9) cause burns to skin, eyes and respiratory system. All solutions should be poured under the hood.

✔  3,5-Dinitrobenzoyl chloride (step 14) is a skin, eye and respiratory system irritant and should be treated with adequate ventilation.

✔  Phenyl isocyanate (step 15) is a moisture-sensitive reagent and any contact with skin or eyes should be avoided. The bottle containing the reagent should be sealed (preferably under nitrogen) when not in use.

✔  Both concentrated nitric and sulfuric acids (step 17) are strong acids that can cause severe burns. They should be treated with respect.

**Experimental**

1. Examine the IR spectrum and compare the major absorptions with those tabulated in Tables 9.1 and 9.2. Check for the presence of a double or triple bond, an aromatic ring, a carbonyl group, or a hydroxyl group.

2. If the unknown is a solid, determine its melting point (see Experiment 3.1). If the unknown is a liquid, determine its boiling point via microboiling point (Experiment 3.2) or distillation (Experiment 4.1) and its refractive index (see Experiment 3.3).

3. Perform the sodium fusion test (see Chapter 8) to identify the presence of nitrogen, sulfur, or halogen. *Note: The instructor should make sure the unknown is anhydrous.*

4. Test the solubility of the unknown in water. Add 4–5 drops of the liquid unknown (0.1 to 0.2 g if it is a solid) to 3 mL of water in a test tube. Mix the contents by shaking the test tube. If the unknown is water soluble, it will also be soluble in aqueous HCl, NaOH, and $NaHCO_3$ solutions. Test the water solution with litmus paper. If the unknown is insoluble in water, test its solubility in 5% HCl, 5% NaOH, and 5% $NaHCO_3$ solutions.

5. If the unknown is soluble in water and tests basic to litmus (i.e., changes pink litmus to blue), or if it is insoluble in water but soluble in 5% HCl solution, it is likely to be an amine. If this is the case, perform the Hinsberg test as described in step 6; otherwise proceed with step 8.

6. Place about 10–15 drops of the amine (0.5 to 0.8 g, if a solid) in a test tube. Add 15 drops of benzenesulfonyl chloride and 10–15 drops of 10% NaOH solution. Stopper the test tube and shake it vigorously for a few minutes. Remove the stopper and warm the test tube in a water bath (60°–70°C) for about 1–2 minutes. No reaction indicates a 3° amine. If a precipitate forms in the alkaline solution, add 3–5 mL of a 10% NaOH solution and shake the test tube. If the precipitate does not dissolve, the unknown is a 2° amine. If the alkaline solution is clear, acidify it with a 10% HCl solution. The formation of a precipitate indicates that the amine is 1°. In the case of 1° and 2° amines, the precipitate (a benzenesulfonamide) can serve as a derivative. Isolate the sulfonamide by suction (vacuum) filtration, and recrystallize it from 95% ethanol-water. Compare your melting point with that of the amines listed in Table 24.1. It is often difficult to isolate the few crystals that are necessary for the melting point determination. In this case, repeat the experiment using two- to three-fold quantities. If the amine is a primary amine proceed with the diazotization procedure in step 7. A tertiary amine may be identified by its reaction with picric acid to form a picrate derivative. This procedure, however, will not be described in this chapter because of the dangerous explosive properties of the *dry* (not wet) picric acid.

7. *Note: Wear plastic gloves when performing this step. If the amine is 2°, the reaction will yield N-nitrosoamines that are carcinogenic compounds.* In a test tube, dissolve 0.2 to 0.5 g of the unknown amine in 5 mL of 10% HCl solution. Cool the solution in an ice/water bath and slowly add while stirring, a solution of 0.5 g of sodium nitrite in 2 mL of water. Primary aliphatic amines evolve nitrogen gas, while secondary aromatic and aliphatic amines produce oily, yellow N-nitrosoamines. In a second test tube, dissolve 0.5 g of phenol in 8 mL of a 5% NaOH solution, and then pour this solution slowly into the first test tube. The formation of a bright (usually) red color confirms a primary aromatic amine.

8. If the unknown is insoluble in water *and* in 5% HCl, but is soluble in 5% $NaHCO_3$ solution, it is probably a carboxylic acid. If it is not soluble in aqueous (5%) $NaHCO_3$ solution, proceed with step 11. Refer to section

**Table 24.1** **Amines**-Unknowns selected from this list

| Name | Formula | Type | BP [MP] (°C) | Benzene-sulfonamide MP (°C) |
|------|---------|------|--------------|------------------------------|
| *1° and 2° amines* | | | | |
| Isopropylamine | $(CH_3)_2CHNH_2$ | 1° | 35 | 3 |
| *n*-Propylamine | | | 49 | 36 |
| *sec*-Butylamine | | | 63 | 70 |
| Isobutylamine | | | 69 | 53 |
| Diethylamine | $(C_2H_5)_2NH$ | 2° | 55 | 42 |
| Piperidine | | | 105 | 93 |
| di-*n*-Propylamine | | | 110 | 51 |
| *n*-Hexylamine | | | 128 | 96 |
| Morpholine | | | 130 | 118 |
| Cyclohexylamine | | | 134 | 89 |
| Aniline | | | 183 | 112 |
| N-Methylaniline | | | 192 | 79 |
| *p*-Toluidine | | | 200 [45] | 120 |
| *p*-Nitroaniline | | | 245 [147] | 139 |

16.1 for the procedure to determine the equivalent weight of the acid. If the unknown is a monocarboxylic acid, the equivalent weight is equal to the molecular weight. Prepare an amide (step 9) or anilide derivative (step 10) to identify the structure of the unknown acid.

9. In a 25-mL round-bottomed flask, place 1 g of the unknown carboxylic acid and 3–4 mL of freshly distilled thionyl chloride, and reflux in a fume hood for about 30 minutes. Cool in ice, and slowly add 10 mL of ice-cold concentrated ammonia. The reaction is vigorous and splattering will occur if the addition is too rapid. Isolate the amide derivative by suction filtration, wash with 10 mL of cold water, and recrystallize from about 10–15 mL ethanol-water (1:1) solution. Determine the MP and match it with the values listed in Table 24.2.

10. To prepare the anilide, proceed as in step 9, but add dropwise 3 mL of freshly distilled aniline (instead of the 10 mL of the cold concentrated ammonia). Stir vigorously. Add enough 5% HCl to dissolve the excess aniline until pH = 4. Isolate the crystals of the anilide derivative by suction (vacuum) filtration, wash with 20 mL of cold water, and recrystallize from 10–15 mL of a 1:1 ethanol-water mixture. Determine the MP and compare it with those of the compounds listed in Table 24.2.

**Table 24.2**  **Carboxylic acids-**Unknowns selected from this list

| Name | Formula | Equivalent weight | BP [MP] (°C) | Amide MP (°C) | Anilide MP(°C) |
|------|---------|-------------------|--------------|---------------|----------------|
| Acetic | CH$_3$COOH | 60 | 118 | 82 | 114 |
| Propionic | | | 141 | 79 | 106 |
| n-Butyric | | | 163 | 115 | 96 |
| n-Valeric | | | 186 | 106 | 63 |
| Caproic | | | 205 | 100 | 95 |
| Caprylic | | | 239 [16] | 107 | 57 |
| Capric | | | 269 [31] | 108 | 70 |
| Benzoic | | | [121] | 129 | 162 |
| p-Nitrobenzoic | | | [239] | 201 | 211 |
| o-Chlorobenzoic | | | [141] | 141 | 118 |
| p-Anisic | | | [184] | 162 | 169 |
| Sorbic | | | [134] | — | 153 |
| p-Toluic | | | [178] | 159 | 146 |
| Phenylacetic | | | [76] | 157 | 118 |
| Oxalic | (COOH)$_2$ | 45 | [101] | 419 (dec) | 246 |
| Malonic | | | [135] (dec) | 170 | 225 |
| Maleic | | | [135] | 181 | 187 |
| Phthalic | | | [208] (dec) | 219 | 251 |
| Adipic | | | [152] | 220 | 239 |
| Succinic | | | [185] | 260 | 230 |

11. If the unknown is insoluble in both 5% HCl and 5% NaHCO$_3$ solutions, try dissolving it in a 5% aqueous NaOH solution. If it does not dissolve, it is *not* a phenol, and you should proceed with step 13. If it is soluble, perform the FeCl$_3$ test according to step 12.

12. Add 0.5 g of the solid (5–10 drops if the unknown is a liquid) to 1–2 mL of a 5% aqueous FeCl$_3$ solution and mix well. Compare the color to the color obtained by performing the same procedure on the phenols that appear in Table 24.3. Also find the MP of the unknown and compare with those recorded in Table 24.3.

**Table 24.3** **Phenols**-Unknowns selected from this list

| Name | Formula | BP [MP] (°C) | Color with 5% FeCl$_3$ |
|---|---|---|---|
| Phenol | C$_6$H$_5$ — OH | 182  [ 42] | |
| o-Cresol | | 191  [30] | |
| m-Cresol | | 202  [2] | |
| p-Cresol | | 202  [36] | |
| 2-Naphthol | | 283 [123] | |
| Resorcinol | | 280 [110] | |
| Pyrogallol | | 309 [133] | |
| Vanillin | | [81] | |
| p-Methoxyphenol | | 243  [56] | |
| Ethylphenol | | 219  [47] | |
| Hydroquinone | | 286 [170] | |
| Salicylic acid | | [159] | |
| p-Hydroxybenzoic acid | | [214] | |
| p-Chlorophenol | | 217  [43] | |
| p-Nitrophenol | | [114] | |

13. Alcohols containing up to 10 carbon atoms react with ceric ammonium nitrate reagent to give a red color. Add 10 drops of the unknown to a mixture of 1 mL of the ceric ammonium nitrate reagent (See Appendix IV) and 5 mL of water. If the test is negative (no color change) proceed with step 16. If the test is positive, (red-orange complex formation) perform the Jones oxidation, the Lucas test, and the iodoform test. (See Experiment 14.2 for the experimental procedures.)  An alcohol derivative can be prepared according to step 14 (3,5-dinitrobenzoate) or step 15 (phenylurethane).

14. Place 1 g (or 1 mL) of the alcohol, 1 g of fresh 3,5-dinitrobenzoyl chloride, and 5 mL of pyridine in a 25-mL round-bottomed flask, and reflux *in a fume hood* for about 30 minutes. *In the hood,* pour the solution into a small beaker containing 10 mL of 10% aqueous NaHCO$_3$ solution. Place the beaker in an ice bath and stir the mixture to induce crystallization. Collect the precipitate via vacuum filtration, and wash it first with 10 mL of 5% NaHCO$_3$ solution and then with 10 mL of water. Recrystallize from 10–15 mL of a 1:1 ethanol-water solution. Determine the MP and compare with the values listed in Table 24.4.

**Table 24.4**   **Alcohols**-Unknowns selected from this list

| Name | Formula | Type | BP [MP] (°C) | 3,5-DNB MP (°C) | Phenylurethane MP (°C) |
|---|---|---|---|---|---|
| Methanol | CH$_3$OH | 1° | 66 | 109 | 47 |
| Ethanol | | | 78 | 94 | 52 |
| 1-Propanol | | | 97 | 75 | 57 |
| 2-Propanol | | | 82 | 122 | 86 |
| 1-Butanol | | | 118 | 64 | 61 |
| 2-Butanol | | | 99 | 76 | 64 |
| 2-Methyl-1-propanol | | | 108 | 88 | 86 |
| 2-Methyl-2-propanol | | | 83 | 142 | 136 |
| 1-Pentanol | | | 138 | 46 | |
| 2-Pentanol | | | 119 | 62 | |
| 3-Pentanol | | | 116 | 100 | 63 |
| Cyclopentanol | | | 141 | 115 | 46 |
| Cyclohexanol | | | 161 | 113 | |
| 2-Octanol | | | 179 | 32 | 114 |
| Benzyl alcohol | | | 205 | 113 | 76 |
| 1-Phenylethanol | | | 203 | 94 | 92 |
| 2-Phenylethanol | | | 220 | 108 | 80 |
| p-Methoxybenzyl alcohol | | | 259 | | 93 |

15. *In the hood,* add 10 drops of freshly distilled phenyl isocyanate to 0.5 g of the alcohol in a test tube and cork. Heat in a hot water bath for about 10 minutes, and then cool in ice. If no precipitate forms, scratch the side of the test tube with a stirring rod. Vacuum filter the precipitate, and recrystallize from about 10 mL of hexane. Determine the MP and compare with the values listed in Table 24.4.

16. If the unknown is insoluble in water, in 5% HCl, in 5% NaOH and in 5% NaHCO$_3$ solutions, and it shows a carbonyl absorption in the IR spectrum, it is an aldehyde, a ketone, or an ester. A positive 2,4-DNP test (Experiment 17.5) indicates an aldehyde or a ketone. Aldehydes give a positive Tollens' test or Fehling's test (see Experiment 17.5) and show IR absorption peaks at 2750 and 2850 cm$^{-1}$. Both aldehydes and ketones can be identified by the MP of their 2,4-dinitrophenylhydrazone or semicarbazone derivatives (Table 17.2). Acetaldehyde and all

methyl ketones also give a positive iodoform test. If the 2,4-DNP test is negative, the compound is probably an ester.

17. To determine if the unknown is an aromatic compound, perform the nitration test. *In the fume hood,* reflux a mixture of 2 mL concentrated nitric acid, 2 mL concentrated sulfuric acid, and 0.5 g of the unknown for about 15 minutes. Pour the contents while the solution is still hot into a beaker containing 10 g of ice and stir thoroughly to induce crystallization. If a precipitate forms, the solution is definitely an aromatic compound. ***Do not perform this test if the unknown is an alcohol.***

**Questions**

1. Identify a simple chemical test that would help distinguish between the following pairs of compounds. State what you do and what you see.
   a. *p*-ethylbenzaldehyde and ethyl phenyl ketone
   b. *p*-nitrobenzoic acid and *p*-nitrotoluene
   c. *p*-methylacetophenone and ethyl phenyl ketone
   d. *p*-methylaniline and *p*-nitrotoluene
   e. *p*-methylphenol and *p*-methylbenzoic acid

2. Give structures for the following aromatic compounds:
   a. ***A*** ($C_8H_{10}O$) positive 2,4-DNP; positive iodoform.
   b. ***B*** ($C_9H_{10}O_2$) insoluble in NaOH; positive 2,4-DNP; negative iodoform; hot $KMnO_4$ oxidation yields *p*-methoxybenzoic acid.
   c. ***C*** ($C_9H_{10}O_2$) soluble in $NaHCO_3$; oxidation with $KMnO_4$ yields phthalic acid; IR peak at 1690 cm$^{-1}$.
   d. ***D*** ($C_9H_{10}O_2$) soluble in $NaHCO_3$; oxidation with $KMnO_4$ yields phthalic acid; IR peak at 1710 cm$^{-1}$.
   e. ***E*** ($C_{10}H_{12}O_2$) insoluble in $NaHCO_3$; oxidation with $KMnO_4$ yields isophthalic acid; IR peak at 1740 cm$^{-1}$.
   f. ***F*** ($C_8H_{11}N$) soluble in HCl; Hinsberg test precipitate insoluble both in excess base and acid (two possible structures).

Name: _____

Unknown # _____

# DATA SHEET: QUALITATIVE ANALYSIS

*Step 1:* IR Spectrum

| Band (cm⁻¹) | Predicted Assignment | Band (cm⁻¹) | Predicted Assignment |
|---|---|---|---|
| | | | |
| | | | |

*Step 2:* Melting point ____°C; Boiling point ____°C; Refractive Index ____ (at 20°C)

*Step 3:* Sodium fusion test: Nitrogen: present/absent (circle one)
Sulfur: present/absent (circle one)
Halogen: present/absent (circle one)

*Step 4:* Solubility in water: soluble/insoluble (circle one)
If unknown is water soluble : litmus paper test conclusion: acidic/basic/neutral (circle one)
If unknown is water insoluble:
solubility in 5% HCl : soluble/insoluble (circle one)
solubility in 5% NaOH: soluble/insoluble (circle one)
solubility in 5% NaHCO₃: soluble/insoluble (circle one)

*Conclusion:* My compound is acidic/basic/neutral (circle one)

*Step 6:* If the compound is water insoluble but soluble in 5% HCl.
Hinsberg test: My compound is a 1°, 2°, 3° amine (circle one)
MP of benzenesulfonamide _____ °C

*Conclusion:* My compound is _____ (by comparing the MP with the MP of the compounds listed in Table 24.1)

*Step 7:* Reaction with NaNO₂ (reconfirmation test for the conclusion from step 6).

*Observation:* _____

*Conclusion:* _____

*Step 8:* If the compound is water insoluble but soluble in 5% NaHCO₃
Equivalent weight: _____

*Step 9:* Melting point of amide derivative _____ °C

*Conclusion:* My compound is _____ (by comparing the MP with the MP of the compounds listed in Table 24.2)

*Step 10:* Melting point of anilide derivative ____ °C

*Conclusion:* My compound is _____ (by comparing the MP with the MP of the compounds listed in Table 24.2)

*Step 11:* If the compound is water insoluble but soluble in 5% NaOH

*Conclusion:* _____

*Step 12:* Ferric chloride test: Color: _____

*Conclusion:* My compound is _____ (by comparing the MP of the compound [from step 2] and the color obtained with the compounds listed in Table 24.3)

*Step 13:* Ceric ammonium nitrate test:

*Observation:* _____

*Conclusion:* _____

      If Ceric ammonium nitrate test is positive:
            Jones test:      Observation: _____ Conclusion: _____
            Lucas test:      Observation: _____ Conclusion: _____
            Iodoform test:  Observation: _____ Conclusion: _____

*Conclusion:* My compound is a 1°, 2°, 3° (circle one) alcohol that possesses/does not possess (circle one) the group $CH_3CH(OH)$ —

*Step 14:* Melting point of 3,5-dinitrobenzoate derivative _____ °C

*Conclusion:* My compound is _____ (by comparing the MP with the MP of the compounds listed in Table 24.4)

*Step 15:* Melting point of phenylurethane derivative _____ °C

*Conclusion:* My compound is _____ (by comparing the MP with the MP of the compounds listed in Table 24.4)

*Step 16:* If the compound is insoluble in $H_2O$, 5% HCl, 5% NaOH, and 5% $NaHCO_3$, and shows a $\diagup\hspace{-0.5em}C{=}O$ absorption in IR.
    2,4 DNP test: Observation: _____
                 MP of ppt (if formed) _____ °C
   Tollens' test:   Positive/negative (circle one)
   Fehling's test:  Positive/negative (circle one)
   Iodoform test:  Positive/negative (circle one)

*Conclusion:* _____

*Step 17:* Nitration test: Positive/negative (circle one)

*Conclusion:* _____

*Final Conclusion:* Based on the above data, Unknown # _____ is _____

# I

# How to Balance an Oxidation-Reduction Reaction

To carry out a chemical reaction in the lab, the chemicals have to be mixed together in the correct proportions. This can only be accomplished by means of a balanced equation, which in many cases is achieved by inspection. However, balancing by inspection is usually not feasible when complex oxidation-reduction reactions are involved. Such equations consist of an **oxidation** (loss of electrons) and a **reduction** (gain of electrons) occurring simultaneously.

For an equation to be balanced, all atoms and ionic charges on the left- and right-hand sides of the equation have to be equal. The following rules must be followed, after dividing the equation into two half equations:

1. Balance all elements except hydrogen and oxygen.
2. Balance charges with $H^+$ (if in acid medium) or $OH^-$ (if in basic medium).
3. Balance oxygens with water.
4. Balance hydrogens with H — **Remember: uncharged** hydrogens **(H)**, not $H^+$ or $H^-$.
5. Cross-multiply each half-equation by the number of hydrogens added to the other half-equation, and add both equations in such a way that the hydrogens cancel. After this step, only the actual participants in the reaction remain in the balanced equation.

Two examples follow that will directly explain the application of these rules.

## EXAMPLE I

Balance the following reaction in acidic medium:

$$C_2H_6O + Cr_2O_7^{-2} \xrightarrow{H^+} C_2H_4O_2 + Cr^{+3}$$

*Half-equation I*      *Half-equation II*
$$C_2H_6O \longrightarrow C_2H_4O_2 \qquad Cr_2O_7^{-2} \longrightarrow Cr^{+3}$$

1. $C_2H_6O \longrightarrow C_2H_4O_2$       $Cr_2O_7^{-2} \longrightarrow 2Cr^{+3}$
2. $C_2H_6O \longrightarrow C_2H_4O_2$       $Cr_2O_7^{-2} + 8H^+ \longrightarrow 2Cr^{+3}$
3. $C_2H_6O + H_2O \longrightarrow C_2H_4O_2$     $Cr_2O_7^{-2} + 8H^+ \longrightarrow 2Cr^{+3} + 7H_2O$
4. $C_2H_6O + H_2O \longrightarrow C_2H_4O_2 + 4(\mathbf{H})$     $Cr_2O_7^{-2} + 8H^+ + 6(\mathbf{H}) \longrightarrow 2Cr^{+3} + 7H_2O$

       multiply this equation by 3         multiply this equation by 2

*Note:* The *least* common denominator is 12.

5. Adding the two half-equations

$$3C_2H_6O + 3H_2O \longrightarrow 3C_2H_4O_2 + 12(\mathbf{H})$$
$$2Cr_2O_7^{-2} + 16H^+ + 12(\mathbf{H}) \longrightarrow 4Cr^{+3} + 14H_2O$$

$$3C_2H_6O + 2Cr_2O_7^{-2} + 16H^+ \longrightarrow 3C_2H_4O_2 + 4Cr^{+3} + 11H_2O$$

To verify that the equation is really balanced:

|  | C | H | O | Cr | net charge |
|---|---|---|---|---|---|
| Reactants | 6 | 34 | 17 | 4 | $+16 + 2(-2) = 16 - 4 = 12$ |
| Products | 6 | 34 | 17 | 4 | $4(+3) = 12$ |

# EXAMPLE II

Balance the following reaction in basic medium:

$$C_7H_8 + MnO_4^- \xrightarrow{\;OH^-\;} C_7H_5O_2^- + MnO_2$$

*Half-equation I*      *Half-equation II*
$$C_7H_8 \longrightarrow C_7H_5O_2^- \quad MnO_4^- \longrightarrow MnO_2$$

Applying the above mentioned rules

1. $C_7H_8 \longrightarrow C_7H_5O_2^-$          $MnO_4^- \longrightarrow MnO_2$
2. $C_7H_8 + OH^- \longrightarrow C_7H_5O_2^-$    $MnO_4^- \longrightarrow MnO_2 + OH^-$
3. $C_7H_8 + OH^- + H_2O \longrightarrow C_7H_5O_2^-$    $MnO_4^- \longrightarrow MnO_2 + OH^- + H_2O$
4. $\underbrace{C_7H_8 + OH^- + H_2O \longrightarrow C_7H_5O_2^- + 6(\mathbf{H})}$   $\underbrace{MnO_4^- + 3(\mathbf{H}) \longrightarrow MnO_2 + OH^- + H_2O}$

         multiply this equation by 1          multiply this equation by 2

5. Adding the two half-equations

$$C_7H_8 + OH^- + H_2O \longrightarrow C_7H_5O_2^- + 6(\mathbf{H})$$
$$2MnO_4^- + 6(\mathbf{H}) \longrightarrow 2MnO_2 + 2OH^- + 2H_2O$$

$$C_7H_8 + 2MnO_4^- \longrightarrow C_7H_5O_2^- + 2MnO_2 + OH^- + H_2O$$

To verify that the equation is really balanced

|  | C | Mn | O | H | net charge |
|---|---|---|---|---|---|
| Reactants | 7 | 2 | 8 | 8 | $2(-1) = -2$ |
| Products | 7 | 2 | 8 | 8 | $(-1) + (-1) = -2$ |

## PROBLEMS

1. Balance the following equations in acidic medium:

   a. $C_7H_8 + MnO_4^- \longrightarrow C_7H_5O_2^- + Mn^{+2}$

   b. $C_8H_{10} + Cr_2O_7^{-2} \longrightarrow C_8H_6O_4 + Cr^{+3}$

   c. $C_6H_5O_2N + Sn \longrightarrow C_6H_7N + Sn^{+2}$

   d. $C_6H_{10} + HNO_3 \longrightarrow C_6H_{10}O_4 + NO$

   e. $Te + NO_3^- \longrightarrow TeO_2 + NO$

   f. $NO + NO_3^- \longrightarrow N_2O_4$

   g. $ReO_2 + Cl_2 \longrightarrow HReO_4 + Cl^-$

   h. $C_7H_6O + Cr_2O_7^{-2} \longrightarrow C_7H_6O_2 + Cr^{+3}$

   i. $HNO_3 + H_2S \longrightarrow NO + S$

   j. $I_2 + HNO_3 \longrightarrow 2HIO_3 + NO_2$

   k. $CH_3-\langle\bigcirc\rangle-CH_2OH + MnO_4^- \longrightarrow HOOC-\langle\bigcirc\rangle-COOH + Mn^{+2}$

2. Balance the following equations in basic medium:

   a. $C_4H_8 + MnO_4^- \longrightarrow C_4H_{10}O_2 + MnO_2$

   b. $C_7H_8O + MnO_4^- \longrightarrow C_7H_5O_3^- + MnO_2$

   c. $Al + NO_3^- \longrightarrow Al(OH)_4^- + NH_3$

   d. $I_2 + Cl_2 \longrightarrow H_3IO_6^{-2} + Cl^-$

   e. $Br_2 \longrightarrow Br^- + BrO_3^-$

   f. $H_2O_2 + ClO_2 \longrightarrow ClO_2^- + O_2$

   g. $CN^- + MnO_4^- \longrightarrow CNO^- + MnO_2$

   h. $S + HO_2^- \longrightarrow SO_4^{-2} + OH^-$

   i. $Bi(OH)_3 + Sn(OH)_4^{-2} \longrightarrow Bi + Sn(OH)_6^{-2}$

   j. $C_6H_{10}O + MnO_4^- \longrightarrow C_6H_8O_4^{-2} + MnO_2$

   k. $CH_3-\langle\bigcirc\rangle-CH_2OH + MnO_4^- \longrightarrow {}^-OOC-\langle\bigcirc\rangle-COO^- + MnO_2$

# appendix II

# How to Calculate the Percent Yield of a Reaction

The **percent yield** is a measure of the efficiency of a synthetic procedure and is defined as

$$\% \text{ yield} = \frac{\text{moles (or mass) of product actually obtained}}{\text{moles (or mass) of product expected}} \times 100$$

## EXAMPLE 1

Calculate the amount of t-butyl benzene expected from 3.70 g t-butyl chloride and excess benzene according to the following equation:

$$(CH_3)_3C-Cl + C_6H_6 \longrightarrow (CH_3)_3C-C_6H_5 + HCl$$

MW =　　　　92.5　　　　　　　　　　　134

The relative molar ratio in this reaction is 1:1 $\longrightarrow$ 1:1. Thus, from 3.70/92.5 = 0.040 moles of $(CH_3)_3C-Cl$, one would expect 0.040 moles of $(CH_3)_3C-C_6H_5$, which is equal to $134 \times 0.040 = 5.36$ g.

## EXAMPLE 2

If in the above example a student isolated 4.02 g of t-butylbenzene, what is the percent yield? The percent yield is defined as

$$\text{Percent yield} = \frac{\text{Actual yield}}{\text{Theoretical yield}} \times 100$$

Thus, the percent yield for the previous reaction is

$$(4.02/5.36) \times 100 = 75.0\%$$

The percent yield can also be calculated from mole quantities.

Actual yield = (4.02/134) = 0.0300 moles
Theoretical yield = 0.0400 moles
Percent yield = (0.0300/0.0400) × 100 = 75.0%

## EXAMPLE 3

Calculate the percent yield of a reaction that leads to the isolation of 4.0 g of 1,2-dibromophenylethane, $C_6H_5CH(Br)CH_2Br$, from 5.2 g of styrene ($C_6H_5CH=CH_2$) and 6.4 g of bromine ($Br_2$).

$$C_6H_5CH=CH_2 + Br_2 \longrightarrow C_6H_5CH(Br)CH_2Br$$

The equation involves two reactants, one of which is the **limiting reagent;** i.e., the reagent which is not in excess.

|  | $C_6H_5CH=CH_2$ | $+ Br_2 \longrightarrow$ | $C_6H_5CH(Br)CH_2Br$ |
|---|---|---|---|
| mass: | 5.2 g | 6.4 g | 4.0 g |
| MW: | 104 | 160 | 264 |
| # moles: | 0.050 | 0.040 | 0.015 |

Bromine is the limiting reagent since only 0.040 moles are used. As a result, the calculation should be based on bromine. Thus, actual yield = 0.015 moles; theoretical yield = 0.040 moles; and percent yield = (0.015/ 0.040) × 100 = 38%. Notice that the limiting reagent is bromine, although its mass is more than that of styrene. Also, note that the equation must be balanced. If the coefficient for styrene had been 2, *it* would have been the limiting reagent.

## EXAMPLE 4

Assume that 60 grams of acetaldehyde ($CH_3CHO$) is reacted with excess NaOH to form crotonaldehyde.

$$2CH_3CHO \longrightarrow CH_3CH=CHCHO + H_2O$$

Calculate the percent yield if 18 g of crotonaldehyde is isolated. The balanced equation is

|  | $2CH_3-CHO \longrightarrow$ | $CH_3-CH=CHCHO + H_2O$ |
|---|---|---|
|  | limiting reagent |  |
| mass | 60 g (start with) | 18 g (end up with) |
| MW | 44 | 70 |
| # moles | 1.4 | 0.26 |

To calculate the percent yield, we must first calculate the amount of product that is theoretically obtainable. From the equation, 2.0 moles of acetaldehyde yield 1.0 mole of crotonaldehyde. Thus, the number of crotonaldehyde moles expected is equal to 1.4/2.0 = 0.70 moles, and the mass of crotonaldehyde expected is equal to 0.70 × 70 = 49 g. Since the actual yield of product is 18 g (or 0.26 moles),

$$\text{Percent yield} = \frac{18g}{49g} \times 100 = 37\%$$

or

$$\frac{0.26 \text{ mole}}{0.70 \text{ mole}} \times 100 = 37\%$$

## PROBLEMS

Calculate the percent yield for the following reactions:

a.                    $C_6H_6 + Br_2 \longrightarrow C_6H_5Br + HBr$

start with 3.9 g   4.2 g          3.0 g (end up with)
(**Answer:** 73%; $Br_2$ is the limiting reagent)

b.                    $C_6H_6 + 2HNO_3 \longrightarrow C_6H_4(NO_2)_2 + 2H_2O$

start with 3.9 g   4.6 g          2.4 g (end up with)
(**Answer:** 39%; $HNO_3$ is the limiting reagent)

c.    $CH_2 {=\!=} CH_2 + Cl_2 \longrightarrow ClCH_2CH_2Cl$

start with 2.8 g   3.5 g          1.2 g (end up with)
(**Answer:** 24%; $Cl_2$ is the limiting reagent)

d. A two-step synthesis:
*First Step:*

$$C_6H_6 + CH_3I \longrightarrow C_7H_8 + HI$$

start with 3.9 g    9.4 g          4.0 g (end up with)

*Second Step:*

$$C_7H_8 + MnO_4^- \longrightarrow C_7H_6O_2 + MnO_2$$

start with 3.0 g                    2.0 g (end up with)
(**Answer:** first step [87%]; second step [50%];
overall yield = [0.87 × 0.50 × 100 = 44%])

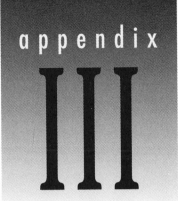

# How to Make Solutions

In preparing laboratory solutions it is important to know how much **solute** is dissolved in a given quantity of **solvent.** The **concentration** of a solution is defined as **the amount of solute dissolved in a given amount of solvent.** In organic chemistry laboratories, the two most common methods for expressing concentrations are **percent by mass** and **molarity.**

## PERCENT BY MASS (w/w)

The **percent by mass** of a solute in a solution is defined as **the number of grams of solute dissolved in 100 g of solution.** Thus, to make a 20% (w/w) solution of NaCl, one needs to dissolve 20 g of NaCl in 80 g of water, so that the resulting final mass is 100 g. This is *not* equal to dissolving 20 g of NaCl in 100 g of water, which would give a total of 120 g of solution. Note also that the molecular weight of solute or solvent, and the density of solvent, need not be known.

## EXAMPLE I

Calculate the percent by mass of a solution made by dissolving 20 g of NaOH (MW 40) in 50 g of water.

$$\% \text{ NaOH} = \frac{\text{mass of NaOH}}{\text{mass of NaOH } + \text{ mass of water}} \times 100$$

Thus,

$$\% \text{ NaOH} = \frac{20}{20+50} \times 100 = 29\%$$

## EXAMPLE II

Calculate the amount of solute present in 20 g of a 2.0% NaOH solution.

$$\% \text{ NaOH} = \frac{\text{mass of NaOH}}{\text{mass NaOH } + \text{ mass water}} \times 100$$

Thus,

$$2.0 = \frac{y}{20} \times 100$$

Solving for $y$

$$2.0 \times 20 = 100y \text{ and } y = 0.40 \text{ g of NaOH}$$

## MOLARITY

**Molarity** is defined as **the number of moles of solute dissolved in one liter of solution.** Thus,

$$\text{Molarity} = (\# \text{ moles})/(\text{volume in liters}), \text{ and}$$
$$\textbf{M} \text{ (molarity in moles/L)} \times \textbf{V} \text{ (in L)} = \textbf{\# moles}$$

and since $\#$ moles = mass/MW

then,

$$\textbf{M} \times \textbf{V} = \textbf{mass/MW}$$

## EXAMPLE III

Calculate the mass of $H_2SO_4$ needed to prepare 2.0 L of 2.0$M$ $H_2SO_4$. How would you make this solution?

Substituting in the formula for molarity

$$\textbf{M} \times \textbf{V} = \text{mass of solute/MW}$$
$$2.0 \times 2.0 = \textbf{x} / 98$$

Solving for **x**

$$\textbf{x} = 2.0 \times 2.0 \times 98 = 3.9 \times 10^2 \text{ g of } H_2SO_4$$

To make this specific solution, one needs to dissolve $3.9 \times 10^2$ g of $H_2SO_4$ in a ***total*** volume of 2 L with water.

## PROBLEMS

1. Calculate the w/w % of the following solutions:
   a. 10 g KCl and 100 g water (**Answer:** 9.1%)
   b. 20 g KCl and 100 g water (**Answer:** 17%)
   c. 50 g $NaNO_2$ and 150 g water (**Answer:** 25%)
   d. 0.40 g KI and 5 g water (**Answer:** 7.4%).
2. How much solute is present in the following:
   a. 50 g of 4.0% NaCl solution (**Answer:** 2.0 g)
   b. 150 g of 9.00% KCl solution (**Answer:** 13.5 g).
3. Determine the molarity of the following solutions:
   a. 1.0 mole of KBr in 200 mL of solution (**Answer:** 5.0$M$)
   b. 0.4 mole of $AgNO_3$ in 300 mL of solution (**Answer:** 1.3$M$)
   c. 100 g of $BaBr_2$ in one liter of solution (**Answer:** 0.337$M$)
   d. 20 g of $CuSO_4 \cdot 5H_2O$ in 2L of solution (**Answer:** 0.040$M$)
   e. 10.0 g of $CH_4O$ in 150 mL of solution (**Answer:** 2.1$M$)
   f. 100 mL of 6.0$M$ HCl and 100 mL of water (**Answer:** 3.0$M$)
   g. 20 mL of 0.2$M$ HCl and 60 mL of water (**Answer:** 0.05$M$)
   h. 50 mL of 2.5$M$ $ZnCl_2$ and 0.45 L of water (**Answer:** 0.25$M$).

4. How much water must be added to 1000 mL of solution containing 49.0 g of $H_2SO_4$ to make the final solution $0.1M$? (**Answer:** 4 L)

5. How many grams of 6.0% (w/w) KCl solution are necessary to yield 80.0 g of KCl? (**Answer:** $1.3 \times 10^3$ g)

6. How would you make 240.0 mL of 4.00% NaCl solution from a stock 0.600% solution? (**Answer:** Reduce 1.6 L of a 0.6% solution to a final volume of 240 mL)

7. Calculate the molarity of each of the following solutions:
   a. 75.0 g of ethanol ($C_2H_6O$) in 450 mL of solution (**Answer:** $3.6M$)
   b. 5.20 g of NaCl in 40 mL of solution (**Answer:** $2.2M$)
   c. 20.0 g of sucrose ($C_{12}H_{22}O_{11}$) in 275 mL of solution (**Answer:** $0.213M$)

8. An aqueous solution labeled 17.5% $HClO_4$ has a density of 1.251 g/mL. What is the molarity of the solution? (**Answer:** $2.19M$)

# Special Reagents

## COMMON LABORATORY REAGENTS

| Reagent | Concentration | Density | Comments |
|---|---|---|---|
| Acetic Acid, glacial | at least 99% | 1.05 | <1% water |
| Ammonium hydroxide, (aqueous ammonia) concentrated | 15$M$ (28%) | 0.90 | |
| Hydrochloric acid, concentrated | 12$M$ (38%) | 1.18 | Colorless |
| Nitric acid, concentrated | 16$M$ (68%) | 1.42 | Colorless |
| Sulfuric acid, concentrated | 18$M$ (36$N$) | 1.84 | <6% water |
| Sulfuric acid, fuming | variable (usually 30% $SO_3$) | | Known as oleum |

## QUALITATIVE ANALYSIS REAGENTS

### Barfoed Reagent
Dissolve 13.3 g of crystallized neutral cupric acetate in 200 mL of 1% aqueous acetic acid solution.

### Benedict's Solution
Add 86.5 g of sodium citrate and 50 g of anhydrous sodium carbonate to 350 mL of distilled water. Heat to dissolve. Cool and filter the solution, and then add a solution of 8.65 g cupric sulfate pentahydrate in 50 mL of water. Dilute the mixture to a total volume of 500 mL. The solution should be clear.

### Bial's Reagent
Add 25–30 drops of 10% ferric chloride solution to 500 g of 30% aqueous HCl solution. To this solution add 1.0 to 1.5 g of orcinol.

### Ceric Ammonium Nitrate Reagent
Dissolve 20 g of Ceric ammonium nitrate $[(NH_4)_2Ce(NO_3)_6]$ in 50 mL of water containing 7 mL of concentrated nitric acid. Use heat to speed up the dissolving process. Keep the cool solution in a tightly stoppered dark bottle.

### 2,4-Dinitrophenylhydrazine Reagent

Dissolve 3 g of 2,4-dinitrophenylhydrazine in 15–20 mL of concentrated sulfuric acid. Add this solution with stirring to a solution of 20 mL water in 70 mL 95% ethanol. Stir the mixture thoroughly and filter. Use the filtrate as the reagent solution.

### Fehling's Solution

Dissolve 34.6 g of pure copper (II) sulfate pentahydrate ($CuSO_4 \cdot 5H_2O$) in 350 mL of distilled water. Add a couple of drops of dilute sulfuric acid and dilute the solution to a total volume of 500 mL. Label this solution as Fehling's solution A.

Dissolve 173 g of Rochelle salt (sodium potassium tartrate) and 65 g of sodium hydroxide pellets in about 350 mL of water, and dilute the solution to a volume of 500 mL. Label this solution as Fehling's solution B. Both solutions A and B should be clear. If not, filter before use.

When performing a test, mix equal volumes of solutions A and B. The solution should be dark blue and clear.

### Hanus Solution

Dissolve 13.2 g iodine in enough glacial acetic acid to make a 1.0L solution with gentle heating. Cool to 15°C and add 5 drops bromine. The solution should be tightly stoppered and protected from sunlight.

### Hopkins-Cole Reagent

Add 5 mL of cold water to 10 g of magnesium in a large conical flask. Next add 250 mL of a cold saturated solution of oxalic acid and stir vigorously. Filter, and then add 25 mL of concentrated acetic acid to the filtrate. Dilute with distilled water to a final volume of 1 liter.

### Iodoform Reagent

Dissolve 15–20 g of iodine in 150–200 mL aqueous 10% potassium iodide solution. Stir and heat if necessary to dissolve the iodine. Keep the solution in a tightly stoppered bottle.

### Lucas Reagent

Dissolve **slowly** 136 g of anhydrous zinc chloride from a freshly opened bottle in 105 g of concentrated hydrochloric acid in an ice bath. In case the zinc chloride is wet, heat in a crucible or porcelain evaporating dish using a Bunsen burner. The solid will melt, and then will solidify. Carefully transfer the solid to a reagent bottle that is immersed in an ice-water bath. Add the hydrochloric acid **slowly**. Prepare the solution in a well-ventilated area (e.g., a fume hood).

### Millon's Reagent

Heat a mixture of 25 g of mercury in 50 mL of concentrated $HNO_3$ until the mercury dissolves. Dilute the resulting solution by adding 100 mL of distilled water.

### Molish Reagent

Dissolve 10 g of α-naphthol in 100 mL of ethanol.

### Seliwanoff's Reagent

Dissolve 0.5–1.0 g resorcinol in 1.0L *6M* HCl.

# appendix V

# Physical Data of the Liquids Used in This Lab Manual

This appendix provides the main physical data for most of the liquids used in this lab manual.

**Suggested References**

Bruno, T. J., and Svoronos, P., *CRC Handbook of Basic Tables for Chemical Analysis.* Boca Raton, FL: CRC Press, 1989.

Lewis, R. J., Sr., *Sax's Dangerous Properties of Industrial Materials.* 8th ed. New York: Van Nostrand Reinhold, 1992.

Patnaik, P., *A Comprehensive Guide to the Hazardous Properties of Chemical Substances.* New York: Van Nostrand Reinhold, 1992.

| No. | Name | Other Name(s) | Formula | MW | MP (°C) | BP (°C) | Refractive Index (20°C) | Dielectric Constant | Density (g/mL) at 20°C | Flash Point (°C) | Auto-ignition Point (°C) | How to Extinguish Flames | No. |
|---|---|---|---|---|---|---|---|---|---|---|---|---|---|
| 1 | Acetic acid | Ethanoic acid | $CH_3COOH$ | 60.1 | 16.7 | 118.1 | 1.3716 | $6.15^{20}$ | 1.049 | 43 | 465 | a,b,c,d | 1 |
| 2 | Acetic anhydride | Acetic acid, anhydride | $(CH_3C{-})_2O$ | 102.1 | −73.1 | 140 | 1.3901 | $20.0^{19}$ | 1.082 | 54 | 390 | a,b,c,d | 2 |
| 3 | Acetone | Dimethyl ketone; propanone | $(CH_3)_2C{=}O$ | 58.1 | −94.6 | 56.5 | 1.3588 | $20.7^{25}$ | 0.797 | −18 | 465 | a,b,c | 3 |
| 4 | Aniline | Aminobenzene; phenylamine | $C_6H_5NH_2$ | 93.1 | −6.2 | 184.4 | 1.5863 | $6.89^{20}$ | 1.022 | 70 | 615 | a,b,c | 4 |
| 5 | Benzaldehyde | Benzene carbaldehyde | $C_6H_5CHO$ | 106.1 | −26 | 179 | 1.544 | $17.8^{20}$ | 1.046 | 64 | 192 | a,b,c,d | 5 |
| 6 | Benzonitrile | Benzoic acid nitrile; cyanobenzene; phenyl cyanide | $C_6H_5C{\equiv}N$ | 103.1 | −12.8 | 191 | 1.5289 |  | 1.246 | 71 |  | a,b,c | 6 |
| 7 | Benzoyl chloride | Benzoic acid, chloride | $C_6H_5CCl$ (=O) | 140.6 | −0.5 | 197 | 1.5537 | $23.0^{20}$ | 1.219 | 72 | — | a,b,(never d) | 7 |
| 8 | Bromine | — | $Br_2$ | 159.8 | −7.3 | 58.7 |  |  | 2.928 | None | None | b(never d) | 8 |
| 9 | Bromobenzene | Phenyl bromide | $C_6H_5Br$ | 157.0 | −30.8 | 156 | 1.5597 | $5.71^{20}$ | 1.495 |  |  |  | 9 |
| 10 | 2,3-Butanediol |  | $CH_3CHCHCH_3$ (OH OH) | 90.1 | 19 | 180 | 1.4306 |  | 1.010 | 85 | 402 | a,b,c | 10 |
| 11 | n-Butyl alcohol | 1-Butanol; n-butanol | $CH_3(CH_2)_2CH_2OH$ | 74.1 | −88.9 | 117.5 | 1.3993 | $17.8^{20}$ | 0.810 | 37 | 365 | a,b,c,d | 11 |
| 12 | sec-Butyl alcohol | 2-Butanol; sec-butanol | $CH_3CHCH_2CH_3$ (OH) | 74.1 | −89 | 99.5 | 1.3954 | $15.8^{25}$ | 0.808 | −10 | 406 | a,b,c,d | 12 |
| 13 | t-Butyl alcohol | 2-Methyl-2-propanol | $(CH_3)_3COH$ | 74.1 | 25.5 | 82.2 | 1.3878 | $10.9^{30}$ | 0.789 | 9 | 478 | a,b,c,d | 13 |
| 14 | n-Butyl chloride | 1-Chlorobutane | $CH_3(CH_2)_3Cl$ | 92.6 | −123.1 | 78.5 | 1.4021 | 1.95 | 0.8865 | −6.7 | 240 | b,c | 14 |
| 15 | t-Butyl chloride | 2-Chloro-2-methyl propane | $(CH_3)_3CCl$ | 92.6 | −25.4 | 51 | 1.3857 |  | 0.842 | 0 |  | a,b,c | 15 |
| 16 | Carbon tetrachloride | Tetrachloromethane | $CCl_4$ | 153.8 | −23.0 | 76.8 | 1.4601 | $2.238^{20}$ | 1.597 | None | None | — | 16 |
| 17 | o-Clorotoluene | 2-Chlorotoluene | $o{-}C_6H_4(Cl)CH_3$ | 126.6 | −35.1 | 159.2 | 1.5268 |  | 1.5268 |  |  | b,c | 17 |
| 18 | Cyclohexane | Hexamethylene | (cyclohexane ring) | 84.2 | 6.5 | 80.7 | 1.4266 | $2.02^{20}$ | 0.779 | −18 | 245 | a,b,c,d | 18 |
| 19 | Cyclohexanol | Adranol; cyclohexyl alcohol | (cyclohexane ring, OH) | 100.2 | 24 | 161.5 | 1.4641 | $15.0^{25}$ | 0.945 | 68 | 300 | a,b,c | 19 |
| 20 | Cyclohexanone | Nadone | (cyclohexane ring, =O) | 98.2 | −45 | 115.6 | 1.4507 | $18.3^{20}$ | 0.948 | 44 | 420 | a,b,c | 20 |

| No. | Name | Other Name(s) | Formula | MW | MP (°C) | BP (°C) | Refractive Index (20°C) | Dielectric Constant | Density (g/mL) at 20°C | Flash Point (°C) | Auto-ignition Point (°C) | How to Extinguish Flames | No. |
|---|---|---|---|---|---|---|---|---|---|---|---|---|---|
| 21 | Cyclohexene | | (structure) | 82.2 | −103.7 | 83 | 1.4465 | $2.22^{25}$ | 0.810 | −5 | 310 | a,b,c | 21 |
| 22 | Cyclopentadiene | 1,3-cyclopentadiene | (structure) | 66.1 | −85 | 42.5 | 1.4440 | | 0.805 | 27 | | a,b,c | 22 |
| 23 | Cyclopentanone | | (structure) | 84.1 | −58.2 | 130.6 | 1.4366 | | 0.951 | 26 | | a,b,c | 23 |
| 24 | Ethanol | Ethyl alcohol | $C_2H_5OH$ | 46.0 | −114 | 78.3 | 1.3611 | $24.3^{25}$ | 0.789 | 13 | 423 | a,b,c | 24 |
| 25 | Ethyl acetate | Ethyl ethanoate | $CH_3COOC_2H_5$ | 88.1 | −83.2 | 77.2 | 1.3723 | $6.02^{25}$ | 0.895 | −4 | 427 | a,b,c | 25 |
| 26 | Ethylbenzene | Phenyl ethane | $C_6H_5C_2H_5$ | 106.2 | −94.9 | 136.2 | 1.4959 | $2.41^{20}$ | 0.867 | 15 | 432 | a,b,c | 26 |
| 27 | Ethylene chloride | 1,2-Dichloroethane; ethylene dichloride | $ClCH_2CH_2Cl$ | 99.0 | −35.4 | 83.5 | 1.4448 | | 1.235 | none | none | — | 27 |
| 28 | Ethyl ether | Diethyl ether; ethoxyethane | $(C_2H_5)_2O$ | 74.1 | −116.2 | 34.6 | 1.3526 | $4.34^{20}$ | 0.714 | −45 | 160 | a,b,c | 28 |
| 29 | Ethyl iodide | Iodoethane | $C_2H_5I$ | 156.0 | −108 | 72 | 0.5133 | | 1.936 | | | b,c | 29 |
| 30 | Formaldehyde (37–50% aqueous) | Methanal | $H_2C{=}O$ | 30.0 | — | 99 | — | — | ~1 | 50 | 430 | a,b,c,d | 30 |
| 31 | Formic acid | Methanoic acid | $HCOOH$ | 46.0 | 8.2 | 100.8 | 1.3714 | $58^{16}$ | 1.227 | 69 | 601 | a,b,c | 31 |
| 32 | Isopropyl alcohol | 2-Propanol; dimethylcarbinol | $(CH_3)_2CHOH$ | 60.1 | −89 | 82.5 | 1.3776 | $18.3^{25}$ | 0.785 | 13 | 455 | a,b,c | 32 |
| 33 | Methyl alcohol | Methanol | $CH_3OH$ | 32.0 | −93.9 | 65 | 1.3288 | $32.6^{25}$ | 0.7914 | 11 | 464 | a,b,d | 33 |
| 34 | Methylethyl Ketone | Butanone; MEK; ethylmethyl ketone | $CH_3\overset{O}{\overset{\|}{C}}C_2H_5$ | 72.1 | −86.4 | 79.6 | 1.3788 | $18.5^{20}$ | 0.8054 | −6 | 516 | a,b,c | 34 |
| 35 | Methyl methacrylate | 2-methylpropanoic acid, methyl ester | $CH_2{=}\underset{CH_3}{C}{-}COOCH_3$ | 100.1 | −48 | 100 | 1.4142 | — | 0.944 | | | b,c | 35 |
| 36 | Methylene chloride | Dichloromethane; Freon 30 | $CH_2Cl_2$ | 84.9 | −96.7 | 39.8 | 1.4242 | $9.08^{20}$ | 1.326 | — | 615 | — | 36 |
| 37 | 1-Pentanol | n-Amyl alcohol; butyl carbinol | $CH_3(CH_2)_3CH_2OH$ | 88.2 | −79 | 137.3 | 1.4101 | $13.9^{25}$ | 0.814 | 58 | | a,b,c | 37 |
| 38 | n-Pentyl nitrite | n-Amyl nitrite | $CH_3(CH_2)_4ONO$ | 117.2 | | 96–99 | 1.3869 | | 0.853 | | 209 | a | 38 |
| 39 | Propanoic acid | Propionic acid | $CH_3CH_2COOH$ | 74.1 | −21.5 | 141.1 | | $18.5^{17}$ | 0.998 | 55 | 513 | a,b,c | 39 |
| 40 | n-Propyl alcohol | 1-Propanol | $CH_3CH_2CH_2OH$ | 60.1 | −126.5 | 97.4 | 1.3850 | $20.1^{25}$ | 0.8035 | 29 | 371 | b | 40 |
| 41 | Thionyl chloride | | $SOCl_2$ | 119.0 | −105 | 79 | | 1.655 | | | | a,b,c | 41 |
| 42 | Toluene | Methylbenzene; phenylmethane | $C_6H_5CH_3$ | 92.2 | −95 | 110.4 | 1.4961 | $2.379^{25}$ | 0.866 | 4 | 535 | a,b,c | 42 |

aAlcohol foam  bCarbon dioxide  cDry chemical extinguisher  dWater mist

appendix

# VI

# Common Drying Agents for Organic Liquids

The table on page 353 lists the most commonly used agents for drying organic liquids and their suggested use. Those squares marked "X" are the best combination of organic family/drying agent. Those marked "never" are the worst combinations, primarily due to possible chemical reactions. For instance, alcohols and sodium metal react vigorously. Consequently, sodium metal should never be used as a drying agent for an alcohol, and another, perhaps $MgSO_4$, should be considered. Those that are blank might be efficient, but are not recommended for use, unless the suggested drying agents are not available. Some combinations do not give efficient results due to complexation (footnoted as *d* in table).

**Suggested References**    Gordon, A. J., and A. R. Ford, *The Chemists Companion: A Handbook for Practical Data, Techniques, and References.* New York: John Wiley, 1972.
Bruno, T. J., and P. D. N. Svoronos, *Handbook of Basic Tables for Chemical Analysis.* Boca Raton, FL: CRC Press, 1989.

# COMMON DRYING AGENTS FOR ORGANIC LIQUIDS

| Organic Family | Na₂CO₃ [a] | K₂CO₃ [a] | MgSO₄ [b] | CaSO₄ [c] | Na₂SO₄ [a] | CaCl₂ | Na | P₂O₅ | NaOH (solid) | KOH (solid) | Quicklime | CaH₂ | LiAlH₄ |
|---|---|---|---|---|---|---|---|---|---|---|---|---|---|
| Alcohols | | X | X | X | | —[d] | Never | Never | Never | Never | X | —[e] | Never |
| Aldehydes | | | X | X | X | —[d] | Never | Never | Never | Never | | —[g] | Never |
| Alkyl halides | | | | | | X | Never | X | Never | Never | | X | Never |
| Amines | | | | | | —[d] | Never | Never | X | X | X | —[f] | Never |
| Anhydrides | | | | | | | | X | | | | | |
| Aryl halides | | | | | | X | Never | X | | | | X | X |
| Carboxylic acids | Never | Never | X | X | X | —[d] | Never | X | Never | Never | Never | Never | Never |
| Esters | | | X | X | X | —[d] | | | Never | Never | | —[g] | Never |
| Ethers | | | | X | Poor | X | X | X | | | | X | X |
| Hydrocarbons, aromatic | X | X | | X | | X | X | X | | | X | X | X |
| Hydrocarbons, saturated | X | X | | X | Poor | X | X | X | | | X | X | X |
| Hydrocarbons, unsaturated | X | X | | | Poor | X | | X | | | | X | X |
| Ketones | X | X | X | X | X | —[d] | Never | Never | Never | Never | Never | —[g] | Never |
| Nitriles | | | | | | | | X | Never | Never | | —[g] | Never |

[a] Excellent in salting out.
[b] Best all-purpose drying agent.
[c] High-capacity, but slow reacting.
[d] Forms complexes.
[e] Lime (common impurity) reacts with acidic hydrogen.
[f] Only for 3° amines (R₃N).
[g] Never if α-hydrogen is present.

# Index

## A

Abbe-Spencer refractometer, 26, 27
Absorbance, 113–19
Absorption bands, infrared, 92, 93
Acetal, formation of, 229–30
Acetaldehyde, 238
Acetamide, 84,100
Acetaminophen, 185–87, 255, 256
Acetanilide, 47, 69, 255–57
Acetic acid, 26, 43, 67, 79, 193, 219–20, 245, 246, 255, 331, 346–49
    esterification of, 78–81, 217–18
Acetic anhydride, 214–15, 255, 321–22, 349
Acetone, 7, 32, 67, 238, 245, 253, 349
Acetonitrile, 32, 67
Acetophenone, 130, 238
Acetylene, synthesis and reactions
    of, 141–45
Acetylsalicylic acid, see Aspirin
Acetylspiraeic acid, 214–15
Achiral molecules, 11
Acid-catalyzed equilibria, 229
Acidic amino acids, 285
Acyl halide, 226, 255, 286
Adaptors, ii
Addition-elimination reactions, 234
Adipic acid, 127, 133–35, 223, 247–48, 331
Adsorption chromatography, 42
Alanine, 284
Alcohols, 78, 175–87, 219, 223, 226, 325, 327, 352
    boiling points of, 183, 333
    densities of, 79
    and derivatives, 332–33
    and esterification, 78–81, 216–18
    from Grignard reaction, 175–76
    and microboiling point determination, 182–83
    oxidation of, 211, 223
    qualitative tests for, 180–83, 326–28, 332
    reduction of, 226
Aldehydes, 175, 211, 223, 226, 229, 234–41, 245, 325, 326, 327, 328
    derivatives of, 236–38
    and ketones, 223–42
    oxidation, 223, 234–36
    physical properties of, 238
    qualitative tests for, 234–41, 328, 331, 333–34
    reaction with ammonia derivatives, 234–38
    reaction with ketones, 243–46
Alder, Kurt, 147
Aldohexose, 273, 275–76, 281
Aldol, 243–46
Aldol condensation, 243–46

Aldopentose, 273, 275–76, 281
Aldose, 273
Alkaloid, 75
Alkanes,
    conformations of n-butane, 9
    qualitative tests for, 143
Alkenes, 14–15, 123–35
    addition stereospecific, 130
    catalytic hydrogenation, 136–40
    oxidation of the C=C bond with cold
        potassium permanganate, 127–29
    oxidation of the C=C bond with hot
        potassium permanganate, 133–35
    reaction wirh bromine, 130–32
    synthesis by dehydration of alcohols,
        123–26
Alkylation, Friedel-Crafts. See Friedel-Crafts
    alkylation
Alkyl halides, 151–73, 175, 352
Alkynes, 141–45
Alumina, eluotropic series of solvents
    for, 43
Amides, 325, 329–30
Amine hydrochloride salt, 325
Amines, 251–59, 325, 329, 352
    from nitro compounds, 251–54
    qualitative analysis of, 325–26, 329–30
Amino acids
    α-, list of, 284–85
    and proteins, 283–303
    qualitative tests for, 290–97
Ammonium hydroxide, 235, 237, 262–63, 346
Amphoteric, 290
n-Amyl alcohol, 113–15, 183
n-Amyl nitrite, 350
Analysis,
    chromatographic. See
        Chromatographic analysis
    qualitative. See Qualitative analysis
Anhydrides, 214, 219, 255, 352
    hydrolysis of, 137–38, 201–02
    reactions of, 219
Anilide preparation, 325, 329–30
Aniline, 195–96, 219–20, 249–50, 255, 256, 330, 349
    acylation of, 255–57
p-Anisaldehyde, 25
Anomers, 273
Anthracene, 201–03
Anthranilic acid, 200, 203
Anti addition to a carbon-carbon double
    bond, 127
    conformations of n-butane, 9
Arabinose, 276–81
Arachidonic acid, 306

Arene, oxidation of side chain of, 197–99, 260–63
Arginine, 285
Aromatic amine, 329
Aromatic compounds, IR substitution
    pattern, 93
    qualitative test, 328
Aromatic diazonium compounds, 258–59, 328–29
Aromatic hydrocarbons, 260, 353
Aromatics, 188–99, 260–63
Arrhenius equation, 166
Aryl halides, 352
Asparagine, 284
Aspartic acid, 285
Aspirin, synthesis of, 214–15
Asymmetry, 11
Axial hydrogen, 10
Azeotropic distillation, 37, 124–25
Azo dyes, formation of, 258–59, 329–30

## B

Balancing, of oxidation-reduction
    reactions, 337–39
Barfoed reagent, 346
Barfoed's test, 276, 277
Barium hydroxide, 247–48
Basic amino acids, 285
Beef fat, 311
Beef tallow and iodine number, 311
Beer-Lambert law, 91, 113
Benedict's solution, 346
Benedict's test, 235, 236, 278,
trans-Benzalacetophenone, 130–32, 249–50
Benzaldehyde, 130–32, 243–44, 245, 349
Benzamide, 20, 84
Benzene, 26, 68
    and its reactions, 188–203
Benzene sulfonamide derivatives, 326, 330
Benzene sulfonyl chloride and amines, 325, 329
Benzenoid compounds, condensed. See
    Polynuclear aromatics
Benzimidazole, 262–63
Benzoic acid, 63–64
    synthesis of, 211–13, 221–22
Benzonitrile, 211–13, 349
Benzophenone, 238
Benzoyl chloride, 349
Benzoyl peroxide, 96, 98
Benzyl acetate, 216
Benzyl alcohol, 333
Benzyl butyrate, 216
Benzyne, 195, 200–02
    addition reactions of, 200
    dimerization of, 201

generation of, 195–200, 202–203
Bial's reagent, 346
Bial's test, 275, 279, 346
Biuret test, 291
Blocking group, 286
Boat conformation, of cyclohexane, 9
Boiling point (BP), 25–26
    determination of (micro), 182
Borate ester, 226
Bromine, 6, 124–26, 130–32, 349
    addition to an alkene, 124–26, 127,
        130–32
p-Bromoacetanilide, 84
Bromobenzene, 349
    nitration of, 188–91
p-Bromobenzenesulfonamide, 84
1-Bromobutane. See n-Butyl bromide
2-Bromobutane. See sec-butyl bromide
Bromoethane, 1H NMR Spectrum of, 107
o-Bromonitrobenzene, 188
p-Bromonitrobenzene, 188
Büchner funnel, ii, 63
Burns, 2
Butane, 6, 9
2,3-Butanediol, 229–31, 349
1-Butanol. See n-Butyl alcohol
2-Butanol. See sec-Butyl alcohol
Butanone, 26, 111–12, 202–03
Butter, 311
n-Butyl alcohol, 79, 183, 349
sec-Butyl alcohol, 14, 79, 183, 349
t-Butyl alcohol, 79, 127–29, 151–52, 154–57,
        183, 192–93, 349
n-Butyl bromide, 164
sec-butyl bromide, 164
n-Butyl chloride, 349
t-Butyl chloride, 151–73, 154–57
    solvolysis of, 154–61
    synthesis of, 151–53
t-Butyl methyl ketone. See Pinacolone
n-Butyric acid, 79, 207

C

Caffeine, 75
    extraction from tea, 75–77
Cahn, R. S., 13, 14
Calcium carbide hydrolysis, 141–42
Calcium chloride, 307
    anhydrous, 352
Calcium hydride, 352
Calcium hydroxide, 141
Calcium sulfate, anhydrous, 352
Canola oil, 311
Capillary filter stick, 72
Capric acid, 207
Caproic acid, 207
Caprylic acid, 207
Carbanions, 243–50
    and unsaturated carbonyls, 243–46,
        249–50
Carbocation, 151, 192
Carbon tetrachloride, 26, 68, 349
Carboxylic acids, 63, 78
    densities of, 79
    and derivatives, 204–22
    equivalent weight determination,
        204–09

reaction with sodium bicarbonate,
        325, 329
Carbohydrates. See Sugars
Carbonation, of Grignard reagent, 176
Carbon tetrachloride, 26, 68
Carbonyl compounds, 223-50. See also
        Unsaturated carbonyls
Carboxylic acids, 63, 78,
    densities of, 79
    and derivatives, 330–31
    equivalent weight determination,
        204–09
    esterification, 78–81, 216–18
    unsaturated, 244
Casein, 290–99
Catalase, 299–302
Celite. See Filter Aid
Ceric ammonium nitrate, 332, 347
Chair conformation, of cyclohexane, 9, 10
Chalcone. See Benzalacetophenone
Chemical shift values, 1H NMR, 107
Chemoselectivity, 251
Chicken fat, 311
Chiral carbons, 11, 13–14
Chiral molecules, Fischer projection
        formulas for, 12–13
p-Chloroacetanilide, 84
Chlorobenzaldehyde, 84, 207
p-Chlorobenzamide, 84
o-Chlorobenzoic acid, synthesis of, 197–99
1-Chlorobutane, 6, 164, 176–78
2-Chlorobutane, 6, 164, 176–78
Chloroform, 26, 68
p-Chlorophenol, 332
2-Chlorotoluene, oxidation of, 197–99
Cholesterol, in fats and oils, 315
    Liebermann-Burchard test for, 321–23
Chromatogram, 51–52
Chromatography, 42–53
    adsorption. See Adsorption
    chromatography column. See Column
        chromatography
    chromatography gas. See Gas
        chromatography (GC)
    gas-liquid. See Gas-liquid
        chromatography (GLC)
    gas-solid. See Gas-solid
        chromatography (GSC)
    gel-permeation. See Gel-permeation
        chromatography
    high-pressure liquid. See High-pressure
        liquid chromatography (HPLC)
    ion-exchange. See Ion-exchange
        chromatography
    thin-layer. See Thin-layer
        chromatography (TLC)
Chromium (VI) oxide, 261
Chromium (III) cation, 223–25
cis-isomers, 10, 11, 14, 15
Citral, steam distillation of, 37–38
Claisen, L., 245
Claisen adaptor, ii
Claisen condensation, 243
Claisen-Schmidt reaction, 245
Coconut oil, 311, 312, 322
Column, distilling. See Distilling column
Column chromatography, 42–44

Common recrystallization solvents, 67–68
Condensed benzenoid compounds, 260
Condenser, ii
Configurational isomers, 10–12, 15
Conformational isomers, 7, 8–10, 15
Conjugated dienes, 147–48
Connecting adaptor, ii
Constant, equilibrium. See Equilibrium
        constant
Constitutional isomers, 6–7
Continuous-wave spectroscopy, 106
Corn oil, 311
Cottonseed oil, 307, 311
Covalent bonds, 5
Cresol, 332
Crop, and second crop, 66
Cupric oxide, 235–36
Cuprous oxide, 235–36
Cuvette, 114
Cyclization reaction, 247–48
Cyclohexane, 7, 9, 10, 43, 68, 349
Cyclohexanediol, from cyclohexene, 127–29
Cyclohexanol, 223–28, 349
    from cyclohexanone, 226–28
    oxidation of, 223–25
Cyclohexanone, 223–28, 349
    from cyclohexanol, 223–25
    reduction of, 226–28
Cyclohexene, 100, 123–29, 133–35, 350
    oxidation to adipic acid. 133–35
    oxidation to 1,2-cyclohexanediol,
        127–29
    synthesis from cyclohexanol, 123–26
    qualitative tests for, 126, 143
Cyclohexylamine, 330
Cyclopentadiene, 147–50, 350
Cyclopentanol, 183, 350
Cyclopentanone, 238, 247–48, 350
    from adipic acid, 247–48
Cymbopogon flexuosus, 37
Cysteine, 284

D

Decarboxylation of adipic acid, 247–48
Dehydrating agents. See Drying Agents
Delta, in 1H NMR Spectra, 106–07
Denaturation of proteins, 290–91, 299–302
Depolymerization-repolymerization, of
        plexiglass, 96, 99
Deshielding, 106
Detector, in GC, 51
Detector, in IR, 91
Deuterated solvent for NMr spectroscopy,
        108
1,2-Diaminobenzene. See o-Phenylene
        diamine
Diastereomers, 12, 13, 15
Diazonium salts, 200, 202, 258, 326–29
Dibenzalacetone, 243–46
Dibromobenzalacetophenone, 130–32
2,3-Dibromobutane, 13
1,2-Dibromocyclohexane, trans, 125
1,2-Dibromocyclopentane, cis- and
        trans-, 10
p-Dichlorobenzene, 7, 63, 100
1,2-Dichloroethane, 202–03
Dichloromethane. See Methylene chloride

Dicyclopentadiene, 143, 148–49
Diels, Otto Paul Hermann, 147
Diels-Alder reaction, 147–50, 201–03
　nes,
　　onjugated. See Conjugated dienes
　　on-conjugated. See Non-conjugated
　　　dienes
　　iophile, 147
　　thylamine, 330
　ethylene glycol diethyl ether, 201, 203
　iethyl ether, 26, 37, 43, 67, 227,
Dimerization of benzyne, 201
1,4-Dimethoxybenzene, 47, 192–94
2,3-Dimethylbutadiene, 108, 232–33
2,3-Dimethyl-2,3-butanediol. See pinacol
3,3-Dimethyl-2-butanone. See pinacolone
4,5-Dimethyldioxolane, synthesis of, 229–30
Dimethyl ether, 7
Dimethylformamide, 26
Dimethyl sulfoxide, 26
Dinitrobenzoate ester (DNB), 327
2,4-Dinitrobenzoyl chloride, 327, 332
2,4-Dinitrobromobenzene, synthesis of,
　　188–91
2,4-Dinitrochlorobenze, reaction with
　　aniline, 195–96
2,4-Dinitrophenylaniline, 195–96
2,4-Dinitrophenyl hydrazine, 236–38,
　　249, 347
1,2-Diols, 232
p-Dioxane, 68
Dipeptides, 283
Diphenylacetylene, 143
Diphenyl sulfone, 84
Diphosphorus pentoxide, 338
Dipolar ion structure. See Zwitterion
Dipole moment, 90
Disaccharides, 264, 276,
Dispersion grating, 91
Distillation, 29–41
　　azeotropic. See Azeotropic distillation
　　fractional. See Fractional distillation
　　micro-. See Micro-distillation
　　simple. See Simple distillation
　　steam. See Steam distillation
　　types of, 29
　　vacuum. See Vacuum distillation
Distilling column, ii
Distribution coefficient, 54–61, 113–15
Drying agents, 37, 351–52

E
E1 (first order) elimination, 123–24
E1CB (carbanion) elimination, 124
E2 (second order) elimination, 123
Eclipsed form, 8, 9, 10
Egg-albumin, 290–95
Electrophilic aromatic substitution, 188–94
　　directing effects of, 189
Elements, cover 3
Eluotropic series of solvents, for alumina
　　and silica, 43
Enantiomers, 11, 12, 13, 15, 130
Enolization, 223, 243–44
Equatorial hydrogens, 10
Equilibrium constant, 78–81
Equivalent weight, 55

Equivalent weight of organic acids, 204–09
Ergosterol, 322
Essential amino acids, 283, 285–86
Esterification, Fischer. See Fischer
　　esterification
Esters, 78, 79
　　as fragrances and flavors, 216
　　saponification, 221–22
　　synthesis of, 78–80, 214–18
Ethanol, 7, 26, 43, 67, 196, 215, 238, 245,
　　246, 307, 350
Ether, 7, 26, 37, 43, 67, 175, 186, 307, 311
　　in Grignard synthesis, 177–79
Ethyl acetate, 26, 43, 55, 56, 67, 350
Ethyl acetoacetate, 115
Ethylbenzene, 350
Ethyl bromide, 1H NMR spectrum of, 107
Ethyl butyrate, 216
Ethylene chloride, 203, 350
Ethyl ether. See Diethyl ether
Ethyl iodide, 185–86, 350
p-Ethylphenol, 332
Ethyl phenylacetate, 216
Extraction
　　and distribution coefficient, 54–56,
　　　113–15
　　and recrystallization, 54–77
　　of caffeine from tea, 75–77
　　separation by, 75–77
E/Z designation, of alkenes, 14–15

F
Fats, 305–23
　　dietary, 311
　　and iodine numbers, 311–19
Fatty acids, 306
Fehling's solution, 235, 347
Fehling's test, 347
Ferric chloride, 307, 331–32
Films and wrappers, IR spectra,
Filter aid. See celite
Filtration,
　　gravity, 63
　　hot, 66–71, 74
　　vacuum, 63–65
First-aid kit, 1
Fischer esterification, 78–80, 216–18
Fischer-Johns apparatus, 20
Fisher projection formulas, 12–13,
　　of monosaccharides, 274
Flammability, of recrystallization solvents,
　　67–68
Flow chart,
　　for recrystallization, 69
　　for qualitative analysis of sugars,
　　　273–81
Formaldehyde, 229–31, 350
Formic acid, 262–63, 350
Fourier transform infrared
　　spectrophotometers, 91
Fractional distillation, 29, 32–35
Frequency, of electromagnetic radiation, 89
Friedel-Crafts alkylation, 192–94
Fructofuranose, 273
Fructose, 264, 273
Fully eclipsed, conformations of
　　n-butane, 9

Functional group isomers, 6–7
Funnels, ii, 56, 70
Furan, 262
Furanose, 273
Furfural, 276

G
Galactaric acid, 278
Galactose, 278
Gas chromatography (GC), 51–53, 225, 227,
　　230, 233, 248
Gas-liquid chromatography (GLC), 51
Gas-solid chromatography (GSC), 51
Gauche, conformations of n-butane, 9
Gel-permeation chromatography, 42
Geometric isomers, 7, 10–11, 15
Geraniol, 39
Gibbs free energy, 78
Glasses (safety glasses), 1
Glassware, ii
Glucopyranose, 274
Glucose, 264, 273–79
Glutamic acid, 285
Glutamine, 284
Glycine, 284
Glycylalanine, 283–89
Glycylglycine, 283–89
Gravity filtration, 63
Grease, 29
Grignard, Victor, 175
Grignard reaction,
　　and acetals, 229
Grignard reagent, carbonation of,
Ground-glass equipment, ii, 3

H
Halogens, sodium fusion test for, 83, 85
Handedness, 11
Hanus solution, 347
Hanus test, 311–15
Heat, and denaturation,
Hefty Bags, IR spectrum of, 97
Hemiacetal, 229
Hemiketal, 229
Heptane, 68
Heterocycles, and polynuclear aromatics,
　　262–63
Hexane, 43, 68
Hickman flask, 40
High-pressure liquid chromatography
　　(HPLC), 42
Hinsberg test, 325, 329
Hirsch funnel, ii
Histidine, 285
Hopkins-Cole test, 293, 347
Hot-stage apparatus, 20, 22
Huckel's rule, 260
Hydrazine, 238
Hydrazone, 238
Hydrobromic acid, 108
Hydrocarbons, classification of, 147
Hydrochloric acid, 151–52, 198, 212, 221,
　　251–54, 301, 346
Hydrogen, generation of, 136, 139
Hydrogen peroxide, decomposition,
　　299–302
Hydrolysis,

of benzonitrile, 211–13,
  of fats, 305–09
  of sucrose, 264–72
*o*-Hydroxybenzoic acid. *See* Salicylic acid
β-Hydroxyketone. *See* aldol
Hydroxylamine, 237, 252

## I

Imidazole, 262
Imides, 219–20
Imines, 238
Infrared (IR), Spectroscopy, 89–104
  and infrared absorption bands, 93
  and infrared spectrum of polystyrene,
    90
Ingold, C. K., 13, 14
Integration of NMR signals, 107–08
Interconversion, of chair forms of
  cyclohexane, 10
Internal alkyne, 141
Iodine monobromide, 311
Iodine numbers, of fats or oils, 311–19
*o*-Iodobenzamide, 84
Iodoform, 180–82, 241, 347
Ion-exchange chromatography, 42
Iron in nitro group reduction, 251
Isoamyl alcohol, 217–18
Isoamyl acetate, synthesis of, 216–18
Isoamyl nitrite, 200–03, 350
Isobutane, 6
Isobutyl propionate, 216
Isoleucine, 284
Isomers, 5
  *cis*. *See* Cis isomers
  configurational. *See* Configurational
    isomers
  conformational. *See* Conformational
    isomers
  constitutional. *See* Constitutional
    isomers
  geometric. *See* Geometric isomers
  optical. *See* Optical isomers
  *ortho, para,* and *meta,* 159
  positional. *See* Positional isomers
  skeletal. *See* Skeletal isomers
  stereoisomers. *See* Stereoisomers
  *trans*. *See* Trans isomers
Isopropyl alcohol, 26, 43, 79, 156–57, 350

## J

Joints, ii
Jones' oxidation, 180, 235

## K

Ketal, formation of, 229–31
Keto-enol equilibrium, 223
β-Ketoester, 244
Ketohexose, 273, 276, 281
Ketones, 72, 235,
  and aldehydes, 223–41
  and derivatives, 236–38
  physical properties of derivatives, 238
  qualitative tests for, 234–41, 328–31,
    333–34
  reaction with alcohols, 229–31
  reduction to secondary alcohols,
    226–28

Ketopentose, 244
Ketose, 244
Kinetics,
  of the acid-catalyzed hydrolysis of
    sucrose, 264–71
  of the solvolysis of *t*-butyl chloride,
    154–61
Knoevenagel condensation, 244

## L

Laboratory notebook, 2–3
Laboratory safety, 1–2
Lactic acid, 11, 12
Lactose, 275
Lactones, 226
Lanosterol, 322
Lard, 311
Lauric acid, 306
Lead acetate, 85, 292
Leaving group, 123
Leucine, 284
Liebermann-Burchard test for cholesterol,
  321–23
Light, velocity of, 89
Ligroin, 68
Limiting reagent, 341
Linoleic acid, 306
Linseed oil, 311, 312, 322
Lipids, 347–49
Liquid solvents, physical data for 67–68
Liquids,
  drying agents for, 351–52
  physical data of, 348–50
Lithium aluminum hydride, 352
Lucas reaction, 180–81
Lucas reagent, 180–81, 347
Lysine, 285

## M

Magnesium chloride, 307
Magnesium metal, in Grignard reactions,
  175–79
Maleic acid, 22
Maleic anhydride, 148–49, 201–03
Maltose, 275
Magnesium sulfate, anhydrous, 352
Manganese dioxide, 128, 133–34, 197–99
Mannose, 274
Margarine, 311
Melting point (MP), determination of,
  19–24
Meso compounds, 13
Metal block apparatus, 19–20, 22
Methanol, 43, 67, 79, 214, 350
Methionine, 284
Methycyclopentane, 7
Methyl acetate, 67
Methyl alcohol. *See* methanol
Methyl anthranilate, 216
Methyl benzoate, saponification of, 221–22
Methyl butyrate, 216
Methyl cellosolve, 67
Methylene chloride, 26, 43, 63, 67, 253, 312,
  316, 321–22, 350
Methylethyl ketone. See Butanone
2-Methyl-2-hexanol, synthesis of, 175–79
Methylmethacrylate, 96, 98, 351

2-Methylnaphthalene, oxidation of, 260–61
2-Methyl-1,4-naphthoquinone, 260–61
Methyl salicylate, 214
Micelles, 305–06
Michael addition, 249–50
Microboiling point, 25
Microdistillation, 40–41
Microrecrystallization, of solid, 72–74
Millon's reagent test, 292, 347
Mixed melting, 21
Mobile phase, 42
Models, molecular. *See* Molecular models
Molarity, 344
Molecular formulas, 5
Molecularity of reaction, 154
Molecular models, and stereochemistry,
  5–18
Molish test, 347
Monochromatic dispersion grating, 91
Monosaccharides, 273–79
Mother liquor, 64, 66
Mucic acid test, 278
Mull method, 95
Multiplet absorption, in NMr spectroscopy,
  107
Mutarotation, 274
Myristic acid, 306

## N

N + 1 rule, in NMr spectroscopy, 107
Naphthalene, 260
Naphtha solvent, 68
*a*-Naphthol, 332
*b*-Naphthol, 258–59, 332
1,4-Naphthoquinone, 260
Neral, 39
Nernst glower, 90
Newman projections, 8, 9
Nichrome wire, 90
Ninhydrin test, 292
Nitration, of bromobenzene, 188–91
Nitric acid, 189–90, 334, 346
Nitriles, 211–13
*p*-Nitroaniline, 258–59
*p*-Nitrobenzene diazonium chloride, 258–59
Nitrogen, socium fusion test, for 83, 85
Nitronium ion, 188
Non-conjugated dienes, 147
Non-reducing carbohydrate, 275,
Non-superimposable isomers, 10
Non-terminal alkyne, 136
Norbornane-*endo*-2,3-dicarboxylic acid,
  *cis*-, 137–39
Norbornene-*endo*-2,3-dicarboxylic acid,
  *cis*-5-, 137–39
Norbornene-*endo*-2,3-dicarboxylic
  anhydride, *cis*-5, 137–39, 148–49
Notebook, 2–3
Nuclear magnetic resonance (NMR). *See*
  Proton Magnetic Resonance
Nuclear spin quantum number, 105
Nuclei, and NMR, 105
Nucleophilic substitution, aliphatic, 151–73
Nucleophilic substitution, aromatic, 195–96
Nucleophilic substitution, definition, 163
Nucleophilic substitution, first order,
  163–67

Nucleophilic substitution, second order, 163–67

305–23
acid, 306, 315
oil, 311, 312
-chain form, of monosaccharides, 274
al isomers, 7, 10, 11–12, 15
um wavelength, in UV-visible spectrophotometry, 113, 114
r, of reaction, 154–56
ganic liquids, drying agents for, 351–52
rganic acids, equivalent weight of, 204–09
Organic solvents, and denaturation, 291
Organometallic compounds, 175
Oxalic acid, 207, 224
Oxidation
    balancing reactions, 337–39
    of cyclohexanol to cyclohexanone, 223–25
    of cyclohexene to adipic acid, 133–35
    of 2-methylnaphthalene, 260–61
    of primary alcohol or aldehyde, 211
    of side chain of arene, 197–99
Oxime, 237
Oxonium ion, 124

**P**

Paladium, on carbon, 136, 138–39
Palmitic acid, 306
Palm oil, 311
Para red, 258–59
Peanut oil, 311
Pellet method, IR, 95–96
Pentane, 43, 68
n-Pentanol. See n-Amyl alcohol
Pentose, 276
Pentyl nitrite. See n-amyl nitrite
Peptide bond, 283
Peptides, synthesis of, 283–91
Percent by mass (w/w), 343–45
Percent transmittance, 90
Percent yield, 340–42
Perhydrocyclopentanophenanthrene ring system, 321
Periodic table of the elements, cover, 3
Perkin condensation, 244
Peroxides, safety hazard, 124, 172
Petroleum ether, 43, 68
Phases in chromatography, 42
Phenacetin, 185–87
Phenanthrene, 260
Phenolphthalein, indicator, 80, 165–66, 205
Phenols, 185, 326, 329–30
    qualitative tests for, 331, 332
Phenylalanine, 284
o-Phenylenediamine, 262
Phenylhydrazine, 237
Phenylhydrazone, 237
Phenylisocyanate, 328, 333
N-Phenylphthalimide, 219–20
Phenylurethane, 333
Pheromones, 217
Phosphoric acid, 123–24, 214–15
Phthalhydrazide, 287–89
Phthalic acid, 219–20, 260

Phthalic anhydride, 219–20, 285–88
Phthaloylglycine, 283–89
Phthaloylglycyl chloride, 283–89
Phthalylglycylglycine, 283–89
Physical data of liquids, 348–50
Pinacol, 108, 232–33
Pinacolone, 108–09, 232–33
Pinacol rearrangement, 232–33
Planck's constant, 89
Plastic films, IR spectra of, 96–97
Plexiglass, depolymerization-repolymerization of, 96, 98
Polymer films, IR spectra of, 96
Polymorphic compounds, 21
Polynuclear aromatics and heterocycles, 260–63
Polypeptides, 283
Polystyrene, 90
Positional isomers, 6
Potassium carbonate, anhydrous, 352
Potassium hydroxide, anhydrous, 352
Potassium iodide, 316,
Potassium permanganate, see unsaturation, permanganate tests for
Prelog, V., 13, 14
Primary amine, 255
Primary structure, of proteins, 290
Prism dispersion grating, 91
Prism, in refractometer, 26, 27
Proline, 284
Propanoic acid, 42–44, 207, 350
1-Propanol. See n-propyl alcohol
Propionaldehyde, 238
Propionic acid. See Propanoic acid
n-Propyl alcohol, 7, 78, 79
Proteins, 290–302
    and amino acids, 283–303
    list of amino acids found in, 284–85
    qualitative tests for, 290–97
Proton magnetic resonance (PMR), 105–12
Pseudo-first-order reactions, 264
Pyranose, 273
Pyridine, 67
Pyridinium hydrogen, sulfate dibromide, 315–19
Pyrimidine, 262
Pyrrole, 262

**Q**

Qualitative analysis, 325–36
    of alcohols, 180–82
    of aldehydes and ketones, 234–41
    of amines, 328–30
    of aminoacids and proteins, 290–97
    of carboxylic acids, 329–31
    of carbohydrates, 273–81
    of hydrocarbons, 143
    of phenols, 331–32
    of unsaturation in lipids, 311–13
Quaternary structure, of proteins, 290
Quicklime, 352

**R**

Racemic mixture, 15
Raoult's law, 33
Rate law, first order, 64

Reaction order, definition and examples, 154
Reactions,
    balancing oxidation-reduction, 337–39
    amino acids and proteins, 290–97
    order of, 154
    and percent yield, 340–42
    pseudo-first-order, 264
    second-order, 154
    and synthesis of acetylene, 141–45
Reagents,
    limiting, 341
    special, 346–47
Recrystallization
    and extraction, 54–77
    flow chart for, 69
    and microrecrystallization, 72–74
    of solids, 66–70
    solvents, 67–68
Reducing disaccharide, 275
Reducing sugars, 275
Reduction
    balancing reactions, 337–39
    of cyclohexanone to cyclohexanol, 226–28
Reflux, setup for, 80
Refractive index, 26–28
Refractometer, 26, 27
Repolymerization, of plexiglass, 96–97
Resonance frequency (MHz) of nuclei, 106
Retention time, gas chromatography, 52
Retro aldol condensation. See reverse aldol condensation
Reverse aldol condensation, 244
Reynolds Oven Cooking Bags, IR spectrum of, 96–97
$R_f$ value in TLC, 45
Round-bottom flasks with standard-taper joints, ii
R/S designation, for chiral carbons, 13–14

**S**

Saccharic acids, 278
Saccharides. See Carbohydrates
Safety, 1–2
Safety glasses, 1
Safflower oil, 311
Sakaguchi reaction test, 293
Salicylic acid, acetylation of, 214–15
    methylation of, 214
Salting out, 218
Salts, and denaturation, 291
Saponification, 221–22, 305–09
Saturated fatty acids, 306
Saturation of nuclear spin states in 1H NMR, 106
Schmidt, J. G., 245
Secondary amine, 255
Secondary structure, of proteins, 290
Second crop, of crystals, 66
Second-order reactions, 264
Seeding, of crystals, 66
Seliwanoff's test, 277, 347
Semicarbazide, 237
Semicarbazone, 237
Separation by extraction, 63–65
Separatory funnels, ii
    handling and support, 56

Serine, 284
Shielding and deshielding effects in NMR spectroscopy, 106
Side chain of arene, oxidation of, 197–99
Silica, eluotropic series of solvents for, 43
Silver nitrate, 152, 164
Silver oxide, 236
Simple distillation, 29, 31
Skeletal isomers, 6
Soap, preparation and properties of, 305–09
Sodium borohydride, 136–37, 140
Sodium carbonate, anhydrous, 352
Sodium dichromate, 180, 223–25
Sodium fusion test, 83–86
Sodium hydroxide, 197–98, 212, 221, 307, 352
Sodium iodide, 164
Sodium metal, 143
Sodium nitrite, 258–59
Sodium sulfate, anhydrous, 352
Sodium thiosulfate, 315–19
Solid
    microrecrystallization of, 72–74
    recrystallization of, 66–70
    sublimation, 75–77
Solubility and qualitative analysis, 325, 329
Solutions, making, 343–45
Solvents,
    common recrystallization solvents, 67–68
    and denaturation, 291
    eluotropic series for alumina, 43
    eluotropic series for silica, 43
Solvolysis, of *t*-butyl chloride, 153–61
Soybean oil, 311
Special reagents, 346–47
Spectrophotometers,
    IR diagram, 91
Spectrophotometric determination of distribution coefficient, 113–15
Spectrophotometry, visible. *See* Visible spectrophotometry
Spectroscopy, 89–121
    continuous-wave. *See* Continuous-wave spectroscopy
    infrared. *See* Infrared spectroscopy
Spectrum, IR, of polystyrene, 90
Spin quantum number, 105
Spin-spin splitting, 107
Spin states, nuclear 105
Spiraeic acid, 214
Staggered form, 8, 10
Standard-taper joint, ii
Stationary phase, 42
Steam distillation, 29, 37–38
Stearic acid, 306
Stereocenters, 11
Stereochemistry, and molecular models, 5–18
Stereoisomers, 6, 7–12, 15
Steroids, 321–22
Stilbene, trans, 47
Structural formulas, 5
Sublimation, 75
    assemblies for, 76
Sucrose, hydrolysis of, 264–69

Sugars, 264–81
    flow chart for qualitative analysis of, 276
    reducing and nonreducing, 275
Sulfanilamide, 218
Sulfur, sodium fusion test for, 83, 85
Sulfuric acid, 80, 189–90, 192–93, 217–18, 321–22, 346
Sulfuric acid, fuming, 346
Sunflower oil, 311
Superimposable structures, 10, 15
Symmetry, 11
Syn, addition to a carbon-carbon double bond, 127, 136–37
Synthesis
    of acetanilide, 255–57
    of acetylene, 141–42
    of adipic acid, from cyclohexene, 133–34
    of *m*-aminoacetophenone from *m*-nitroacetophenone, 251–54
    of aspirin, 214–15
    of benzimidazole, 262–63
    of benzoic acid from benzonitrile, 211–13
    from methylbenzoate, 221–22
    of *t*-butyl chloride, 151–53
    of 2-chlorobenzoic acid, 197–99
    of cyclohexanediol from cyclohexene, 127–29
    of cyclohexanol from cyclohexanone, 226–28
    of cyclohexanone from cyclohexanol, 223–25
    of cyclohexene, from cyclohexanol, 123–26
    of cyclopentanone from adipic acid, 247–48
    of dibenzalacetone, 243–46
    of dibromobenzalacetophenone, 130–31
    of 2,5-*di-t-butyl*-1,4-dimethoxybenzene, 192–94
    of dimethyldioxolane, 229–31
    of 2,4-dinitrobromobenzene, 188–91
    of 2,4-dinitrophenylaniline, 195–96
    of glycylglycine, 283–89
    of isoamyl acetate, 216–18
    of 2-methyl-2-hexanol, 175–79
    of 2-methyl-1,4-naphthoquinone, 260–61
    *cis*-5-norbornane-*endo*-2,3-dicarboxylic acid, 136–41
    *endo*-norbornene-*endo*-2,3-dicarboxylic anhydride, 147–49
    of para red, 258–59
    of phenacetin, 185–87
    of N-phenylphthalimide, 219–22
    of peptides, 283–89
    of pinacolone 232–33
    of triptycene, 200–03

T
Tallow, 311
Terminal alkyne, 141
Terpenes, 39
Tertiary amine, 255
Tertiary structure, of proteins, 290
Tetrahydrofuran, 26

Tetrahydroxyadipic acid. *See* Mucic acid
Thiazole, 262
Thiele-Dennis tube, 19, 21–22
Thin-layer chromatography (TLC), 45–49
Thionyl chloride, 330, 350
Thiophene, 262
Threonine, 285
Tollens' test, 235–36,
Toluene, 26, 56, 61, 68, 350
Toxicity, of recrystallization solvents, 67–68
Trans isomers, 10, 11, 14, 15
Transmittance, 90
Tripeptides, defined, 283
Triptycene, 200–03
Tryptophan, 285
Tylenol, 255
Tyrosine, 285
Tyrosylserine, 247

U
Unsaturated carbonyls, and carbanions, 243–50
Unsaturated fatty acids, 311–19
Unsaturated hydrocarbons, 147
Unsaturation, bromine test for, 124–25, 126, 140, 143, 149
Unsaturation, permanganate test for, 124–25, 126, 140, 149
U-shaped capillary, 72
UV-visible spectroscopy (spectrophotometry) 113–21

V
Vacuum adaptor, ii
Vacuum distillation, 29
Vacuum filtration, 63
Valine, 285
Vanadium pentoxide, 260
Velocity of light, 89
Visible spectrophotometry, techniques, 113–21

W
Water,
    aldol condensation and, 243–46
    reaction with calcium carbide, 141–42
    as a common recrystallization solvent, 67
    and eluotropic series of solvents for alumina and silica, 43
    and equilibrium constant of esterification, 78–81
    and the Grignard reagent, 175
Water-drop technique, 64, 76
Wave number, 90
Wavelength, 89, 90
West condenser, ii
Williamson ether synthesis, 185–187
Wrappers, IR spectrum of, 96

X
Xanthoproteic test, 255–56, 261

Z
Zwitterions, 255